MODULAR MATHEMATICS

Module B: Pure Maths 2

Second Edition

By the same authors

MODULE A: PURE MATHS 1
MODULE C: STATISTICS 1
MODULE E: MECHANICS 1
MODULE F: MECHANICS 2
CORE MATHS FOR A-LEVEL
MATHEMATICS — THE CORE COURSE FOR A-LEVEL
FURTHER PURE MATHEMATICS — with C. Rourke
MATHEMATICS — MECHANICS AND PROBABILITY
FURTHER MECHANICS AND PROBABILITY
APPLIED MATHEMATICS I
APPLIED MATHEMATICS II
PURE MATHEMATICS I
PURE MATHEMATICS II

MODULAR MATHEMATICS

Module B: Pure Maths 2

Second Edition

L. Bostock, B.Sc.

S. Chandler, B.Sc.

Stanley Thornes (Publishers) Ltd

First published in 1991 by Stanley Thornes (Publishers) Ltd, Ellenborough House, Wellington Street, CHELTENHAM GL50 1YW

Second edition 1995

98 99 00 / 10 9 8 7 6 5 4 3

A catalogue record of this book is available from the British Library.

ISBN 0–7487–1775–7

Cover photograph by Martyn Chillmaid
Photograph on page 64 reproduced
courtesy of Images Colour Library

Typeset by Tech-Set, Gateshead, Tyne & Wear.
Printed and bound by T J Press

CONTENTS

vi

PREFACE TO THE SECOND EDITION

This is the second book in a series of modular mathematics courses for students wishing to gain academic qualifications beyond GCSE. An AS-level subject in mathematics requires two modules and four modules are needed for an A-level.

The P2 syllabus of the London Modular Mathematics has been changed to comply with the new subject core for A and AS Level and this second edition of Module B has been extensively revised to cover the work now required. Together with the second edition of Module A, most courses for AS Pure Mathematics are catered for, as well as the pure mathematics contents of a variety of A-level Mathematics subjects.

The contents of this book assume knowledge of the topics in Module A. Students who are taking statistics as a separate module will cover the work in the last three chapters of this book in that module and can safely omit chapters 18, 19 and 20. There may also be an overlap between some of the trigonometry in this book and work you may have done for GCSE. Chapters containing such topics do, however, cover some questions that are more sophisticated than those likely to be met at Key Stage 4. Before assuming that these more demanding problems can be tackled, we suggest that the exercises at the ends of these chapters are used, except for the starred questions, to identify any weaknesses. Further work to deal with any troublespots can then be carried out using the relevant sections of the chapter.

The starred questions in some exercises go beyond the minimum requirement for this syllabus, and some need ideas that extend beyond the scope of the syllabus. Others illustrate some interesting, but more difficult, applications of the topic while others are simply harder problems. They are there to stretch those of you who like being challenged.

The exercises in the chapters all start with straightforward questions designed to build up confidence. More sophisticated questions of the examination type are given in consolidation sections that appear at regular intervals throughout the book. These are intended for use later on, to give practice when confidence has been built up. The consolidation sections also include a summary of the work in preceding chapters and a set of multiple choice questions which probe understanding in a way that other questions cannot do.

There are many computer programs that aid the understanding of mathematics. In particular, a good graph-drawing package is invaluable for investigating graphical aspects of functions. Graphics calculators are also very useful and some of them now available are capable of numerical solution of equations and numerical differentiation and integration. In a few places we have indicated where such aids can be used effectively, but this should be regarded as a minimum application of the use of technology. Some computer programs, such as *Mathematica*, are capable of handling a wide variety of symbolic algebraic processes. Two publications that we recommend in this connection are: *Graphics Calculators in the Mathematics Classroom* and *Spreadsheets: Exploring their Potential in Secondary Mathematics*. These are published by The Mathematical Association and are available from the Association at 259 London Road, Leicester, LE2 3BE, UK.

We are grateful to the following examination boards for permission to reproduce questions from their past examination papers. Part questions are indicated by the suffix p. Questions from specimen papers are indicated by the suffix s, and it should be noted that these questions have not been subjected to the rigorous checking and moderation procedure by the Boards that their examination questions undergo. It should also be noted that the answers to all past examination questions are our responsibility.

University of London Examinations and Assessment Council (ULEAC)
The Associated Examining Board (AEB)
Welsh Joint Examination Council (WJEC)
University of Cambridge Local Examination Syndicate (UCLES)
Oxford University Delegacy of Local Examinations (UODLE)
Oxford and Cambridge Schools Examination Board (OCSEB, MEI)

L. Bostock
1995   S. Chandler

NOTATION

$=$	is equal to	$\propto$	is proportional to
$\equiv$	is identical to	$\rightarrow$	maps to
$\approx$	is approximately equal to*	$\Rightarrow$	implies
$>$	is greater than	$\Leftrightarrow$	implies and is implied by
$\geqslant$	is greater than or equal to	$\in$	is a member of
$<$	is less than	$:$	is such that
$\leqslant$	is less than or equal to	$[a, b]$	the interval $a \leqslant x \leqslant b$
∞	infinity; infinitely large		

A stroke through a symbol negates it, e.g. $\neq$ means 'is not equal to'

ABBREVIATIONS

$\parallel$	is parallel to	$-$ve	negative
$+$ve	positive	w.r.t.	with respect to

USEFUL FORMULAE

For a cone with base radius r, height h and slant height l

$$\text{volume} = \tfrac{1}{3}\pi r^2 h \qquad \text{curved surface area} = \pi r l$$

For a sphere of radius r

$$\text{volume} = \tfrac{4}{3}\pi r^3 \qquad \text{surface area} = 4\pi r^2$$

For any pyramid with height h and base area a

$$\text{volume} = \tfrac{1}{3}ah$$

*Practical problems rarely have exact answers. Where numerical answers are given they are corrected to two or three decimal places depending on their content, e.g. π is 3.142 correct to 3 dp and although we write $\pi = 3.142$ it is understood that this is not an exact value. We reserve the symbol $\approx$ for those cases where the approximation being made is part of the method used.

COMPUTER PROGRAM REFERENCES

Where computer programs or calculators can be helpful, they are identified by marginal symbols in the following way.

INSTRUCTIONS FOR ANSWERING MULTIPLE CHOICE EXERCISES

These exercises are included in each consolidation section. The questions are set in groups; the answering techniques are different for each group and are classified as follows:

TYPE I

These questions consist of a problem followed by several alternative answers, only *one* of which is correct.

Write down the letter corresponding to the correct answer.

TYPE II

A single statement is made. Write T if it is true and F if it is false.

CHAPTER 1

ALGEBRA

In *Module A* we saw that the value of a fraction is unaltered if the *numerator and denominator* are multiplied or divided by the same number,

e.g. $\dfrac{\frac{1}{2}a+b}{a+2b} = \dfrac{a+2b}{2(a+2b)} = \dfrac{1}{2}$

We also saw that fractions are multiplied by taking the product of the numerators and the product of the denominators,

e.g. $\dfrac{e^2}{1+e} \times \dfrac{e}{1-e} = \dfrac{e^3}{(1+e)(1-e)}$

and, to divide by a fraction, we multiply by its reciprocal,

e.g. $\dfrac{1+x}{1-x} \div \dfrac{2x}{1+x} = \dfrac{1+x}{1-x} \times \dfrac{1+x}{2x} = \dfrac{(1+x)^2}{2x(1-x)}$

and $\dfrac{\sin x+1}{\cos x} \div \cos x = \dfrac{\sin x+1}{\cos x} \times \dfrac{1}{\cos x} = \dfrac{\sin x+1}{\cos^2 x}$

ADDITION AND SUBTRACTION OF FRACTIONS

Before fractions can be added or subtracted, they must be expressed with the same denominator, i.e. we have to find a common denominator. Then the numerators can be added or subtracted,

e.g. $\dfrac{2}{p} + \dfrac{3}{q} = \dfrac{2q}{pq} + \dfrac{3p}{pq} = \dfrac{2q+3p}{pq}$

Example 1a

Simplify $\quad x - \dfrac{1}{x}$

$$x - \dfrac{1}{x} = \dfrac{x}{1} - \dfrac{1}{x} = \dfrac{x^2}{x} - \dfrac{1}{x} = \dfrac{x^2-1}{x} = \dfrac{(x-1)(x+1)}{x}$$

1

Simplify

1. $\dfrac{1}{a} - \dfrac{1}{b}$

2. $\dfrac{1}{3x} + \dfrac{1}{5x}$

3. $\dfrac{1}{p} - \dfrac{1}{q}$

4. $\dfrac{1}{2x} + \dfrac{3}{5x}$

5. $x + \dfrac{1}{x}$

6. $\dfrac{x}{y} - \dfrac{y}{x}$

7. $2p - \dfrac{1}{p}$

8. $\dfrac{x}{3} + \dfrac{x+1}{4}$

9. $\frac{1}{2}(x-1) + \frac{1}{3}(x+1)$

10. $\dfrac{x+2}{5} - \dfrac{2x-1}{3}$

11. $\dfrac{1}{\sin A} + \dfrac{1}{\sin B}$

12. $\dfrac{1}{\cos A} + \dfrac{1}{\sin A}$

13. $3x + \dfrac{1}{4x}$

14. $x - \dfrac{2}{2x+1}$

15. $x + 1 + \dfrac{1}{x+1}$

16. $1 + \dfrac{1}{x} + \dfrac{1}{2x}$

17. $1 - x + \dfrac{1}{x}$

18. $\dfrac{1}{n} + \dfrac{1}{n^2}$

19. $\dfrac{x}{a^2} + \dfrac{x}{b^2}$

20. $1 + \dfrac{1}{a} + \dfrac{1}{a+1}$

21. $\dfrac{e^x}{x} + \dfrac{1}{2}e^x$

22. $\dfrac{1}{(\cos A + 1)} - \dfrac{1}{(\sin A + 1)}$

Example 1b

Simplify $\dfrac{2}{x+2} - \dfrac{x-4}{2x^2+x-6}$

$$\frac{2}{x+2} - \frac{x-4}{2x^2+x-6} = \frac{2}{x+2} - \frac{x-4}{(x+2)(2x-3)}$$

$$= \frac{2(2x-3)}{(x+2)(2x-3)} - \frac{x-4}{(x+2)(2x-3)}$$

$$= \frac{2(2x-3)-(x-4)}{(x+2)(2x-3)}$$

$$= \frac{4x-6-x+4}{(x+2)(2x-3)}$$

$$= \frac{3x-2}{(x+2)(2x-3)}$$

EXERCISE 1b

Simplify

1. $\dfrac{1}{x+1} + \dfrac{1}{x-1}$

2. $\dfrac{1}{x+1} + \dfrac{1}{x-2}$

3. $\dfrac{4}{x+2} + \dfrac{3}{x+3}$

4. $\dfrac{1}{x^2-1} + \dfrac{1}{x+1}$

5. $\dfrac{2}{a^2-1} - \dfrac{3}{a-1}$

6. $\dfrac{1}{x^2+2x+1} + \dfrac{1}{x+1}$

7. $\dfrac{3}{4x^2+4x+1} - \dfrac{2}{2x+1}$

8. $\dfrac{2}{x^2+5x+4} - \dfrac{3}{x+1}$

9. $\dfrac{4}{(x+1)^2} + \dfrac{2}{x+1}$

10. $\dfrac{3}{(x+2)^2} - \dfrac{1}{x+4}$

11. $\dfrac{1}{2(x-1)} + \dfrac{2}{3(x+4)}$

12. $\dfrac{7}{5(x+2)} - \dfrac{2}{x+4}$

13. $\dfrac{4}{3(x+2)} - \dfrac{3}{2(3x-5)}$

14. $\dfrac{3}{x+1} - \dfrac{2}{x-2} + \dfrac{4}{x+3}$

15. $\dfrac{1}{x+1} - \dfrac{2}{x+2} + \dfrac{3}{x+3}$

16. $\dfrac{x+2}{(x+1)^2} - \dfrac{1}{x}$

17. $\dfrac{4t}{t^2+2t+1} + \dfrac{3}{t+1}$

18. $\dfrac{2t}{t^2+1} - \dfrac{t^2+1}{t^2-1}$

19. $\dfrac{1}{y^2-x^2} + \dfrac{3}{y+x}$

20. $1 + \dfrac{1}{n} + \dfrac{1}{n+1} + \dfrac{1}{n+2}$

21. $\dfrac{1}{e^2-1} - \dfrac{e}{e+1}$

22. $\dfrac{\sin A}{\sin A - 1} - \dfrac{\cos A}{\cos A - 1}$

DIVISION OF ONE POLYNOMIAL BY ANOTHER POLYNOMIAL

Long division can be used to divide say $x^3 + 4x^2 - 7$ by $x^2 - 3$ ($x^3 + 4x^2 - 7$ is called the *dividend* and $x^2 - 3$ is called the *divisor*.)

$$
\begin{array}{r}
x + 4 \\
x^2 - 3 \overline{\smash{)}\ x^3 + 4x^2 \qquad - 79} \\
\underline{x^3 \qquad\quad - 3x} \\
4x^2 + 3x - 7 \\
\underline{4x^2 \qquad - 12} \\
+ 3x + 5
\end{array}
$$

Divide x^3 by x^2: it goes in x times.

Multiply the divisor by x and then subtract it from the dividend.

The result is the new dividend; repeat the process until the dividend is not divisible by x^2.

The number over the division line is the *quotient*, and what is left is called the *remainder*.

The relationship between the divisor, the dividend, the quotient and the remainder can be expressed as

$$x^3 + 4x^2 - 7 \equiv (x+4)(x^2-3) + 3x + 5$$

IMPROPER FRACTIONS

When the highest power of x in the numerator of a fraction is greater than or *equal* to the highest power of x in the denominator, the fraction is called *improper*.

For example, $\dfrac{x^3 + 4x^2 - 7}{x^2 - 3}$ and $\dfrac{x^2 + 7}{x^2 - 2}$ are improper fractions.

Improper fractions can be expressed in a form in which any fractions are proper, by dividing the numerator by the denominator,

i.e. $\dfrac{x^3 + 4x^2 - 7}{x^2 - 3}$ can be written as $x + 4 + \dfrac{3x + 5}{x^2 - 3}$

Long division is not always necessary; it is often simpler to rearrange the numerator by adding and subtracting appropriate terms. This is illustrated in the following worked example.

Example 1c

Express $\dfrac{x^3 + 5x^2 - 3x}{x^3 + 1}$ in a form without an improper fraction.

$$\frac{x^3 + 5x^2 - 3x}{x^3 + 1} \equiv \frac{x^3 + 1 + 5x^2 - 3x - 1}{x^3 + 1} \equiv \frac{x^3 + 1}{x^3 + 1} + \frac{5x^2 - 3x - 1}{x^3 + 1}$$

$$\equiv 1 + \frac{5x^2 - 3x - 1}{x^3 + 1}$$

Notice that we add 1 to x^3 so that the expression can be split into two fractions, one of which divides out exactly. It is important to realise that, having added 1 to the numerator we also have to subtract 1 so that the value of the numerator is not altered.
After some practice, it is possible to do the intermediate steps mentally.

EXERCISE 1c

1. Find the quotient and the remainder for each of the following divisions.

(a) $(x^3 + x^2 - 3x + 6) \div (x^2 + 3)$

(b) $(x^4 - 5x^2 + 2) \div (x + 1)$

(c) $(2x^3 - 4x^2 + 3x - 1) \div (x^2 - 1)$

(d) $(3x^3 - 5) \div (x - 2)$

(e) $(x^5 - 5x^2 + 1) \div (x^3 + 1)$

(f) $(2x^3 - 5x^2 + 6x + 2) \div (x - 3)$

(g) $(x^2 - 7x + 2) \div (x + 3)$

(h) $(5x^3 - x^2 + 1) \div (x^2 - 1)$

(i) $(3x^2 - 7) \div (x^2 + 1)$

(j) $(4x^3 - 9x + 1) \div (2x - 1)$

2. Express the following fractions as the sum of a polynomial and a proper fraction.

(a) $\dfrac{x+4}{x+1}$ (b) $\dfrac{2x}{x-2}$ (c) $\dfrac{x^2+3}{x^2-1}$

(d) $\dfrac{x^2}{x-2}$ (e) $\dfrac{x^2+3x}{x-4}$ (f) $\dfrac{x^2-4}{x(x+1)}$

THE REMAINDER THEOREM

When $f(x) = x^3 - 7x^2 + 6x - 2$ is divided by $x - 2$, we get a quotient and a remainder. The relationship between these quantities can be written as

$$f(x) = x^3 - 7x^2 + 6x - 2 \equiv (\text{quotient})(x-2) + \text{remainder}$$

Now substituting 2 for x eliminates the term containing the quotient, giving

$$f(2) = \text{remainder}$$

This is a particular illustration of the more general case, namely if a polynomial $f(x)$, is divided by $(x-a)$ then

$$f(x) \equiv (\text{quotient})(x-a) + \text{remainder}$$

$\Rightarrow$ $f(a) \equiv \text{remainder}$

This result is called the *remainder theorem* and can be summarised as

when a polynomial $f(x)$ is divided by $(x-a)$, the remainder is $f(a)$

Examples 1d

1. Find the remainder when

   (a) $x^3 - 2x^2 + 6$ is divided by $x + 3$
   (b) $6x^2 - 7x + 2$ is divided by $2x - 1$

(a) When $f(x) = x^3 - 2x^2 + 6$ is divided by $x + 3$, the remainder is

$$f(-3) = (-3)^3 - 2(-3)^2 + 6 = -39$$

(b) If $f(x) = 6x^2 - 7x + 2$, then

$$f(x) = (2x-1)(\text{quotient}) + \text{remainder}$$

$\Rightarrow$ $\text{remainder} = f\left(\tfrac{1}{2}\right) = 0$

Note that as the remainder is zero, $2x - 1$ is a factor of $f(x)$

The Factor Theorem

This is a special case of the remainder theorem because if $x - a$ is a factor of a polynomial $f(x)$ then there is no remainder when $f(x)$ is divided by $x - a$,

i.e. $f(a) = 0$

This result, which is called the factor theorem, states that

$$\text{if, for a polynomial } f(x), \ f(a) = 0$$
$$\text{then } x - a \text{ is a factor of } f(x)$$

The factor theorem is very helpful when factorising cubics or higher degree polynomials.

Examples 1d (continued)

2. Factorise $x^3 - x^2 + 2x - 8$

$$f(x) \equiv x^3 - x^2 + 2x - 8$$

We will test for factors of the form $x - a$ by finding $f(a)$ for various values of a. Note that, as the factors of 8 and 1, 2, 4 and 8, the values we choose for a must belong to the set $\{\pm 1, \ \pm 2, \ \pm 4, \ \pm 8, \}$.

$f(1) = 1 - 1 + 2 - 8 \neq 0$, so $x - 1$ is not a factor of $f(x)$

$f(-1) = -1 - 1 - 2 - 8 \neq 0$, so $x + 1$ is not a factor of $f(x)$

$f(2) = 8 - 4 + 4 - 8 = 0$, therefore $(x - 2)$ is a factor of $f(x)$

Now that a factor has been found, it should be taken out; this can be done by inspection or by long division.

$$x^3 - x^2 + 2x - 8 = (x - 2)(x^2 + x + 4)$$

and $x^2 + x + 4$ does not factorise.

Therefore $x^3 - x^2 + 2x - 8 = (x - 2)(x^2 + x + 4)$

The Factors of $a^3 - b^3$ and $a^3 + b^3$

$a^3 - b^3 = 0$ when $a = b$, hence $a - b$ is a factor of $a^3 - b^3$

Therefore $$a^3 - b^3 \equiv (a - b)(a^2 + ab + b^2)$$

in particular $$x^3 - 1 \equiv (x - 1)(x^2 + x + 1)$$

Also, $a^3 + b^3 = 0$ when $a = -b$, so $a + b$ is a factor of $a^3 + b^3$

Therefore $\qquad a^3 + b^3 \equiv (a + b)(a^2 - ab + b^2)$

in particular $\qquad x^3 + 1 \equiv (x + 1)(x^2 - x + 1)$

EXERCISE 1d

1. Find the remainder when the following functions are divided by the linear factors indicated.

   (a) $x^3 - 2x + 4$, $x - 1$ $\qquad$ (b) $x^3 + 3x^2 - 6x + 2$, $x + 2$

   (c) $2x^3 - x^2 + 2$, $x - 3$ $\qquad$ (d) $x^4 - 3x^3 + 5x$, $2x - 1$

   (e) $9x^5 - 5x^2$, $3x + 1$ $\qquad$ (f) $x^3 - 2x^2 + 6$, $x - a$

   (g) $x^2 + ax + b$, $x + c$ $\qquad$ (h) $x^4 - 2x + 1$, $ax - 1$

2. Determine whether the following linear functions are factors of the given polynomials

   (a) $x^3 - 7x + 6$, $x - 1$ $\qquad$ (b) $2x^2 + 3x - 4$, $x + 1$

   (c) $x^3 - 6x^2 + 6x - 2$, $x - 2$ $\qquad$ (d) $x^3 - 27$, $x - 3$

   (e) $2x^4 - x^3 - 1$, $2x - 1$ $\qquad$ (f) $x^3 + ax^2 - a^2x - a^3$, $x + a$

3. Factorise the following functions as far as possible.

   (a) $x^3 + 2x^2 - x - 2$ $\qquad$ (b) $x^3 - x^2 - x - 2$

   (c) $x^4 - 1$ $\qquad$ (d) $x^3 + 3x^2 + 3x + 2$

   (e) $2x^3 - x^2 + 2x - 1$ $\qquad$ (f) $27x^3 - 1$

   (g) $x^3 + a^3$ $\qquad$ (h) $x^3 - y^3$

4. If $x^2 - 7x + a$ has a remainder 1 when divided by $x + 1$, find a.

5. If $x - 2$ is a factor of $ax^2 - 12x + 4$, find a.

6. One solution of the equation $x^2 + ax + 2 = 0$ is $x = 1$, find a.

7. One root of the equation $x^2 - 3x + a = 0$ is 2. Find the other root.

SOLUTION OF CUBIC EQUATIONS

Consider the general cubic function $f : x \rightarrow ax^3 + bx^2 + cx + d$.
From the shape of the graph we know that $f(x)$ takes *all* real values from $-\infty$ to ∞, so there is at least one value of x for which $f(x) = 0$,

i.e. **a cubic equation has at least one real root.**

If this root is rational, i.e. of the form $\dfrac{a}{b}$ where a and b are integers, then we can find it by using the factor theorem.

Examples 1e

1. Solve the equation $x^3 - 2x^2 - x + 2 = 0$

Possible factors of $x^3 - 2x^2 - x + 2$ are $(x \pm 1)$ and $(x \pm 2)$

Using the factor theorem with $f(x) = x^3 - 2x^2 - x + 2$ gives

$f(1) = 1 - 2 - 1 + 2 = 0$ so $(x - 1)$ is a factor of $f(x)$

Hence $x^3 - 2x^2 - x + 2 = (x - 1)(x^2 - x - 2)$

$$= (x - 1)(x + 1)(x - 2)$$

$\therefore \qquad x^3 - 2x^2 - x + 2 = 0$ when $x = 1, -1$ or 2

The factor and remainder theorems can be useful when dealing with some problems involving quadratic equations.

2. The equation $f(x) = 0$ has a repeated root, where $f(x) = 4x^2 + px + q$.
When $f(x)$ is divided by $x + 1$ the remainder is 1. Find the values of p and q.

$$f(-1) = 4 - p + q = 1 \quad \Rightarrow \quad q = p - 3 \qquad [1]$$

If $4x^2 + px + q = 0$ has a repeated root then '$b^2 - 4ac$' $= 0$

i.e. $\qquad p^2 - 16q = 0 \qquad\qquad\qquad [2]$

Solving equations [1] and [2] simultaneously gives

$$p^2 - 16(p - 3) = 0 \quad \Rightarrow \quad p^2 - 16p + 48 = 0$$

$$\Rightarrow \quad (p - 12)(p - 4) = 0$$

$\therefore \qquad$ either $p = 12$ and $q = 9$ or $p = 4$ and $q = 1$

1. Factorise $2x^3 - x^2 - 2x + 1$. Hence find the values of x for which $2x^3 - x^2 - 2x + 1 = 0$.

2. Given that $f(x) = x^3 - x^2 - x - 2$ show that $f(x) = 0$ has only one root.

3. Find the value of p for which $x = \frac{1}{2}$ is a solution of the equation $4x^2 - px + 3 = 0$.

4. Show that the x coordinates of the points of intersection of the curves

$$y = \frac{1}{x} \quad \text{and} \quad x^2 + 4y = 5$$

satisfy the equation $x^3 - 5x + 4 = 0$.

Solve this equation.

5. Factorise $x^3 - 4x^2 + x + 6$
Hence sketch the curve $y = x^3 - 4x^2 + x + 6$

6. A function f is defined by

$$f(x) = 5x^3 - px^2 + x - q$$

When $f(x)$ is divided by $x - 2$, the remainder is 3. Given that $(x - 1)$ is a factor of $f(x)$

(a) find p and q

(b) find the number of real roots of the equation
$$5x^3 - px^2 + x - q = 0$$

7. The polynomial $x^3 + px^2 + q$ has factors $(x - 2)$ and $(x + 4)$.

(a) Find two equations relating p and q.
Solve these equations to find p and q.

(b) Solve the equation
$$x^3 + px^2 + q = 0$$

Approximate Solution of Equations

We have shown that a cubic equation has at least one real root but if that root is not rational the factor theorem does not help in finding it. We can, however, find an approximate value for such a root by drawing sketch graphs. If the zoom facility on a graphics calculator is used, the root can be found as accurately as required.

Consider the equation $x^3 + 2x^2 + 5x - 1 = 0$

The equation can be written as $x^3 = 1 - 5x - 2x^2$ so a sketch of the curves $y = x^3$ and $y = 1 - 5x - 2x^2$ shows the number of roots.

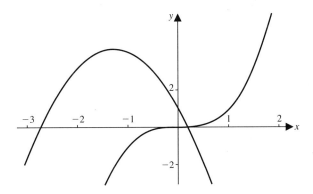

From the sketch we see that there is only one root and it is near the origin.

We can locate the root a little more accurately by using the fact that, if a root of $f(x) = 0$ lies between $x = a$ and $x = b$, then $f(a)$ and $f(b)$ are opposite in sign.

When $f(x) = x^3 + 2x^2 + 5x - 1$,

$$f(0) = -1 < 0 \quad \text{and} \quad f(0.5) = (0.5)^3 + 2(0.5)^2 + 5(0.5) - 1 > 0$$

Therefore the root of the equation $x^3 + 2x^2 + 5x - 1 = 0$ lies between 0 and 0.5. Although we can continue using this method to narrow down the interval in which the root lies, there are disadvantages; several steps are needed to get a root correct to even one decimal place and it is not easy to keep track of the calculations.

APPROXIMATE SOLUTION OF EQUATIONS BY ITERATION

Iteration methods generate a sequence of numbers which converge to the root of the equation. The advantages of iteration are that a sequence of numbers is easy to keep track of and, if the sequence converges quickly, the number of steps needed to get an answer accurate to 3 significant figures, say, can be quite small. There are several iterative methods but we will look at just one.

Consider the recurrence relation $u_{n+1} = g(u_n)$.
For a given value of u_1, this generates the sequence $u_1, u_2, u_3, \ldots$
If the sequence converges to α, then both u_n and u_{n+1} approach the value α as n increases.

This means that, as $n \to \infty$, $u_{n+1} = g(u_n) \to \alpha = g(\alpha)$
i.e. the sequence generated by $u_{n+1} = g(u_n)$ converges to a root of the equation $x = g(x)$.

Therefore if x_n is an approximate value of a root of the equation $x = g(x)$, the sequence generated by $x_{n+1} = g(x_n)$, will, *provided that it converges*, generate better and better approximations to that root.

Returning to the equation $x^3 + 2x^2 + 5x - 1 = 0$

and arranging it as $$x = (1 - x^3 - 2x^2)/5$$

we can use the iteration formula $x_{n+1} = (1 - x_n{}^3 - 2x_n{}^2)/5$ with $x_1 = 0$ to produce the sequence

$0, \quad 0.2, \quad 0.1824\ldots, \quad 0.1854\ldots, \quad 0.1854\ldots, \quad \ldots$

After two iterations, we see that the root is 0.2 correct to 1 significant figure. Another two iterations gives the root as 0.185 correct to 3 significant figures.

The example demonstrates the main advantage of iteration methods; once the iteration formulae is obtained, the method only involves feeding numbers into the formula. This is a process that is ideally suited to calculators and computers.

If you have a calculator with an ANS button, (i.e. a button which feeds in the answer to the previous calculation) you can generate the terms as follows:

enter 0 (the first approximation for the root), press EXE,

enter $(1 - \text{ANS}^3 - 2\text{ANS}^2) \div 5$
(i.e. the iteration formula with ANS replacing x_n), press EXE,

then continue to press EXE and successive terms of the sequence appear in the display.

A computer spreadsheet also will easily generate successive approximations. The following illustration shows how this can be done.

The first approximation is entered in cell A1. The iteration formula, with A1 replacing x_n is entered in cell A2. The 'fill' command is then used to replicate this down the column, so generating the sequence.

(The 'fill' command, or the equivalent, is the one that automatically copies the formula, replacing cell A1 by cell A2, A2 by A3, and so on.)

There is one main disadvantage of this method; not all (and sometimes none) of the possible rearrangements of an equation to the form $x = g(x)$ result in a convergent sequence. The conditions for this method to succeed are investigated in the next section.

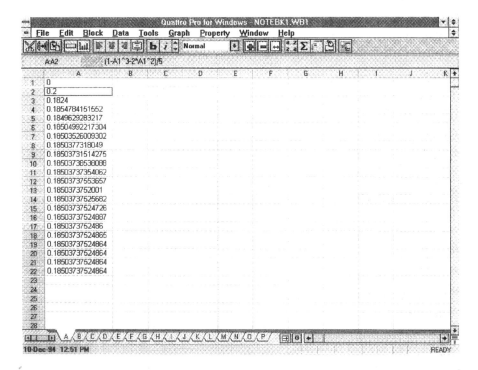

Conditions for Convergence

The roots of the equation $x = g(x)$ are the values of x at the points of intersection of the line $y = x$ and the curve $y = g(x)$.

Taking x_1 as a first approximation to a root α then in the diagram,
        A is the point on the *curve* where $x = x_1$, $y = g(x_1)$
        B is the point where $x = \alpha$, $y = g(x_1)$
        C is the point on the *line* where $x = x_2$, $y = g(x_1)$

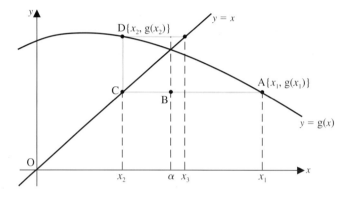

If, in the region of α, the slope of $y = g(x)$ is less steep than that of the line $y = x$, i.e. provided that $|g'(x)| < 1$

then $CB < BA$

so x_2 is closer to α than is x_1 and x_2 is a better approximation to α

But C is on the line $y = x$

therefore $x_2 = g(x_1)$

Now taking the point D on the curve where $x = x_2$, $y = g(x_2)$ and repeating the argument above we find that x_3 is a better approximation to α than is x_2

where $x_3 = g(x_2)$

This process can be repeated as often as necessary to achieve the required degree of accuracy.

Hence   **the iteration formula $x_{n+1} = g(x_n)$ converges
to a root α of the equation $x = g(x)$
provided that $|g'(x)| < 1$ near to $x = \alpha$**

The rate at which these approximations converge to α depends on the value of $|g'(x)|$ near α. The smaller $|g'(x)|$ is, the more rapid is the convergence.

It should be noted that this method fails if $|g'(x)| > 1$ near α.
The following diagrams illustrate some of the factors which determine the success, or otherwise, of this method.

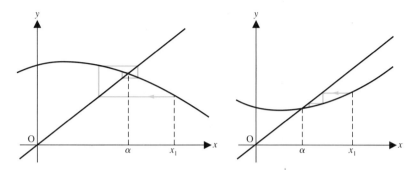

Rapid rate of convergence $(|g'(x)|$ small$)$.

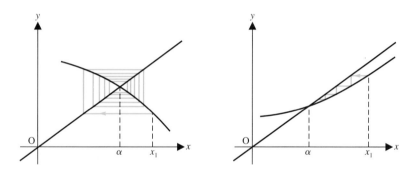

Slow rate of convergence $(|g'(x)| < 1$ but close to $1)$.

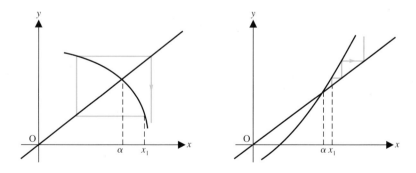

Divergence, i.e. failure, $(|g'(x)| > 1)$.

Example 1f

The equation $x^3 + 3x^2 - 7 = 0$ can be arranged as $x = \sqrt{\dfrac{7 - x^3}{3}}$. Given that there is a root between 1 and 2, and using the given arrangement as the basis of an iteration formula, find this root correct to 2 significant figures.

Using $\quad x_{n+1} = \sqrt{\dfrac{7 - x_n^3}{3}} \quad$ with $\quad x_1 = 1 \quad$ gives

$$x_2 = \sqrt{\frac{7 - 1^3}{3}} = 1.414\ldots$$

$$x_3 = \sqrt{\frac{7 - (1.41\ldots)^3}{3}} = 1.179\ldots$$

$$x_4 = \sqrt{\frac{7 - (1.17\ldots)^3}{3}} = 1.336\ldots$$

$$x_5 = \sqrt{\frac{7 - (1.33\ldots)^3}{3}} = 1.239\ldots$$

$$x_6 = \sqrt{\frac{7 - (1.23\ldots)^3}{3}} = 1.303\ldots$$

$$x_7 = \sqrt{\frac{7 - (1.30\ldots)^3}{3}} = 1.263\ldots$$

Therefore $x = 1.3$ correct to 2 significant figures.

Note that, although this process can be done automatically with an appropriate calculator, it is important that, when a question asks for the intermediate values, all stages in the working are written down.

EXERCISE 1f

Show that each of the following equations has a root between 0 and 1. Taking 0 as the first approximation for this root, and using the iteration formula given three times, find a better approximation for this root, giving answers corrected to 4 significant figures.

1. $x^3 - x^2 + 10x - 2 = 0; \quad x_{n+1} = (2 + x_n^2 - x_n^3)/10$

2. $3x^3 - 2x^2 - 9x + 2 = 0$; $x_{n+1} = (3x_n^3 - 2x_n^2 + 2)/9$

3. $2x^3 + x^2 + 6x - 1 = 0$; $x_{n+1} = (1 - 2x_n^3 - x_n^2)/6$

4. $x^2 + 8x - 8 = 0$; $x_{n+1} = (8 - x_n^2)/8$

5. $5x^2 - x^3 - 1 = 0$; $x_{n+1} = \sqrt{\dfrac{1}{5 - x_n}}$

6. Show that the equation $e^x = 2x + 1$ has a root between 1 and 2.
   Using the iteration formula $x_{n+1} = \ln(2x_n + 1)$ five times find an approximate value for this root, corrected to 3 significant figures.

7. (a) Find, to the nearest integer, the larger root of the equation $x^2 - 7x + 2 = 0$.
   (b) By trying various rearrangements of the equation $x^2 - 7x + 2 = 0$, find an iteration formula that converges to the larger root. Use it to find this root correct to 3 significant figures.

8. Repeat question 7 with the equation $x^3 - 3x + 1 = 0$.

MIXED EXERCISE 1

1. Simplify   (a) $\dfrac{2p}{r} - \dfrac{3}{p}$ (b) $x + \dfrac{3}{4x - 1}$

2. Factorise $x^3 + x^2 - 5x - 2$

3. Use the iteration formula $x_{n+1} = \dfrac{6}{x_n^2 + 5x_n}$ with $x_1 = -1$ to find a root
   of the equation $x^3 + 5x^2 - 6 = 0$ correct to 2 significant figures.

4. Simplify $\dfrac{1}{x + 1} + \dfrac{1}{2x - 1} + \dfrac{1}{x}$

5. Express $\dfrac{x^3 - 2x^2 + 3x - 1}{x - 1}$ as the sum of a polynomial and a proper fraction.

6. Find the larger root of the equation $x^2 - 3x + 1 = 0$ correct to 2 significant
   figures using the iteration given by $x_{n+1} = 3 - \dfrac{1}{x_n}$ with $x_1 = 2$.

7. What is the remainder when $(x - 1)$ is divided into $x^5 - 3$?

8. Simplify

   (a) $\dfrac{2x}{x^2 - 1} + \dfrac{x^2 - 2x}{x^2 - 2x + 1}$ (b) $1 + \dfrac{1}{2x} + \dfrac{1}{x - 1}$ (c) $\dfrac{1}{a} + \dfrac{1}{b} + \dfrac{1}{c}$

9. Find the root of the equation $2x = \cos x$ correct to 2 decimal places using the iteration formula $x_{n+1} = \frac{1}{2} \cos x_n$ with $x_1 = 0.5$.

10. Factorise   (a) $x^4 - 1$ (b) $81x^5 - 625x$.

11. When $x^3 - ax^2 + 5x - 7$ is divided by $x - 3$, the remainder is 2. Find the value of a.

12. Show that $(a + 2b)$ is a factor of $2a^3 + 5a^2b - ab^2 - 6b^3$. Hence factorise $2a^3 + 5a^2b - ab^2 - 6b^3$ completely.

*13. A box with a square cross-section is to be made from a square sheet of metal of side x cm so that the height of the box is 2 cm less than its width and the volume of the box is 16 cm^3. Find the value of x.

*14. (a) Show that $\sin x = 1 - x$ has one root.

   (b) Use a suitable iteration formula to find this root correct to 3 significant figures. This takes several iterations so you will need to use appropriate technical help.

*15. (a) By drawing *sketch* graphs show that the equation $x^3 - 2x + 1 = 0$ probably has one root between -1 and -2 and possibly one or two further roots greater than zero.

   (b) Using pencil and paper methods only, confirm that there is a root between -1 and -2 and investigate the number of positive roots. If there are any positive roots, give intervals, of width half a unit, in which they lie.

   (c) Use iteration to find the value of the negative root correct to 3 significant figures. State the iteration formula used.

CHAPTER 2

COORDINATE GEOMETRY

In *Module A* we saw that, for any two points $A(x_1, y_1)$ and $B(x_2, y_2)$ the length of the line joining A and B is

$$\sqrt{[(x_2 - x_1)^2 + (y_2 - y_1)^2]}$$

and the gradient of the line through A and B is

$$\frac{y_2 - y_1}{x_2 - x_1}$$

THE MIDPOINT OF A LINE JOINING TWO GIVEN POINTS

Consider the line joining the points $A(1, 1)$ and $B(5, 3)$.

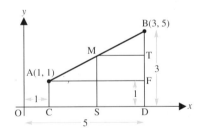

M is the midpoint of AB, hence S is the midpoint of CD.

Therefore the x coordinate of M is given by $OC + \frac{1}{2}CD$
i.e. by $1 + \frac{1}{2}(5 - 1) = 3$.

Now $1 + \frac{1}{2}(5 - 1) = \frac{1}{2}(1 + 5)$ which is the average of the x coordinates of A and B.

Similarly, T is the midpoint of BF so, from the diagram,
the y coordinate of M is $1 + \frac{1}{2}(3 - 1) = \frac{1}{2}(1 + 3)$
which is the average of the y coordinates of A and B.

The same argument can be applied to the general points $A(x_1, y_1)$ and $B(x_2, y_2)$ to show that

> **the midpoint of the line joining $A(x_1, y_1)$ and $B(x_2, y_2)$**
> **has coordinates $[\frac{1}{2}(x_1 + x_2), \frac{1}{2}(y_1 + y_2)]$,**
> **i.e. the averages of the coordinates of A and B**

The next worked example shows that this formula holds when some of the coordinates are negative.

Example 2a

Find the coordinates of the midpoint of the line joining $A(-3, -2)$ and $B(1, 3)$.

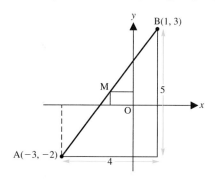

The coordinates of M are

$$[\tfrac{1}{2}(x_1 + x_2), \tfrac{1}{2}(y_1 + y_2)] = [\tfrac{1}{2}(-3 + 1), \tfrac{1}{2}(-2 + 3)] = (-1, \tfrac{1}{2})$$

Alternatively, from the diagram, M is half-way from A to B horizontally and vertically,

i.e. at M $\quad x = -3 + \tfrac{1}{2}(4) = -1 \quad$ and $\quad y = -2 + \tfrac{1}{2}(5) = \tfrac{1}{2}$

This confirms that the formula works when some of the coordinates are negative.

EXERCISE 2a

1. Find the coordinates of the midpoint of the line joining

   (a) $A(1, 2)$ and $B(4, 6)$ (b) $C(3, 1)$ and $D(2, 0)$

   (c) $J(4, 2)$ and $K(2, 5)$ (d) $P(a, b)$ and $Q(b, a)$

2. Find the coordinates of the midpoint of the line joining

   (a) $A(-1, -4)$ and $B(2, 6)$ (b) $S(0, 0)$ and $T(-1, -2)$

   (c) $E(-1, -4)$ and $F(-3, -2)$ (d) $P(t, 2t)$ and $Q\left(\dfrac{1}{t}, t^2\right)$

3. Find the coordinates of the midpoint of the line from the point $(4, -8)$ to the origin.

4. A, B and C are the points $(7, 3)$, $(-4, 1)$ and $(-3, -2)$ respectively.
   (a) Show that $\triangle ABC$ is isosceles.
   (b) Find the midpoint of BC.
   (c) Find the area of $\triangle ABC$.

5. The vertices of a triangle are $A(0, 2)$, $B(1, 5)$ and $C(-1, 4)$. Find
   (a) the perimeter of the triangle,
   (b) the coordinates of D where D is the midpoint of BC,
   (c) the length of AD.

6. M is the midpoint of the line joining A to B. The coordinates of A and M are $(5, 7)$ and $(0, 2)$ respectively. Find the coordinates of B.

7. Repeat question 6 for the points $A(-2, 5)$ and $M(-1, -3)$.

8. In triangle OAB, where O is the origin, A is the point $(6, 2)$ and the median through A cuts OB at $D(-1, 4)$. Find the coordinates of B.

The Equation of a Straight Line

In *Module A* we saw that the equation of a straight line can be written in the form

$$y = mx + c$$

where m is the gradient and c is the intercept of the line on the y-axis.

We also saw that if the gradient, m, of a line and the coordinates (x_1, y_1) of a point on it are known, the equation of the line can be found using

$$y - y_1 = m(x - x_1)$$

Further, when two points on the line, (x_1, y_1) and (x_2, y_2), are known, the equation of the line is given by

$$y - y_1 = \left(\frac{y_2 - y_1}{x_2 - x_1}\right)(x - x_1)$$

PARALLEL LINES

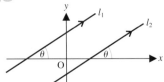

If l_1 and l_2 are parallel lines, they are equally inclined to the x-axis, i.e.

parallel lines have equal gradients.

PERPENDICULAR LINES

Consider the perpendicular lines AB and CD whose gradients are m_1 and m_2 respectively.

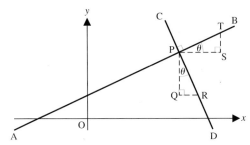

If AB makes an angle θ with the x-axis then CD makes an angle θ with the y-axis. Therefore triangles PQR and PST are similar.

Now the gradient of AB is $\dfrac{ST}{PS} = m_1$

and the gradient of CD is $\dfrac{-PQ}{QR} = m_2$,   i.e. $\dfrac{PQ}{QR} = -m_2$

But $\dfrac{ST}{PS} = \dfrac{QR}{PQ}$ ($\triangle$s PQR and PST are similar)

therefore $m_1 = -\dfrac{1}{m_2}$ or $m_1 m_2 = -1$

i.e.   **the product of the gradients of perpendicular lines is -1,**

or, if one line has gradient m,
any line perpendicular to it has gradient $-\dfrac{1}{m}$

Problems In Coordinate Geometry

We now look at a miscellaneous selection of problems in coordinate geometry. A clear, large and reasonably accurate diagram, showing all the given information, will often suggest the most direct method for solving a particular problem.

Examples 2b

1. The points $A(2, 1)$, $B(5, 1)$ and $C(4, 7)$ lie on the circumference of a circle. Find the coordinates of the centre of the circle.

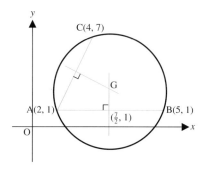

The centre of a circle lies on the perpendicular bisector of a chord. So we need the equations of the perpendicular bisectors of AB and AC and then if we find the point where these lines intersect, that point is the centre of the circle.

From the diagram, the perpendicular bisector of AB is the line $x = \frac{7}{2}$

The midpoint of AC is the point $D(3, 4)$

The gradient of AC is $\dfrac{7-1}{4-2} = 3$

$\therefore$ the gradient of the perpendicular bisector of AC is $-\frac{1}{3}$

The equation of the line through D and G is $y - 4 = -\frac{1}{3}(x - 3)$

$$\Rightarrow \quad 3y + x = 15$$

When $x = \frac{7}{2}$, $3y + \frac{7}{2} = 15 \quad \Rightarrow \quad y = \frac{23}{6}$

Therefore G is the point $\left(\frac{7}{2}, \frac{23}{6}\right)$

2. One vertex of the parallelogram OABC is at the origin. The equation of the side OA is $y = 3x$, the equation of AB is $y = x + 5$ and B is the point $(6, 11)$. Find the coordinates of the points A and C.

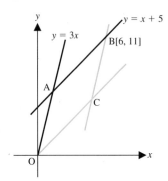

A is the point of intersection of the lines $y = 3x$ and $y = x + 5$

Solving these equations simultaneously gives $x = \frac{5}{2}, y = \frac{15}{2}$

∴   A is the point $\left(\frac{5}{2}, \frac{15}{2}\right)$

C is the point of intersection of the lines OC and BC.

The line OC is parallel to AB.

∴   the gradient of OC is 1 and OC goes through the origin,

therefore the equation of OC is $y = x$ [1]

The line BC is parallel to OA, therefore its gradient is 3; and it passes through $(6, 11)$,

∴   the equation of BC is $y - 11 = 3(x - 6)$ $\Rightarrow$ $y = 3x - 7$ [2]

Solving [1] and [2] simultaneously gives $x = \frac{7}{2}, y = \frac{7}{2}$

These values give the coordinates of C,

∴   C is the point $\left(\frac{7}{2}, \frac{7}{2}\right)$

EXERCISE 2b

1. Determine whether AB and CD are parallel, perpendicular or neither.

   (a) $A(0, -1)$, $B(1, 1)$, $C(1, 5)$, $D(-1, 1)$

   (b) $A(1, 1)$, $B(3, 2)$, $C(-1, 1)$, $D(0, 1)$

   (c) $A(3, 3)$, $B(-3, 1)$, $C(-1, -1)$, $D(1, -7)$

   (d) $A(2, -5)$, $B(0, 1)$, $C(-2, 2)$, $D(3, -7)$

   (e) $A(2, 6)$, $B(-1, -9)$, $C(2, 11)$, $D(0, 1)$

2. Find the equation of the line through $(1, 5)$ that is perpendicular to the line $y = 5 - 2x$.

3. Find the equation of the line through the point $(-1, 3)$ that is

   (a) parallel   (b) perpendicular

   to the line $2x - y + 3 = 0$.

4. Write down the equation of the perpendicular bisector of the line joining the points $(2, -3)$ and $(-\frac{1}{2}, \frac{7}{2})$.

5. Write down the equation of the perpendicular bisector of the line joining the points (s, t), $(2s, 3t)$.

6. The coordinates of the points A, B and C are $(3, 1)$, $(-1, 4)$ and $(-2, -3)$ respectively. Find the equation of the line perpendicular to AB and passing through the midpoint of AC.

7. The line l, whose equation is $3x - 2y + 4 = 0$, cuts the x-axis at A and the y-axis at B. Find the equation of the line perpendicular to l which passes through the midpoint of AB.

8. A circle goes through the points $A(1, 3)$, $B(-1, 5)$ and $C(3, -1)$. Find the coordinates of the centre of the circle and the radius of the circle.

9. The vertices of a triangle are $A(2, 1)$, $B(5, 3)$ and $C(-2, -7)$. By finding the equation of the line through the midpoints of AB and AC, prove that this line is parallel to BC.

10. The line joining $A(1, 4)$ and $B(5, -3)$ is a diameter of a circle. The point (p, q) is on the circumference of the circle. Find a relationship between p and q.

THE LOCATION OF A POINT IN SPACE

We saw in *Module A* that any point P in a plane can be located by giving its distances from a fixed point O, in each of two perpendicular directions. These distances are the Cartesian coordinates of the point.

Now we consider locating a point in three-dimensional space.

If we have a fixed point, O, then any other point can be located by giving its distances from O in each of *three* mutually perpendicular directions, i.e. we need *three* coordinates to locate a point in space. So we use the familiar *x*- and *y*-axes, together with a third axis O*z*.
Then any point has coordinates (x, y, z) relative to the origin O.

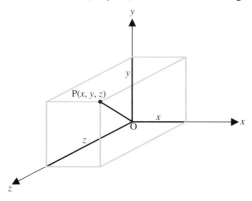

For example, if P is the point $(2, 3, 5)$ then P is
2 units from O in the direction O*x*
3 units from O in the direction O*y*
5 units from O in the direction O*z*.

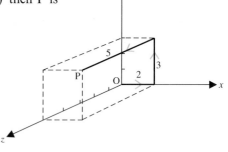

The Distance of a Point from the Origin

Consider the point $A(4, 2, 2)$

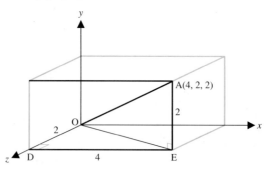

From the diagram, in $\triangle ODE$ we have $OD = 2$, $DE = 4$ and $\angle D = 90°$

Using Pythagoras' Theorem in $\triangle ODE$ gives $OE = \sqrt{2^2 + 4^2}$

Also, in $\triangle OAE$, $AE = 2$, $\angle E = 90°$ and OE is now known, so using Pythagoras' theorem gives

$$OA = \sqrt{2^2 + 4^2 + 2^2} = \sqrt{24}$$

The same argument can be applied to any point $P(x, y, z)$ giving

$$OP = \sqrt{x^2 + y^2 + z^2}$$

EXERCISE 2c

1. Write down the coordinates of the vertices of the cuboid when

   (a) $OG = 5$, $OD = 2$ and $OE = 6$

   (b) B is the point $(3, 2, 1)$

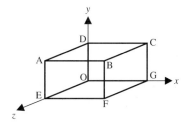

2. Write down the coordinates of the vertices of the cuboid when

   (a) O is the midpoint of DA, $DA = 8$, $AB = 2$ and $AE = 3$

   (b) F is the point $(5, 6, 4)$ and G is the point $(-4, 6, 4)$

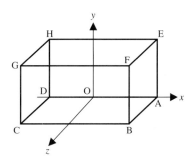

3. B is the point $(3, 5, 2)$

   (a) Write down the coordinates of the other vertices of the cuboid.

   (b) Find the length of
      (i) OA (ii) OB (iii) AF

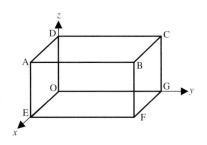

THE DISTANCE BETWEEN TWO POINTS

Now consider the points $A(3, 2, 1)$ and $B(6, 5, 4)$

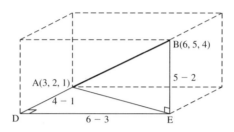

A is 3 units from O in the direction Ox and B is 6 units from O in the direction Ox, therefore D is $6 - 3$ units from A in the direction Ox, i.e. $DE = 6 - 3$
Similarly $EB = 5 - 2$ and $AD = 4 - 1$

Using Pythagoras' theorem in $\triangle ADE$ gives $AE = \sqrt{(6 - 3)^2 + (4 - 1)^2}$
Then applying Pythagoras' Theorem to $\triangle AEB$ gives

$$AB = \sqrt{(6 - 3)^2 + (4 - 1)^2 + (5 - 2)^2} = \sqrt{27} = 3\sqrt{3}$$

The same argument can be applied to the general points $A(x_1, y_1, z_1)$ and $B(x_2, y_2, z_2)$ to show that

**the distance between the points $A(x_1, y_1, z_1)$ and $B(x_2, y_2, z_2)$
is given by**

$$AB = \sqrt{(x_2 - x_1)^2 + (y_2 - y_1)^2 + (z_2 - z_1)^2}$$

Note that we have chosen to draw the axes so that the x- and y-axes are in the conventional positions for work in two dimensions. In three dimensions however, the axes are sometimes drawn with different orientations so check any diagrams given carefully.

1. U is the point $(3, 7, 4)$

   (a) Write down the coordinates of the
       points (i) P (ii) Q (iii) V

   (b) Write down the coordinates of the
       midpoint of (i) OR (ii) SU

   (c) Find the length
       of (i) OT (ii) OU (iii) PV

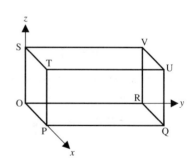

2. A, B and C are the points $(2, 0, 0)$, $(2, 4, 0)$ and $(2, 4, 6)$ respectively.

   (a) Draw a diagram showing these points clearly.

   (b) Find the distance between B and C.

3. Repeat question 2 for the points A$(-1, 0, 0)$, B$(-1, 2, 3)$ and C$(2, -2, 5)$

4. Use the information in the diagram to find

   (a) ∠UTV

   (b) the area of the face STVW.

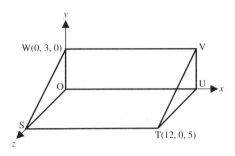

5*. The diagram shows a computer generated plot of a surface. (The equation of this surface is $z = \sin xy$.)

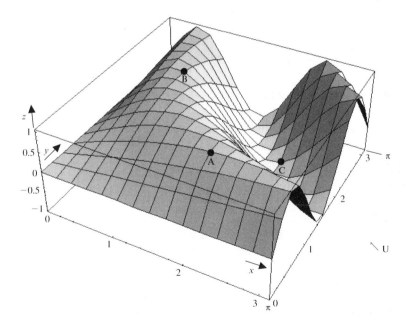

   (a) Find the coordinates of the points marked A, B and C.

   (b) Find, correct to 3 significant figures, the distance AB

   (c) Write down the equation of the curve showing on the face marked U.

CHAPTER 3

SINE AND COSINE FORMULAE

TRIGONOMETRIC RATIOS OF OBTUSE ANGLES

In this chapter we are going to find relationships between the sides and angles of any triangle, i.e. a triangle that does not contain a right-angle. Such triangles include those containing an obtuse angle, so we start with a reminder about the sine and cosine ratios of an obtuse angle. We established the following relationships in *Module A*.

For *any* angle θ, $\sin(\pi - \theta) = \sin\theta$

$$\cos(\pi - \theta) = -\cos\theta$$

When A is an obtuse angle, measured in degrees, these relationships become

$$\mathbf{\sin A = \sin(180° - A)}$$

$$\mathbf{\cos A = -\cos(180° - A)}$$

For example, $\sin 145° = \sin 35°$ and $\cos 145° = -\cos 35°$

FINDING UNKNOWN SIDES AND ANGLES IN A TRIANGLE

Triangles are involved in many practical measurements (e.g. surveying) so it is important to be able to make calculations from limited data about a triangle.

Although a triangle has three sides and three angles, it is not necessary to know all of these in order to define a particular triangle. If enough information about a triangle is known, the remaining sides and angles can be calculated. This is called *solving* the triangle and it requires the use of one of a number of formulae.

The two relationships that are used most frequently are the sine rule and the cosine rule.

When working with a triangle ABC the side opposite to $\angle A$ is denoted by a, the side opposite to $\angle B$ by b and so on.

THE SINE RULE

In a triangle ABC, $\quad \dfrac{a}{\sin A} = \dfrac{b}{\sin B} = \dfrac{c}{\sin C}$

This relationship can be proved quite simply and the proof is given below.

Proof

Consider a triangle ABC in which there is no right angle.

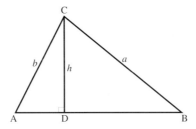

A line drawn from C, perpendicular to AB, divides triangle ABC into two right-angled triangles, CDA and CDB.

In $\triangle$CDA $\quad \sin A = h/b \quad \Rightarrow \quad h = b \sin A$

In $\triangle$CDB $\quad \sin B = h/a \quad \Rightarrow \quad h = a \sin B$

Therefore $\quad a \sin B = b \sin A$

i.e. $\quad \dfrac{a}{\sin A} = \dfrac{b}{\sin B}$

We could equally well have divided $\triangle$ABC into two right-angled triangles by drawing the perpendicular from A to BC (or from B to AC). This would have led to a similar result,

i.e. $\quad \dfrac{b}{\sin B} = \dfrac{c}{\sin C}$

By combining the two results we produce the sine rule,

$$\dfrac{a}{\sin A} = \dfrac{b}{\sin B} = \dfrac{c}{\sin C}$$

Note that this proof is equally valid when $\triangle$ABC contains an obtuse angle.

Suppose that $\angle A$ is obtuse.

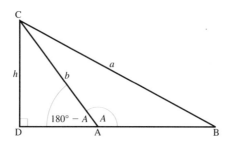

This time $h = b \sin (180° - A)$ but, as $\sin (180° - A) = \sin A$, we see that once again $h = b \sin A$.

In all other respects the proof given above is unaltered, showing that the sine rule applies to any triangle.

Using the Sine Rule

$$\frac{a}{\sin A} = \frac{b}{\sin B} = \frac{c}{\sin C}$$

This rule is made up of three separate fractions, only two of which can be used at a time. We select the two which contain three known quantities and only one unknown.

Note that, when the sine rule is being used to find an unknown angle, it is more conveniently written in the form

$$\frac{\sin A}{a} = \frac{\sin B}{b} = \frac{\sin C}{C}$$

Examples 3a

1. In $\triangle ABC$, $BC = 5$ cm, $A = 43°$ and $B = 61°$. Find the length of AC.

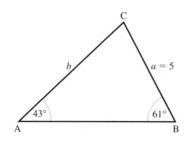

$\angle A$, $\angle B$ and a are known and b is required, so the two fractions we select from the sine rule are

$$\frac{a}{\sin A} = \frac{b}{\sin B}$$

i.e. $$\frac{5}{\sin 43°} = \frac{b}{\sin 61°}$$

$\Rightarrow$ $$b = \frac{5 \sin 61°}{\sin 43°} = 6.412$$

Therefore $AC = 6.41$ cm correct to 3 sf

2. In ABC, $AC = 17$ cm, $\angle A = 105°$ and $\angle B = 33°$. Find AB.

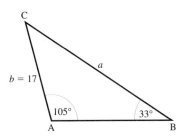

The two sides involved are b and c, so before the sine rule can be used we must find C.

$$\angle A + \angle B + \angle C = 180° \quad \Rightarrow \quad \angle C = 42°$$

Now from the sine rule we can use

$$\frac{b}{\sin B} = \frac{c}{\sin C}$$

$\Rightarrow$ $$\frac{17}{\sin 33°} = \frac{c}{\sin 42°}$$

i.e. $$c = \frac{17 \times 0.6691}{0.5446} = 20.88$$

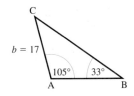

Therefore $AB = 20.9$ cm correct to 3 sf

EXERCISE 3a

1. In $\triangle ABC$, $AB = 9$ cm, $\angle A = 51°$ and $\angle C = 39°$
   Find BC.

2. In $\triangle XYZ$, $\angle X = 27°$, $YZ = 6.5$ cm and $\angle Y = 73°$
   Find ZX.

3. In $\triangle PQR$, $\angle R = 52°$, $\angle Q = 79°$ and $PR = 12.7$ cm.
   Find PQ.

4. In $\triangle$ABC, AC = 9.1 cm, $\angle$A = 59° and $\angle$B = 62°.
   Find BC.

5. In $\triangle$DEF, DE = 174 cm, $\angle$D = 48° and $\angle$F = 56°.
   Find EF.

6. In $\triangle$XYZ, $\angle$X = 130°, $\angle$Y = 21° and XZ = 53 cm.
   Find YZ.

7. In $\triangle$PQR, $\angle$Q = 37°, $\angle$R = 101° and PR = 4.3 cm.
   Find PQ.

8. In $\triangle$ABC, BC = 73 cm, $\angle$A = 54° and $\angle$C = 99°.
   Find AB.

9. In $\triangle$LMN, LN = 637 cm, $\angle$M = 128° and $\angle$N = 46°.
   Find LM.

10. In $\triangle$XYZ, XY = 92 cm, $\angle$X = 59° and $\angle$Y = 81°.
    Find XZ.

11. In $\triangle$PQR, $\angle$Q = 64°, $\angle$R = 38° and PR = 15 cm.
    Find QR.

12. In $\triangle$ABC, AB = 24 cm, $\angle$A = 132° and $\angle$C = 22°.
    Find AC.

13. In $\triangle$XYZ, $\angle$X = 49°, XY = 98 cm and $\angle$Z = 100°.
    Find XZ.

14. In $\triangle$ABC, AB = 10 cm, BC = 9.1 cm and AC = 17 cm.
    Can you use the sine rule to find $\angle$A? If you answer YES, write down the two
    parts of the sine rule that you would use. If you answer NO, give your reason.

The Ambiguous Case

Consider a triangle specified by two sides and one angle.

If the angle is between the two sides there is only one possible triangle,
e.g.

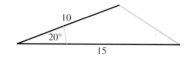

If, however, the angle is not between the two given sides it is sometimes possible to draw two triangles from the given data.

Consider, for example, a triangle ABC in which $\angle A = 20°$, $b = 10$ and $a = 8$

The two triangles with this specification are shown in the diagram; in one of them B is an acute angle, while in the other one, B is obtuse. Note also that the length of the side AB is different in the two cases.

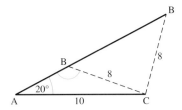

Therefore, when two sides and an angle of a triangle are given, there may be two possible cases, one of which involves an obtuse angle.

The following worked examples illustrate how this situation may arise in *calculations* when two sides and a non-included angle of a triangle are given.

1. In the triangle ABC, find C given that AB = 5 cm, BC = 3 cm and A = 35°.

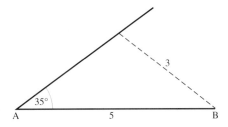

We know a, c and $\angle A$ so the sine rule can be used to find $\angle C$.

As we are looking for an angle, the form we use is

$$\frac{\sin A}{a} = \frac{\sin C}{c} \quad \Rightarrow \quad \frac{\sin 35°}{3} = \frac{\sin C}{5}$$

Hence $\sin C = \dfrac{5 \times 0.5736}{3} = 0.9560$

One angle whose sine is 0.9560 is $73°$ but there is also an obtuse angle with the same sine, i.e. $107°$.

If $\quad C = 107,\quad$ then $\quad A + C = 107 + 35 = 142$

$\Rightarrow\quad B = 180 - 142 = 38$

So in this case $\angle C = 107°$ *is* an acceptable solution and we have two possible triangles.

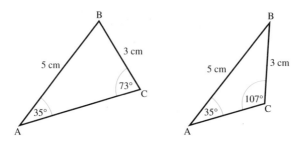

Therefore $\angle C$ is either $73°$ or $107°$.

The reader should *not* assume that there are *always* two possible angles when the sine rule is used to find a second angle in a triangle. The next example shows that this is not so.

Examples 3b (continued)

2. In the triangle XYZ, $\angle Y = 41°$, $XZ = 11$ cm and $YZ = 8$ cm. Find $\angle X$.

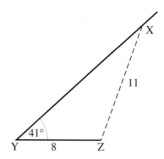

Using the part of the sine rule that involves x, y, $\angle X$ and $\angle Y$ we have

$$\frac{\sin X}{x} = \frac{\sin Y}{y} \quad \Rightarrow \quad \frac{\sin X}{8} = \frac{\sin 41°}{11}$$

Hence $\quad \sin X = \dfrac{8 \times 0.6561}{11} = 0.4771$

The two angles with a sine of 0.4771 are $28°$ and $152°$

Checking to see whether 152° is a possible value for ∠X we see that

$$\angle X + \angle Y = 152° + 41° = 193°$$

This is greater than 180°, so it is not possible for the angle X to have the value 152°.

In this case then, there is only one possible triangle containing the given data, i.e. the triangle in which ∠X = 28°

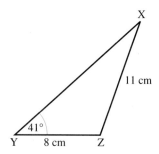

It is interesting to see how the different situations that arose in the two examples above can be illustrated by the construction of the triangles with the given data.

When AB = 5 cm, BC = 3 cm and ∠A = 35° we have

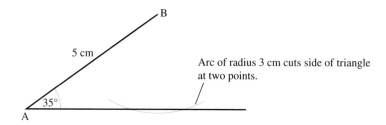

Arc of radius 3 cm cuts side of triangle at two points.

When XZ = 11 cm, YZ = 8 cm and ∠Y = 41° we have

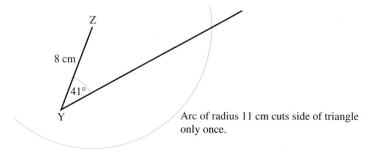

Arc of radius 11 cm cuts side of triangle only once.

In each of the following questions, find the angle indicated by the question mark, giving two values in those cases where there are two possible triangles. Illustrate your solution to each question.

	AB	BC	CA	∠A	∠B	∠C
1.		2.9 cm	6.1 cm	?	40°	
2.	5.7 cm		2.3 cm		20°	?
3.	21 cm	36 cm		29.5°		?
4.		2.7 cm	3.8 cm	?	54°	
5.	4.6 cm		7.1 cm		?	33°
6.	9 cm	7 cm		?		40°

THE COSINE RULE

When solving a triangle, the sine rule cannot be used unless the data given includes one side and the angle opposite to that side. If, for example, a, b and C are given then in the sine rule we have

$$\frac{a}{\sin A} = \frac{b}{\sin B} = \frac{c}{\sin C}$$

and it is clear that no pair of fractions contains only one unknown quantity.

Some other method is therefore needed in such circumstances and the one we use is called the *cosine rule*. This rule states that

$$a^2 = b^2 + c^2 - 2bc \cos A$$

The proof of the cosine rule is given here for readers who are interested.

Proof

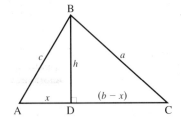

Let ABC be a non-right-angled triangle in which BD is drawn perpendicular to AC. Taking x as the length of AD, the length of CD is $(b - x)$. Then, using h as the length of BD, we can use Pythagoras' theorem to find h in each of the right-angled triangles BDA and BDC, i.e.

$$h^2 = c^2 - x^2 \quad \text{and} \quad h^2 = a^2 - (b - x)^2$$

Therefore $\qquad c^2 - x^2 = a^2 - (b - x)^2$

$\Rightarrow \qquad\qquad c^2 - x^2 = a^2 - b^2 + 2bx - x^2$

$\Rightarrow \qquad\qquad\quad a^2 = b^2 + c^2 - 2bx$

But $\qquad x = c \cos A$

Therefore $\qquad a^2 = b^2 + c^2 - 2bc \cos A$

The proof is equally valid for an obtuse-angled triangle, as is shown below.

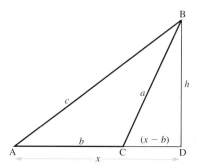

In this case the length of CD is $(x - b)$, so in $\triangle$BCD we have

$$h^2 = a^2 - (x - b)^2 = a^2 - x^2 + 2bx - b^2$$

This is identical to the expression found for h^2 above.

The remainder of the proof above is unchanged so we have now proved that, in *any* triangle,

$$a^2 = b^2 + c^2 - 2bc \cos A$$

When the altitude is drawn from A or from C similar expressions for the other sides of a triangle are obtained, i.e.

$$b^2 = c^2 + a^2 - 2ca \cos B$$

and

$$c^2 = a^2 + b^2 - 2ab \cos C$$

Examples 3c

1. In △ABC, BC = 7 cm, AC = 9 cm and C = 61°. Find AB.

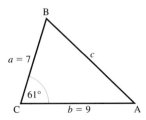

Using the cosine rule, starting with c^2, we have

$$c^2 = a^2 + b^2 - 2ab \cos C$$

$$\Rightarrow \quad c^2 = 7^2 + 9^2 - (2)(7)(9)(0.4848)$$

$$\Rightarrow \quad c = 8.302$$

Hence AB = 8.30 cm correct to 3 sf

2. XYZ is a triangle in which ∠Y = 121°, XY = 14 cm and YZ = 26.9 cm. Find XZ.

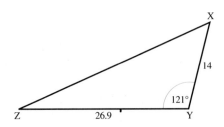

Using $y^2 = z^2 + x^2 - 2zx \cos Y$ gives

$$y^2 = (14)^2 + (26.9)^2 - (2)(14)(26.9)(-0.5150)$$

Note that, because Y is an obtuse angle, it has a negative cosine. Extra care therefore has to be taken with the sign of the term $-2zx \cos Y$. The best way to avoid mistakes is to enclose the cosine in brackets as shown.

Hence $y^2 = 1307.51 \quad \Rightarrow \quad y = 36.16$

Therefore XZ = 36.2 cm correct to 3 sf.

In each question use the data given for △PQR to find the length of the third side.

	PQ	QR	RP	P	Q	R
1.		8 cm	4.6 cm			39°
2.	11.7 cm		9.2 cm	75°		
3.	29 cm	37 cm			109°	
4.		2.1 cm	3.2 cm			97°
5.	135 cm		98 cm	48°		
6.	4.7 cm	8.1 cm			138°	
7.		44 cm	62 cm			72°
8.	19.4 cm		12.6 cm	167°		

Using the Cosine Rule to Find an Angle

So far the cosine rule has been used only to find an unknown side of a triangle. When we want to find an unknown angle, it is advisable to rearrange the formula to some extent.

The version of the cosine rule that starts with c^2,

i.e.
$$c^2 = a^2 + b^2 - 2ab \cos C$$

can be written as
$$2ab \cos C = a^2 + b^2 - c^2$$

and further as
$$\cos C = \frac{a^2 + b^2 - c^2}{2ab}$$

The reader should find this last form quite easy to remember if it is noted that the side opposite to the angle being found, c^2 in this case, appears only once as the last term in the formula. Some readers however may prefer to work from the basic cosine formula for all calculations, carrying out any necessary manipulation in each problem as it arises.

Examples 3d

1. If, in $\triangle ABC$, $a = 9$, $b = 16$ and $c = 11$, find, to the nearest degree, the largest angle in the triangle

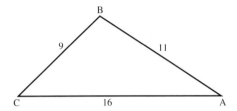

The largest angle in a triangle is opposite to the longest side, so in this question we are looking for angle B and we use

$$\cos B = \frac{c^2 + a^2 - b^2}{2ca}$$

$$= \frac{121 + 81 - 256}{(2)(11)(9)} = -0.2727$$

The negative sign shows that $\angle B$ is obtuse.

Hence $B = 106°$ and this is the largest angle in $\triangle ABC$.

2. The sides a, b, c of a triangle ABC are in the ratio $3 : 6 : 5$. Find the smallest angle in the triangle.

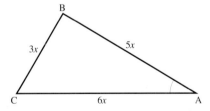

The actual lengths of the sides are not necessarily 3, 6 and 5 units so we represent them by $3x$, $6x$ and $5x$. The smallest angle is A (opposite to the smallest side).

$$\cos A = \frac{b^2 + c^2 - a^2}{2bc}$$

$$= \frac{36x^2 + 25x^2 - 9x^2}{60x^2} = \frac{52}{60} = 0.8667$$

Therefore the smallest angle in $\triangle ABC$ is $30°$.

1. In $\triangle XYZ$, $XY = 34$ cm, $YZ = 29$ cm and $ZX = 21$ cm.
   Find the smallest angle in the triangle.

2. In $\triangle PQR$, $PQ = 1.3$ cm, $QR = 1.8$ cm and $RP = 1.5$ cm.
   Find $\angle Q$.

3. In $\triangle ABC$, $AB = 51$ cm, $BC = 37$ cm and $CA = 44$ cm.
   Find $\angle A$.

4. Find the largest angle in $\triangle XYZ$ given that $x = 91$, $y = 77$ and $z = 43$.

5. What is the size of   (a) the smallest,   (b) the largest angle
   in $\triangle ABC$ if $a = 13$, $b = 18$ and $c = 7$?

6. In $\triangle PQR$ the sides PQ, QR and RP are in the ratio $2:1:2$
   Find $\angle P$.

7. ABCD is a quadrilateral in which $AB = 5$ cm, $BC = 8$ cm, $CD = 11$ cm,
   $DA = 9$ cm and angle $ABC = 120°$. Find the length of AC and the size of the
   angle ADC.

GENERAL TRIANGLE CALCULATIONS

If three independent facts are given about the sides and/or angles of a triangle
and further facts are required, a choice must be made between using the sine
rule or the cosine rule for the first step.

As the sine rule is easier to work out, it is preferred to the cosine rule whenever
the given facts make this possible, i.e. whenever an angle and the opposite side are
known.   (Remember that if two angles are given, then the third angle is also
known.)

The cosine rule is used only when the sine rule is not suitable and it is never
necessary to use it more than once in solving a triangle.

Suppose, for example, that the triangle PQR given below is to be solved.

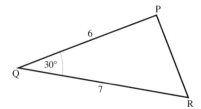

Only one angle is known and the side opposite to it is not given. We must therefore use the cosine rule first to find the length of PR.

Once we know q as well as $\angle Q$, the sine rule can be used to find either of the remaining angles, the third angle then following from the sum of the angles in the triangle.

EXERCISE 3e

Each of the following questions refers to a triangle ABC. Fill in the blank spaces in the table.

	∠A	∠B	∠C	a	b	c
1.		80°	50°			68 cm
2.			112°	15.7 cm	13 cm	
3.	41°	69°		12.3 cm		
4.	58°				131 cm	87 cm
5.		49°	94°		206 cm	
6.	115°		31°			21 cm
7.	59°	78°		17 cm		
8.		48°	80°		31.3 cm	
9.	77°				19 cm	24 cm
10.		125°		14 cm		20 cm

11. A tower stands on level ground. From a point P on the ground, the angle of elevation of the top of the tower is 26°. Another point Q is 3 m vertically above P and from this point the angle of elevation of the top of the tower is 21°. Find the height of the tower.

12. A survey of a triangular field, bounded by straight fences, found the three sides to be of lengths 100 m, 80 m and 65 m. Find the angles between the boundary fences.

MIXED EXERCISE 3

1. Find the value, between $0°$ and $180°$, of $\angle A$ if

   (a) $\cos A = -\cos 64°$ (b) $\sin 94° = \sin A$

2. If $\angle X$ is acute and $\sin X = \frac{7}{25}$, find $\cos(180° - X)$

3. Given that $\sin A = \frac{5}{8}$, find $\tan A$ in surd form if

   (a) $\angle A$ is acute   (b) $\angle A$ is obtuse

4. Find, in surd form, $\sin \theta$ and $\cos \theta$, given

   (a)   (b)

   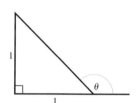 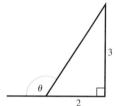

5. Given that $\sin X = \frac{12}{13}$ and X is obtuse, find $\cos X$.

6. In $\triangle ABC$, $BC = 11$ cm, $\angle B = 53°$ and $\angle A = 76°$, find AC.

7. In $\triangle PQR$, $p = 3$, $q = 5$ and $R = 69°$, find r.

8. In $\triangle XYZ$, $XY = 8$ cm, $YZ = 7$ cm and $ZX = 10$ cm, find $\angle Y$.

9. In $\triangle ABC$, $AB = 7$ cm, $BC = 6$ cm and $\angle A = 44°$, find all possible values of $\angle ACB$.

10. Find the angles of a triangle whose sides are in the ratio $2:4:5$.

11. Use the cosine formula, $\cos A = \dfrac{b^2 + c^2 - a^2}{2bc}$, to show that

   (a) $\angle A$ is acute if $a^2 < b^2 + c^2$
   (b) $\angle A$ is obtuse if $a^2 > b^2 + c^2$.

CHAPTER 4

TRIANGLES

THE AREA OF A TRIANGLE

The simplest way to find the area of a triangle is to use the formula

$$\text{Area} = \tfrac{1}{2}\,\text{base} \times \text{perpendicular height}$$

Clearly this is of immediate use only when the perpendicular height is known. It can be adapted, however, to cover other cases.

Consider the triangle shown below, in which b, c, and $\angle A$ are known.

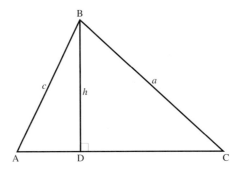

The line BD, drawn from B perpendicular to AC, is the height, h, of the triangle, so the area of the triangle is $\tfrac{1}{2}bh$.

In the triangle ADB, $\qquad \sin A = \dfrac{h}{c} \quad \Rightarrow \quad h = c \sin A$

Therefore the area of triangle ABC is

$$\tfrac{1}{2}bc \, \sin A$$

Drawing the perpendicular heights from A to C give similar expressions, i.e.

$$\text{Area of triangle ABC} = \tfrac{1}{2}ab \sin C = \tfrac{1}{2}ac \sin B$$

Each of these formulae can be expressed in the 'easy to remember' form

Area $= \frac{1}{2}$ product of two sides $\times$ sine of included angle

Find the area of triangle PQR,
given that $P = 65°$, $Q = 79°$
and $PQ = 30$ cm.

The given facts do not include two sides and the included angle so we must first find another side. To do this the sine rule can be used and we need angle R.

$$\angle R = 180° - 65° - 79° = 36°$$

From the sine rule, $\qquad \dfrac{p}{\sin P} = \dfrac{r}{\sin R}$

$\Rightarrow \qquad\qquad\qquad p = \dfrac{30 \times \sin 65°}{\sin 36°} = 46.26$

i.e. $\qquad$ QR $= 46.3$ cm (correct to 3 sf).

Now we can use area PQR $= \frac{1}{2} pr \sin Q$

$\qquad \frac{1}{2} pr \sin Q = \frac{1}{2} \times 46.26 \times 30 \times \sin 79 = 681.2$

So the area of triangle PQR is 681 cm^2 (correct to 3 sf).

Find the area of each triangle given in Questions 1 to 5.

1. $\triangle XYZ$; $XY = 180$ cm, $YZ = 145$ cm, $\angle Y = 70°$

2. $\triangle ABC$; $AB = 75$ cm, $AC = 66$ cm, $\angle A = 62°$

3. $\triangle PQR$; $QR = 69$ cm, $PR = 49$ cm, $\angle R = 85°$

4. $\triangle XYZ$; $x = 30$, $y = 40$, $\angle Z = 49°$

5. $\triangle PQR$; $p = 9$, $r = 11$, $\angle Q = 120°$

6. In triangle ABC, $AB = 6$ cm, $BC = 7$ cm and $CA = 9$ cm.
   Find $\angle A$ and the area of the triangle.

7. $\triangle PQR$ is such that $\angle P = 60°$, $\angle R = 50°$ and $QR = 12$ cm.
   Find PQ and the area of the triangle.

8. In $\triangle XYZ$, $XY = 150$ cm, $YZ = 185$ cm and the area is $11\,000$ cm^2.
   Find $\angle Y$ and XZ.

9. The area of triangle ABC is 36.4 cm^2. Given that $AC = 14$ cm and
   $\angle A = 98°$, find AB.

PROBLEMS

Many practical problems which involve distances and angles can be illustrated by a diagram. Often, however, this diagram contains too many lines, dimensions, etc. to be clear enough to work from. In these cases we can draw a second figure by extracting a triangle (or triangles) in which three facts about sides and/or angles are known. The various methods given in Chapter 3 can then be used to analyse this triangle and so to solve the problem.

Examples 4b

1. Two boats, P and Q, are 300 m apart. The base, A, of a lighthouse is in line with PQ. From the top, B, of the lighthouse the angles of depression of P and Q are found to be 35° and 48°. Write down the values of the angles BQA, PBQ and BPQ and find, correct to the nearest metre, the height of the lighthouse.

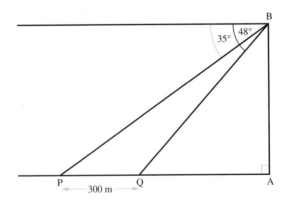

$\angle$ BQA $= 48°$ (alternate angles)

$\angle$ PBQ $= 48° - 35° = 13°$

$\angle$ BPQ $= 35°$ (alternate angles)

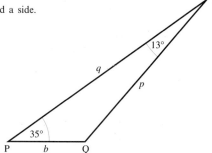

Now we can extract $\triangle$PBQ, knowing two angles and a side.

From the sine rule,

$$\frac{p}{\sin P} = \frac{b}{\sin B}$$

$\therefore$
$$p = \frac{(300)(\sin 35°)}{\sin 13°}$$

$$= 764.9$$

We can now use the right-angled triangle ABQ

$$h = p \sin 48°$$

$$= (764.9)(\sin 48°)$$

$$= 568.4$$

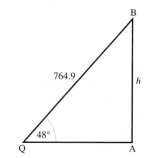

Correct to the nearest metre the height of the light-house is 568 m.

2. A traveller pitches camp in a desert. He knows that there is an oasis in the distance, but cannot see it. Wishing to know how far away it is, he measures 250 m due north from his starting point, A, to a point B where he can see the oasis, O, and finds that its bearing is 276°. He then measures a further 250 m due north to point C from which the bearing of the oasis is 260°. Find how far from the oasis he has camped.

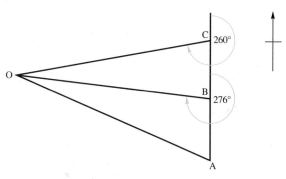

$\angle$OCB $= 260° - 180° = 80°$ and $\angle$OBC $= 360° - 276° = 84°$

As two angles and a side are known, $\triangle OBC$ can be used.

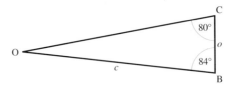

$$\angle BOC = 180° - 80° - 84° = 16°$$

From the sine rule, $\quad \dfrac{c}{\sin\ C} = \dfrac{o}{\sin\ O}$

$\Rightarrow \qquad\qquad\qquad c = \dfrac{250 \times \sin\ 80°}{\sin\ 16°} = 893.2$

Now in $\triangle ABO$, $\angle ABO = 276° - 180° = 96°$ and we also know OB and AB. As two sides and the included angle are known, it is the cosine rule that must be used.

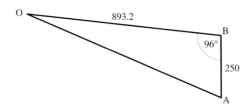

$$OA^2 = OB^2 + AB^2 - 2 \times OB \times AB \times \cos\ ABO$$

$$= (893.2)^2 + (250)^2 - 2 \times 893.2 \times 250 \times \cos\ 96°$$

$$= 906\,989$$

$\Rightarrow \qquad OA = 952.4$

To the nearest metre the initial distance from the oasis was 952 m.

EXERCISE 4b

1. In a quadrilateral PQRS, PQ = 6 cm, QR = 7 cm, RS = 9 cm, $\angle PQR = 115°$ and $\angle PRS = 80°$. Find the length of PR. Considering it as split into two separate triangles, find the area of the quadrilateral PQRS.

2. A light aircraft flies from an airfield, A, a distance of 50 km on a bearing of 049° to a town, B. The pilot then changes course and flies on a bearing of 172° to a landing strip, C, 68 km from B. How far is the landing strip from the airfield?

3. In a surveying exercise, P and Q are two points on land which is inaccessible. To find the distance PQ, a line AB of length 300 metres is marked out so that P and Q are on opposite sides of AB. The directions of P and Q relative to the line AB are then measured and are shown in the diagram. Calculate the length of PQ. (Hint. Find AP and AQ.)

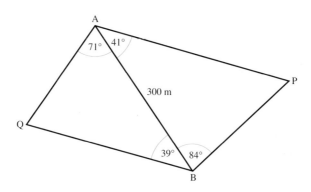

4.

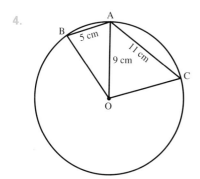

AB, of length 5 cm, and AC, of length 11 cm, are two chords of a circle with centre O and radius 9 cm. Find each of the angles BAO and CAO and hence calculate the area of the triangle ABC.

5. The diagram shows the cross section of a beam of length 2 m.

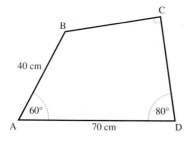

Calculate

(a) the length of BD

(b) the angle ADB

(c) the length of CD

(d) the area of the cross section

(e) the volume of the beam.

THREE-DIMENSIONAL PROBLEMS

One of the difficulties which many people experience with this topic, arises when attempting to illustrate a three-dimensional situation on a two-dimensional diagram. The following hints may help in producing a clear representation of the 3-D problem from which appropriate calculations can be made.

- Vertical lines should be drawn vertically on the page.

- Lines in the East–West direction should be drawn horizontally on the page. North–South lines are shown as inclined at an acute angle to the East direction.

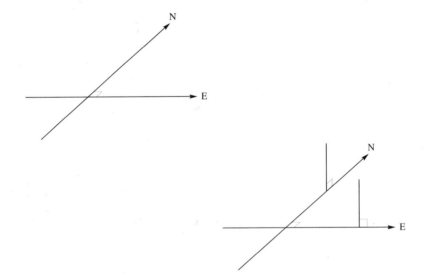

- All angles that are 90° in three dimensions should be marked as right angles on the diagram, particularly those that do not *appear* to be 90°.

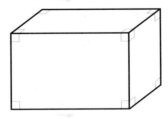

- Perspective drawing is rarely used, so parallel lines are drawn parallel in the diagram.

● When viewing a 3-D object, some of its sides are usually not visible. It is helpful to indicate these by broken lines.

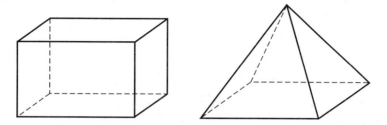

● In a situation involving two points in the foreground and an object in the background it is usually clearer to draw the object *between* the two points.

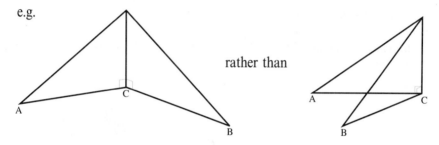

● It is often helpful to draw a separate diagram showing each individual triangle in which calculations are needed.

The following facts and definitions should also be known.

● Two non-parallel planes meet in a line called the common line.

● A line that is perpendicular to a plane is also perpendicular to every line in that plane

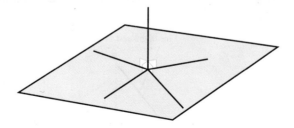

and if a line is perpendicular to two non-parallel lines in a plane, then it is perpendicular to the plane.

● The angle between a line and a plane is defined as the angle between that line and its projection on the plane. (Its projection can be thought of as the shadow of the line cast on the plane by a beam of light shining at right-angles to the plane.)

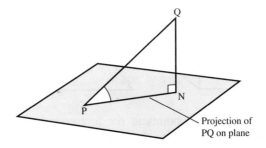

Projection of
PQ on plane

● The angle between two planes is defined as follows. From any point A, on the common line of two planes P_1 and P_2, lines AB and AC are drawn, one in each plane, perpendicular to the common line. Then angle BAC is the angle between the two planes.

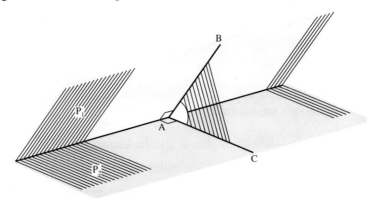

● If one of the planes, P_2 say, is horizontal, then AB is called *a line of greatest slope* of plane P_1.

Line of greatest slope

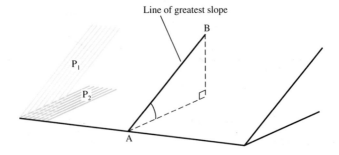

1. The diagram shows a cube of side 6 cm. M is the midpoint of AB. Find

(a) the length of MC

(b) the length of MR

(c) to the nearest degree, the angle between MR and the plane ABCD.

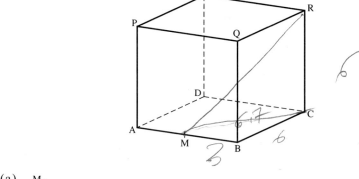

(a)

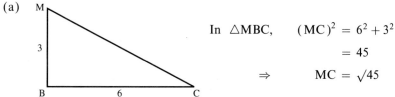

In $\triangle$MBC, $(MC)^2 = 6^2 + 3^2$

$= 45$

$\Rightarrow \qquad MC = \sqrt{45}$

Therefore the length of MC is 6.71 cm (correct to 3 sf).

(b)

In $\triangle$MCR, $(MR)^2 = (MC)^2 + 6^2$

$= 45 + 36$

$= 81$

$\Rightarrow \qquad MR = 9$

Therefore the length of MR is 9 cm.

To find the angle between RM and the plane ABCD, we need the projection of RM on the plane. As RC is perpendicular to ABCD, it follows that CM is the projection of RM on that plane. So the angle we are looking for is RMC.

(c)

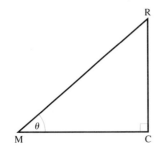

In $\triangle$RMC, $\sin \theta = \frac{6}{9} = 0.6667$

$\Rightarrow$ $\theta = 41.8°$

Therefore, to the nearest degree, the angle between RM and the plane ABCD is 42°.

2. The base AB of an isosceles triangle ABC is horizontal. The plane containing the triangle is inclined to the horizontal at 54°. If the angle ACB is 48°, find the angle between AC and the horizontal plane.

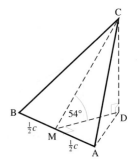

AB is the line common to the horizontal plane and the plane of the triangle, and M is its midpoint. Because the triangle is isosceles, CM is perpendicular to AB. So CM is a line of greatest slope and therefore makes an angle of 54° with its projection, MD, on the horizontal plane.

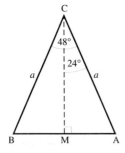

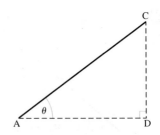

Let the length of AC and BC be a

In $\triangle$CMA,   CM $= a \cos 24°$

In $\triangle$CDM,   CD $=$ CM $\sin 54°$

$= (a \cos 24°)(\sin 54°)$

The angle between AC and the horizontal plane is the angle between AC and its projection AD on that plane. If this angle is θ then

$$\sin\theta = \frac{CD}{CA} = \frac{(a)(\cos 24°)(\sin 54°)}{a}$$

$\Rightarrow \qquad \theta = 47.7°$

EXERCISE 4c

1.

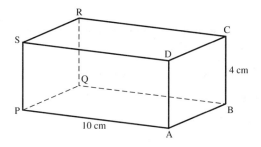

In the cuboid shown above, ABCD is a square of side 4 cm and PA = 10 cm. Find the length of

(a) AC   (b) AS   (c) AQ   (d) a diagonal of the cuboid.

2.

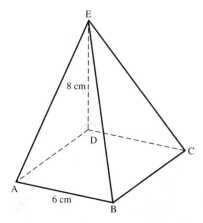

The pyramid ABCDE has a square base of side 6 cm. E is 8 cm vertically above D. Calculate

(a) the lengths of AD, BD and BE

(b) the angle between AE and the plane ABCD

(c) the angle between BE and the plane ABCD

(d) the angle between the planes EBC and ABCD.

3.

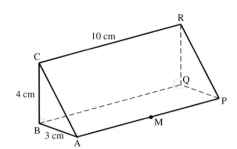

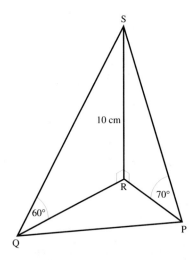

Given the triangular prism in the diagram, in which M is the midpoint of AP, find the following lengths and angles.

(a) RM   (b) RA   (c) QA

(d) the angle between RA and the plane ABQP

(e) the angle between RM and the plane ABQP.

4. Given a regular tetrahedron (i.e. a pyramid where each face is an equilateral triangle), find the cosine of the angle between two of the faces.

5.

Three points, P, Q and R lie in a plane. The line RS is perpendicular to the plane and is of length 10 cm. If angle SPR = 70°, angle SQR = 60° and PQ = 7 cm, calculate each of the angles in triangle PQR.

6. Find the angle between two diagonals of a cube.

7.

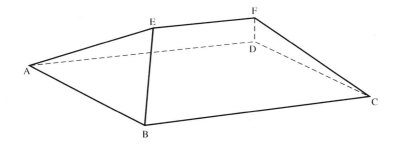

The diagram shows a roof whose base is a rectangle of length 15 m and width 9 m. Each end face is an isosceles triangle inclined to the horizontal at an angle α, and each long face is a trapezium inclined to the horizontal at an angle β. If $\tan \alpha = 2$ and $\tan \beta = \frac{4}{3}$, calculate

(a) the height of the ridge (EF) above the base

(b) the length of the ridge

(c) the angle between AE and the horizontal

(d) the total surface area of the roof.

8.

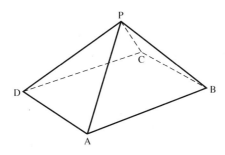

The diagram shows a solid figure in which ABCD is a horizontal rectangle. AB = 13 cm, BC = 8 cm, AP = DP = 9 cm and BP = CP = 7 cm. Calculate

(a) the length of AC

(b) the height of P above the plane ABCD

(c) the angle between AP and the horizontal

(d) the angle between the faces APB and ABCD.

9. An aircraft is noted simultaneously by three observers, A, B and C, stationed in a horizontal straight line. AB and BC are each 200 m and the noted angles of elevation of the aircraft from A and C are 25° and 40° respectively. What is the height of the aircraft? Find also the angle of elevation of the aircraft from B.

10.

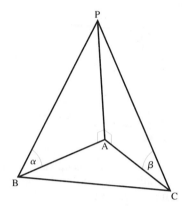

ABC is a horizontal triangle in which BC = 10 m. P is a point 12 m vertically above A. The angles of elevation of P from B and C are α and β, where tan $\alpha = 1$ and tan $\beta = \frac{6}{7}$. Find the angle between the planes PBC and ABC.

11. The coordinates of the points A, B and C are (2, 1, 5), (−3, 4, 0) and (3, −1, −2) respectively. Find (a) angle ABC (b) the area of △ABC.

Harder Problems

Certain problems in three dimensions are rather more demanding than those seen up to now. An example of such a problem is given below.

Example 4d

A, B and C are points on a horizontal line such that AB = 60 m and BC = 30 m. The angles of elevation, from A, B and C respectively, of the top of a clock tower are α, β, and γ, where tan $\alpha = \frac{1}{13}$, tan $\beta = \frac{1}{15}$ and tan $\gamma = \frac{1}{20}$. The foot of the tower is at the same level as A, B and C. Find the height of the tower.

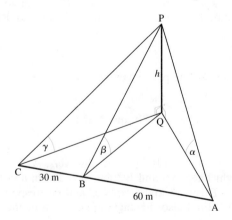

If the height of the tower, PQ, is h then

$$h = AQ \tan \alpha = QB \tan \beta = QC \tan \gamma$$

i.e. $\qquad h = \dfrac{QA}{13} \quad = \dfrac{QB}{15} \quad = \dfrac{QC}{20}$

$\Rightarrow \qquad QA = 13h, \quad QB = 15h, \quad QC = 20h$

Now considering the base triangle ABCQ, we have

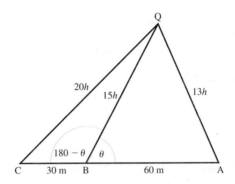

Using the cosine rule in $\triangle ABQ$ gives

$$\cos \theta = \frac{(60)^2 + (15h)^2 - (13h)^2}{2(60)(15h)}$$

and using the cosine rule in $\triangle CBQ$ gives

$$\cos(180° - \theta) = -\cos \theta = \frac{(30)^2 + (15h)^2 - (20h)^2}{2(30)(15h)}$$

$\therefore \qquad \dfrac{(60)^2 + (15h)^2 - (13h)^2}{2(60)(15h)} = \dfrac{(20h)^2 - (30)^2 - (15h)^2}{2(30)(15h)}$

$\Rightarrow \qquad\qquad 3600 + 56h^2 = 2(175h^2 - 900)$

$\Rightarrow \qquad\qquad\qquad 5400 = 294h^2$

Hence $\qquad\qquad\qquad\quad h = 4.285$

The height of the clock tower is 4.29 m (correct to 3 sf).

MIXED EXERCISE 4 $\quad$ 1, 3, 5, 7.

1. In $\triangle PQR$, PQ $= 11$ cm, PR $= 14$ cm and QPR $= 100°$. Find the area of the triangle.

2. The area of ABC is 9 cm². If AB $=$ AC $= 6$ cm, find sin A. Are there two possible triangles? Give a reason for your answer.

3.

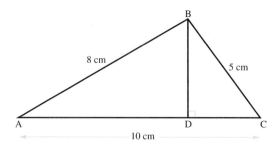

Given the information in the diagram,

(a)   find ∠ABC

(b)   find the area of △ABC

(c)   *hence* find the length of BD.

4. Triangle **PQR**, in which **PRQ** = 120°, lies in a horizontal plane and X is a point 6 cm vertically above R. If **XQR** = 45° and **XPR** = 60°, find the lengths of the sides of △PQR. Find also the area of this triangle.

5.

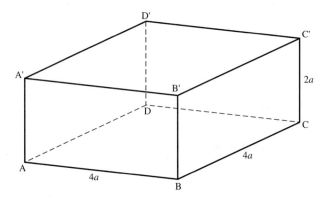

Given the cuboid shown in the diagram, find

(a)   the angle between AC and the plane ABB′A′

(b)   the angle between the planes ACD′ and ABCD.

6. Two rectangular panels, **ABCD** and **ABEF**, each measure 1.5 m by 2 m. They are hinged along the edge AB which is 2 m long. If the angle between their planes is 60°, find the angle between the diagonals AC and AE.

7. A river running due east has straight parallel banks. A vertical post stands with its base, P, on the north bank of the river. On the south bank are two surveyors, A who is to the east, and B who is to the west of the post. A and B are at a distance $\frac{2}{7}a$ apart and the angle APB is 150°. The angles of elevation from A and B of the top, Q, of the post are 45° and 30°. Find, in terms of a, the width of the river and the height of the post.

*8. ABCD is a tetrahedron whose horizontal base ABC is an equilateral triangle. The angle between each pair of slant edges is θ where $\tan \theta = \frac{5}{12}$ and the length of these edges is a. Find the height of D above ABC.

*9. Two lamp standards are each of height h m. They are d m apart on level ground where a man of height t m is also standing. Each light casts a shadow of the man on the ground. Prove that, no matter where the man stands, the distance between the ends of his two shadows is $\dfrac{dt}{h-t}$ m. (Hint. Look for similar triangles.)

*10. The roof of a south-facing house slopes down at an angle α to the horizontal. A gulley at the end of the roof is in the direction θ east of north. If the gulley is inclined to the horizontal at an angle β, show that $\tan \beta = \tan \alpha \cos \theta$.

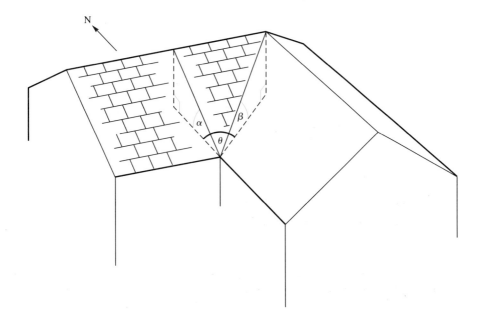

*11.

The distances and angles of elevation of A and B were measured from P.
A is 3550 m from P at an angle of elevation of 50°.
B is 2236 m from P at an angle of elevation of 28°.
Assuming that P is at sea level, find

(a)   the height of A above sea level,

(b)   the height of B above sea level,

(c)   the horizontal distance of A from B.

CONSOLIDATION A

SUMMARY

The Remainder Theorem

When a polynomial, $f(x)$, is divided by $(x-a)$, the remainder is equal to $f(a)$.

The Factor Theorem

If $f(a) = 0$ then $(x-a)$ is a factor of $f(x)$.

Coordinate Geometry

The coordinates of the midpoint of the line joining (x_1, y_1) to (x_2, y_2) are

$$(\tfrac{1}{2}\{x_1 + x_2\}, \tfrac{1}{2}\{y_1 + y_2\})$$

Parallel lines have equal gradients.

When two lines are perpendicular, the product of their gradients is -1.

The distance between two points (x_1, y_1, z_1) and (x_2, y_2, z_2) in three dimensions is $\sqrt{(x_1 - x_2)^2 + (y_1 - y_2)^2 + (z_1 - z_2)^2}$

Trigonometry

$$\sin\theta = \sin(180° - \theta)$$
$$\cos\theta = -\cos(180° - \theta)$$
$$\tan\theta = -\tan(180° - \theta)$$

In any triangle ABC,

$$\frac{a}{\sin A} = \frac{b}{\sin B} = \frac{c}{\sin C} \qquad (\text{sine formula})$$

$$\left.\begin{array}{l} a^2 = b^2 + c^2 - 2bc\cos A \\ b^2 = a^2 + c^2 - 2ac\cos B \\ c^2 = a^2 + b^2 - 2ab\cos C \end{array}\right\} \quad (\text{cosine formula})$$

The area of $\triangle ABC$ is $\tfrac{1}{2}ab\sin C$ or $\tfrac{1}{2}bc\sin A$ or $\tfrac{1}{2}ac\sin B$

TYPE I

1. The midpoint of the line joining $(3, -3)$ to $(3, -7)$ is

 A $(0, 5)$ **B** $(3, -5)$ **C** $(0, -5)$ **D** $(3, 5)$ **E** $(6, -10)$

2. The value of k for which $x - 1$ a factor of $4x^3 - 3x^2 - kx + 2$ is

 A -1 **B** 0 **C** 1 **D** 2 **E** 3

3. ABCDE is a right square-based pyramid (i.e. E is vertically above the centre of the base). The angle between the face BEC and the base ABCD is

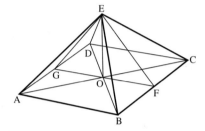

 A F **B** B **C** C **D** O **E** D

4. The equation of the line through the origin and parallel to $4x - y + 5 = 0$ is

 A $4x - y = 0$ **B** $4y - x + 5 = 0$ **C** $y = 4x - 5$
 D $4y - x = 0$ **E** $x - 4y + 1 = 0$

5. $x^3 - 3x^2 + 2x - 6$ has a factor

 A $x - 3$ **B** $x - 2$ **C** $x - 4$ **D** $x + 3$ **E** $x + 2$

6. In $\triangle ABC$

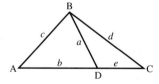

 A $\sin A = \dfrac{a \sin B}{b}$ **B** $\sin B = \dfrac{b \sin C}{c}$ **C** $\sin C = \dfrac{c \sin A}{a}$

 D $\sin B = \dfrac{(b+e) \sin A}{a}$ **E** $\sin A = \dfrac{d \sin C}{c}$

7. $x^3 - 3x^2 + 6x - 2$ has remainder 2 when divided by

 A $x - 1$ **B** $x + 1$ **C** x **D** $x + 2$ **E** $2x - 1$

8. The lines $y = 2x - 4$ and $3x - y = 8$

A are perpendicular   B both pass through the origin
C intersect at $(8, -4)$ D are parallel
E form two sides of a triangle

TYPE II

9. A root of the equation $x^2 + 2x - 1 = 0$ can be found from the iteration $x_{n+1} = \frac{1}{2}(1 - x_n^2)$ starting with $x_0 = 1$

10. The equation of a line l is $y = 2x - 1$

The line through the origin perpendicular to l is $y + 2x = 0$

11. In triangle ABC

$$\frac{a}{\sin A} = \frac{b}{\sin \gamma}$$

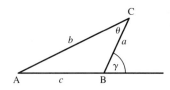

12. If in triangle ABC, angle A is $30°$ then $\sin B = b/2a$

MISCELLANEOUS EXERCISE A $4, 7, 11, 14, 17$

1.

In the triangle ABC, $AC = 3$ cm, $BC = 2$ cm, $\angle BAC = \theta$ and $\angle ABC = 2\theta$. Calculate the value of θ correct to the nearest tenth of a degree.

Hence find the size of the angle ACB and, without further calculation, explain why the length of AB is greater than 2 cm.   (NEAB)

2. The points P, Q and R have coordinates $(2, 4)$, $(7, -2)$ and $(6, 2)$ respectively. Find the equation of the straight line l which is perpendicular to the line PQ and which passes through the midpoint of PR.   (AEB)$_s$

3. Given that $(x - 2)$ and $(x + 2)$ are each factors of $x^3 + ax^3 + bx - 4$, find the values of a and b.
For these values of a and b, find the other linear factor of $x^3 + ax^3 + bx - 4$.   (UCLES)$_s$

4. Given that $f(x) \equiv 3 + 4x - x^4$, show that the equation $f(x) = 0$ has a root $x = a$, where a is in the interval $1 \leqslant a \leqslant 2$.

It may be assumed that if x_n is an approximation to a, then a better approximation is given by x_{n+1}, where

$$x_{n+1} = (3 + 4x_n)^{1/4}$$

Starting with $x_0 = 1.75$, use this result twice to obtain the value of a to 2 decimal places.   (ULEAC)

5.

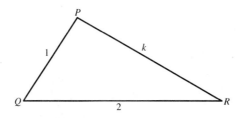

The diagram shows triangle PQR, in which $PQ = 1$ unit, $QR = 2$ units and $RP = k$ units. Express $\cos R$ in terms of k.

Given that $\cos R < \frac{7}{8}$, show that $2k^2 - 7k + 6 < 0$. Find the set of values of k satisfying this inequality.   (UCLES)

6. Show that $(x - 2)$ is a factor of $x^3 - 9x^2 + 26x - 24$.

Find the set of values of x for which $x^3 - 9x^2 + 26x - 24 < 0$.   (AEB)$_s$

7. (a) Show that the equation $x^3 + 3x^2 - 7 = 0$ may be rearranged into the form
$$x = \sqrt{\left(\frac{a}{x+b}\right)}, \text{ and state the values of } a \text{ and } b.$$

   (b) Hence, using the iteration formula
$$x_{n+1} = \sqrt{\left(\frac{a}{x_n + b}\right)}$$

   with $x_0 = 1$, together with your values of a and b, find the approximate solution x_4 of the equation, giving your answer to an appropriate degree of accuracy. Show your intermediate answers for x_1, x_2 and x_3 clearly in your solution.   (ULEAC)$_s$

8. The points $O(0, 0)$, $A(3, 4)$ and $B(8, 4)$ are three vertices of a rhombus OABC.

   (a) Find the coordinates of C.

   (b) Show that the diagonals OB and AC bisect each other at right-angles.

9.

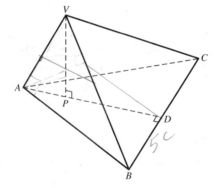

A walker at point P can see the spires of St Mary's (M) and St George's (G) on the bearings of 340° and 015° respectively. She then walks 500 m due North to a point Q and notes the bearings as 315° to St Mary's and 030° to St George's.

Taking the measurements as exact and assuming they are all made in the same horizontal plane,

(a) copy the diagram and fill in the information given,

(b) calculate the distance PM to the nearest metre,

(c) Given that the distance PG is 966 m, calculate MG, the distance between the spires, to the nearest metre. (MEI)ₛ

10. The polynomial $x^3 + ax + b$ has $x - 1$ and $x - 3$ as two of its factors. Use this fact to write down two equations for a and b, and solve them. Hence find the third factor. (OCSEB)ₛ

11.

The horizontal base of tetrahedron $VABC$ is an equilateral triangle of side 10 cm. The vertex V is such that $VA = 8$ cm and $VB = VC = 10$ cm. The mid-point of BC is D and the point P is on the line AD vertically below V as shown in the diagram.

Calculate, giving your answers to 1 decimal place

(a) the angle between VA and the horizontal,

(b) the height of V above the plane ABC,

(c) the angle between VB and the horizontal,

(d) the angle between the planes VBC and ABC. (ULEAC)ₛ

12. Show that both $(x - \sqrt{3})$ and $(x + \sqrt{3})$ are factors of $x^4 + x^3 - x^2 - 3x - 6$.

    Hence write down on quadratic factor of $x^4 + x^3 - x^2 - 3x - 6$, and find a second quadratic factor of this polynomial. (UCLES)

13.

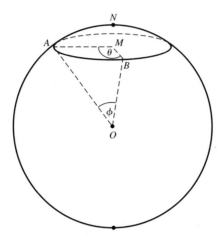

The diagram represents the Earth modelled as a sphere with centre O and radius $2a$. The points A and B are situated on the Earth's surface and are on the same circle of latitude, which has centre M and radius a. The angle $AMB = \theta$ and the angle $AOB = \phi$. Using the cosine rule, or otherwise, prove that

$$4 \cos \phi = 3 + \cos \theta.$$

Given that $a = 3200$ km and $\theta = 130°$, calculate, to three significant figures,

(a) the arc length AB of the circle with centre M and radius a,

(b) the arc length AB of the circle with centre O and radius $2a$. (NEAB)

14. $f(x) \equiv x^3 + 2x^2 - 11x - 12$.

(a) Show that $(x + 1)$ is a factor of $f(x)$.

(b) Solve the equation $f(x) = 0$. (ULEAC)

15. The cubic function f is given by

$$f(x) = x^3 + ax^2 - 4x + b$$

where a, b are constants.

Given that $(x - 2)$ is a factor of $f(x)$ and that a remainder of 6 is obtained when $f(x)$ is divided by $(x + 1)$, find the values of a and b. (WJEC)

16. The sequence given by the iteration formula

$$x_{n+1} = 2(1 + e^{-x_n}),$$

with $x_1 = 0$, converges to a. Find a correct to 3 decimal places, and state an equation of which a is a root. (UCLES)

17.

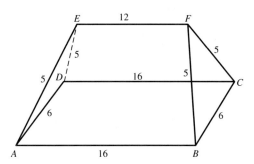

A tent is erected as shown. The base ABCD is rectangular and horizontal and the top edge EF is also horizontal.

The lengths, in metres, of the edges are

$$AE = BF = CF = DE = 5, \quad AB = CD = 16, \quad AD = BC = 6,$$

$$EF = 12.$$

(a) Calculate the size of $\angle ADE$, giving your answer to the nearest degree.

(b) Show that the vertical height of EF above the base $ABCD$ is $2\sqrt{3}$ m.

Calculate, to the nearest degree, the size of the acute angle between

(c) the face ADE and the horizontal,

(d) the edge AE and the horizontal. (ULEAC)

18. Show that the equation $x^3 - x^2 - 2 = 0$ can be arranged in the form $x = \sqrt[3]{(f(x))}$ where $f(x)$ is a quadratic function.

Use an iteration of the form $x_{n+1} = g(x_n)$ based on this rearrangement and, with $x_1 = 1.5$, to find x_2 and x_3, giving your answers to 3 decimal places.

(AEB)$_{\text{sp}}$

CHAPTER 5

LOGARITHMS

Exponentials

The power or index to which a number is raised can also be called the exponent. In the case of 3^{-2}, the exponent is -2; 3 is called the base.

The function given by $f(x) = a^x$ is an exponential function and, as we saw in *Module A*, such functions are useful for modelling situations where a quantity grows or decays by a constant factor over equal time intervals, i.e. undergoes exponential growth or decay. We will be considering such situations again in a later chapter.

When the base is the number e ($\approx 2.718\ldots$), the function $f : x \to e^x$ is called *the* exponential function.

We also defined the inverse of the exponential function as $f^{-1} : x \to \ln x$ where 'ln' means 'the power to which e is raised to give' and 'ln' is called 'the logarithm to the base e'.

LOGARITHMS

The same argument applies when the base is any positive number.

Consider the function $f : x \to 10^x$.

The graph of the function shows that it is a one–one mapping, therefore the inverse function f^{-1} exists.

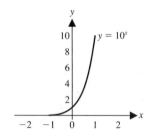

The equation of the curve representing f is $y = 10^x$. The curve $y = f^{-1}(x)$ is obtained by reflecting $y = 10^x$ in the line $y = x$.

The equation of the reflected curve is given by $10^y = x$, i.e. y is the power to which 10 is raised to give x.

Using 'log$_{10}$' to mean 'the power to which 10 is raised to give', we have

$$y = \log_{10} x$$

('log$_{10}$' is an abbreviation of 'the logarithm to the base 10')

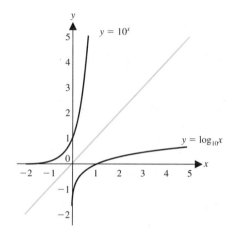

When $x = 100$, $\log_{10} 100$ means the power to which 10 must be raised to give 100. Now $10^2 = 100$, therefore $\log_{10} 100 = 2$,

i.e. $\qquad 10^2 = 100 \quad \Leftrightarrow \quad \log_{10} 100 = 2$

The same argument can be applied to any positive base, i.e. $\log_2 8$ means the power to which 2 must be raised to give 8. Since $2^3 = 8$ we have $\log_2 8 = 3$,

i.e. $\qquad\qquad 2^3 = 8 \quad \Leftrightarrow \quad \log_2 8 = 3$

and $\qquad\qquad 3^4 = 81 \quad \Leftrightarrow \quad \log_3 81 = 4,$

Similarly $\quad \log_5 25 = 2 \quad \Leftrightarrow \quad 5^2 = 25,$ and so on.

In general

$$a^c = b \quad \Leftrightarrow \quad \log_a b = c$$

Note that the symbol $\Leftrightarrow$ means each fact implies the other.

Example 5a

(a) Write $\log_2 64 = 6$ in index form.

(b) Write $5^3 = 125$ in logarithmic form.

(c) Complete the statement $2^{-3} = ?$ and then write it in logarithmic form.

(a) If $\log_2 64 = 6$ then the base is 2, the number is 64 and the power (i.e. the log) is 6

$$\log_2 64 = 6 \quad \Rightarrow \quad 2^6 = 64$$

(b) If $5^3 = 125$ then the base is 5, the log (i.e. the power) is 3 and the number is 125

$$5^3 = 125 \quad \Rightarrow \quad \log_5 125 = 3$$

(c) $2^{-3} = \frac{1}{8}$

The base is 2, the power (log) is -3 and the number is $\frac{1}{8}$

$$2^{-3} = \frac{1}{8} \quad \Rightarrow \quad \log_2\left(\frac{1}{8}\right) = -3$$

EXERCISE 5a

Convert each of the following facts to logarithmic form.

1. $10^3 = 1000$

2. $2^4 = 16$

3. $10^4 = 10\,000$

4. $3^2 = 9$

5. $4^2 = 16$

6. $5^2 = 25$

7. $10^{-2} = 0.01$

8. $9^{1/2} = 3$

9. $5^0 = 1$

10. $4^{1/2} = 2$

11. $12^0 = 1$

12. $8^{1/3} = 2$

13. $p = q^2$

14. $x^y = 2$

15. $p^q = r$

Convert each of the following facts to index form.

16. $\log_{10} 100\,000 = 5$

17. $\log_4 64 = 3$

18. $\log_{10} 10 = 1$

19. $\log_2 4 = 2$

20. $\log_2 32 = 5$

21. $\log_{10} 1000 = 3$

22. $\log_5 1 = 0$

23. $\log_3 9 = 2$

24. $\log_4 16 = 2$

25. $\log_3 27 = 3$

26. $\log_{36} 6 = \frac{1}{2}$

27. $\log_a 1 = 0$

28. $\log_x y = z$

29. $\log_a 5 = b$

30. $\log_p q = r$

Evaluating Logarithms

It is generally easier to solve a simple equation in index form than in log form so we often use an index equation in order to evaluate a logarithm. For example to evaluate $\log_{49} 7$ we can say

if $\qquad x = \log_{49} 7$ then $49^x = 7 \quad \Rightarrow \quad x = \frac{1}{2}$

therefore $\qquad\qquad\qquad\qquad \log_{49} 7 = \frac{1}{2}$

In particular, for any base b,

if $\qquad x = \log_b 1$ then $b^x = 1 \quad \Rightarrow \quad x = 0$

i.e. **the logarithm to any base of 1 is zero.**

EXERCISE 5b

Evaluate

1. $\log_2 4$
2. $\log_{10} 1\,000\,000$
3. $\log_2 64$
4. $\log_3 81$

5. $\log_8 64$
6. $\log_4 64$
7. $\log_9 3$
8. $\log_{1/2} 4$

9. $\log_{10} 0.1$
10. $\log_{121} 11$
11. $\log_5 1$
12. $\log_2 2$

13. $\log_{64} 4$
14. $\log_{99} 1$
15. $\log_{27} 3$
16. $\log_a a^3$

THE LAWS OF LOGARITHMS

When working with indices, we found certain rules that powers obey in the multiplication and division of numbers. Because logarithm is just another word for index or power, it is to be expected that logarithms too obey certain laws and these we are now going to investigate.

Consider $\qquad x = \log_a b$ and $\quad y = \log_a c$

$\Rightarrow \qquad\qquad a^x = b \qquad$ and $\quad a^y = c$

$\qquad\qquad$ Now $\quad bc = (a^x)(a^y)$

$\Rightarrow \qquad\qquad bc = a^{x+y}$

Therefore $\qquad \log_a bc = x + y$

i.e. $\qquad\qquad \log_a bc = \log_a b + \log_a c$

This is the first law of logarithms and, as a can represent *any* base, this law applies to the log of *any* product *provided that the same base is used for all the logarithms in the formula.*

Using x and y again, a law for the log of a fraction can be found.

$$\frac{b}{c} = \frac{a^x}{a^y} \quad \Rightarrow \quad \frac{b}{c} = a^{x-y}$$

Therefore $\log_a (b/c) = x - y$

i.e. $\log_a (b/c) = \log_a b - \log_a c$

A third law allows us to deal with an expression of the type $\log_a b^n$

Using $x = \log_a b^n \quad \Rightarrow \quad a^x = b^n$

i.e. $a^{x/n} = b$

Therefore $\dfrac{x}{n} = \log_a b \quad \Rightarrow \quad x = n \log_a b$

i.e. $\log_a b^n = n \log_a b$

So we now have the three most important laws of logarithms. Because they are true for *any* base it is unnecessary to include a base in the formula but

in each of these laws every logarithm must be to the same base

$$\log bc = \log b + \log c$$
$$\log b/c = \log b - \log c$$
$$\log b^n = n \log b$$

Examples 5c

1. Express $\log pq^2 \sqrt{r}$ in terms of $\log p$, $\log q$ and $\log r$

$$\log pq^2 \sqrt{r} = \log p + \log q^2 + \log \sqrt{r}$$
$$= \log p + 2 \log q + \tfrac{1}{2} \log r$$

2. Simplify $3 \log p + n \log qq - 4 \log r$

$$3 \log p + n \log q - 4 \log r = \log p^3 + \log q^n - \log r^4$$
$$= \log \frac{p^3 q^n}{r^4}$$

EXERCISE 5c

Express in terms of $\log p$, $\log q$, and $\log r$

1. $\log pq$ 2. $\log pqr$ 3. $\log p/q$ 4. $\log pq/r$

5. $\log p/qr$ 6. $\log p^2 q$ 7. $\log q/r^2$ 8. $\log p\sqrt{q}$

9. $\log p^2 q^3 / r$

10. $\log \sqrt{(q/r)}$

11. $\log q^n$

12. $\log p^n q^m$

13. $\log 2pq$

14. $\log \frac{1}{2} pq$

15. $\log 2pq^2$

16. $\log \dfrac{p}{2q}$

Simplify

17. $\log p + \log q$

18. $2 \log p + \log q$

19. $\log q - \log r$

20. $3 \log q + 4 \log p$

21. $n \log p - \log q$

22. $\log p + 2 \log q - 3 \log r$

23. $\log p - \log 2$

24. $2 \log p - p \log 2$

25. $\log p + \log q - \log 3$

26. $\log (p + 2) - \log (q - 2)$

Changing the Base of a Logarithm

Suppose that $x = \log_a c$ and that we wish to express x as a logarithm to the base b

$$\log_a c = x \quad \Rightarrow \quad c = a^x$$

Now taking logs to the base b gives

$$\log_b c = x \log_b a \quad \Rightarrow \quad x = \frac{\log_b c}{\log_b a}$$

i.e.
$$\log_a c = \frac{\log_b c}{\log_b a}$$

In the special case when $c = b$, i.e. when $\log_b c = 1$, this relationship becomes

$$\log_a b = \frac{1}{\log_b a}$$

Evaluating Logarithms

Scientific calculators usually have two sets of logarithms stored in them. These are logs to the base 10 (common logarithms) and logs to the base e (natural logarithms); the respective keys are marked log and ln.

Example 5d

Evaluate $\log_2 3$

First we can change the base to 10, then use a calculator.

$$\log_2 3 = \frac{\log_{10} 3}{\log_{10} 2}$$

$$= 1.58 \text{ correct to 3 sf}$$

Alternatively we can change the base to e.

$$\log_2 3 = \frac{\ln 3}{\ln 2} = 1.58 \text{ correct to 3 sf}$$

EXERCISE 5d

1. Use a calculator to evaluate, correct to 3 significant figures,

   (a) $\log_{10} 4$ (b) $\ln 5$ (c) $\log_{10} 0.2$

   (d) $\ln 0.2$ (e) $\ln 1.7$ (f) $\log_{10} 3.86$

2. Change the base of each logarithm to 10

   (a) $\log_3 8$ (b) $\log_5 10$ (c) $\log_{100} 5$

3. Change the base of each logarithm to e

   (a) $\log_4 3$ (b) $\log_7 2$ (c) $\log_{0.1} 8.2$

4. Change the base of each logarithm to a.

   (a) $\log_x y$ (b) $\log_x a$ (c) $\log_y 8$

5. Find, correct to three significant figures, the value of

   (a) $\log_8 12$ (b) $\log_5 100$ (c) $\log_3 15$

   (d) $\log_{0.1} 2.6$ (e) $\log_{0.5} 3.2$ (f) $\log_{250} 107$

SOLUTION OF EQUATIONS OF THE FORM $a^x = b$

When the unknown quantity forms part of an index, taking logs will often transform the index into a factor.

For example, if $5^x = 10$ then taking logs of both sides gives

$$x \log 5 = \log 10 \quad \Rightarrow \quad x = \frac{1}{\log 5}$$

Similarly, $2^{3x-1} = 5$ becomes $(3x - 1) \log 2 = \log 5$

$$\Rightarrow \qquad 3x - 1 = \frac{\log 5}{\log 2} = 2.321\ldots$$

$$\Rightarrow \qquad x = \tfrac{1}{3}(2.321\ldots + 1) = 1.11 \ (3 \text{ sf})$$

Solve the equations.

1. $3^x = 6$ 2. $5^x = 4$ 3. $1.2^x = 7.3$ 4. $5.8^x = 278$

5. $2^{2x} = 5$ 6. $3^{x-1} = 7$ 7. $3(2^x) = 8$ 8. $5^{3x-1} = 50$

9. $4^{2x+1} = 3$ 10. $5^x(5^{x-1}) = 10$ 11. $2^x(2^{2x+1}) = 3$ 12. $e^{3x} = 5$

Harder Exponential and Logarithmic Equations

Taking logs leads to a solution when the terms with x in the exponent can be expressed as a single term but this is not always possible.

Consider, for example, the equation

$$2^{2x} + 3(2^x) - 4 = 0$$

Now $2^{2x} + 3(2^x)$ cannot be simplified into a single term but, as 2^{2x} is $(2^x)^2$, using $y = 2^x$ gives $y^2 + 3y - 4 = 0$

This is a quadratic equation which can now be solved.

When an equation contains logarithms involving the unknown, first make sure that all logs are to the same base, then check to see if a simple substitution will reduce the equation to a recognisable form. Sometimes the best policy is to remove the logarithms. Equations of each of these types are considered in the following worked examples.

Examples 5f

1. Solve the equation $\log_3 x - 4\log_x 3 + 3 = 0$

The two log terms have different bases so we begin by changing the base of the log in the second term to 3. The given equation then becomes

$$\log_3 x - \frac{4}{\log_3 x} + 3 = 0$$

$$\Rightarrow \quad (\log_3 x)^2 - 4 + 3\log_3 x = 0$$

Substituting y for $\log_3 x$ gives $y^2 - 4 + 3y = 0 \quad \Rightarrow \quad y^2 + 3y - 4 = 0$

$$\therefore \qquad\qquad (y+4)(y-1) = 0 \quad \Rightarrow \quad y = 1 \quad \text{or} \quad -4$$

i.e. $\quad \log_3 x = 1 \quad \text{or} \quad \log_3 x = -4 \quad \Rightarrow \quad x = 3 \quad \text{or} \quad x = 3^{-4} = \frac{1}{81}$

2. Solve for x and y the equations

$$yx = 16 \quad \text{and} \quad \log_2 x - 2\log_2 y = 1$$

$$yx = 16 \qquad\qquad\qquad\qquad\qquad [1]$$

$$\log_2 x - 2\log_2 y = 1 \qquad\qquad\qquad [2]$$

Using the laws of logs, equation [2] can be written as

$$\log_2 \frac{x}{y^2} = 1 \quad \Rightarrow \quad \frac{x}{y^2} = 2, \quad \text{i.e.} \quad x = 2y^2$$

Substituting $2y^2$ for x in equation [1] gives

$$2y^3 = 16 \quad \Rightarrow \quad y^3 = 8$$

$$\therefore \qquad y = 2 \quad \text{and, from [1],} \quad x = 8$$

EXERCISE 5f) HomeWork) Friday

Solve the equations

1. $2(2^{2x}) - 5(2^x) + 2 = 0$

2. $3^{2x+1} - 26(3^x) - 9 = 0$

3. $4^x - 6(2^x) - 16 = 0$

4. $\log_2 x + \log_x 2 = 2$

5. $\log_2 x = \log_4(x+6)$

6. $4\log_3 x = \log_x 3$

7. Solve the equations

(a) $3^{2x} = 10$

(b) $4^x + 2^x - 6 = 0$

(c) $(2^x)(3^x) = \frac{1}{36}$

(d) $2\log_2 x = \log_x 2$

(e) $3\log_4 x - \log_4 6 = 16$

(f) $\log_x 1.2 = 2.1$

Solve the equations simultaneously.

*8. $2 \lg y = \lg 2 + \lg x$ and $2^y = 4^x$ ($\lg$ means $\log_{10}$)

*9. $\log_x y = 2$ and $xy = 8$

*10. $\log_3 x = y = \log_9 (2x - 1)$

*11 $\lg (x + y) = 0$ and $2 \lg x = \lg (y + 1)$

*12. (a) Change the base of $\log_{10} x$ to e.

(b) Describe the transformation that maps the curve

$$y = \ln x \quad \text{to} \quad y = \log_{10} x$$

(c) Hence *sketch* the curves $y = \ln x$ and $y = \log_{10} x$ on the same set of axes.

(d) If f is the inverse of g, where $g : x \to \log_3 x$, give the equation of the curve $y = f(x)$.

(e) Sketch the curve $y = \log_a x$, where $a > 1$, marking the point where the curve cuts the x-axis.

*13. Investigate the behaviour of $\log_a x$ as x increases, $(x > 0)$, when

(a) $a > 1$ (b) $a < 1$

Justify your conclusions.

CHAPTER 6

TRIGONOMETRY

THE TRIGONOMETRIC FUNCTIONS

The graphs and main properties of the basic trigonometric functions are as follows.

memory

The sine function, $f : \theta \rightarrow \sin \theta$
    f is periodic with period 2π,
    f is odd, i.e. $-\sin \theta = \sin (-\theta)$,
    the range of f is $-1 \leqslant \sin \theta \leqslant 1$

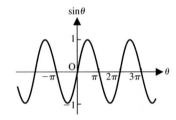

The cosine function, $f : \theta \rightarrow \cos \theta$
    f is periodic with period 2π,
    f is even, i.e. $\cos \theta = \cos (-\theta)$,
    the range of f is $-1 \leqslant \cos \theta \leqslant 1$

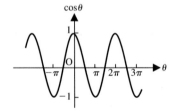

The tangent function, $f : \theta \rightarrow \tan \theta$
    f is periodic with period π,
    f is odd, i.e. $-\tan \theta = \tan (-\theta)$,
    the range of f is all real values.

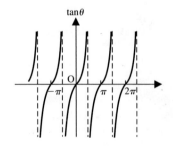

THE RECIPROCAL TRIGONOMETRIC FUNCTIONS

The reciprocals of the three main trig functions have their own names and are sometimes referred to as the *minor* trig ratios.

memory

$$\frac{1}{\sin \theta} = \text{cosec } \theta \qquad \frac{1}{\cos \theta} = \text{sec } \theta \qquad \frac{1}{\tan \theta} = \text{cot } \theta$$

The names given above are abbreviations of cosecant, secant and cotangent respectively.

The graph of $f(\theta) = \text{cosec } \theta$ is given below.

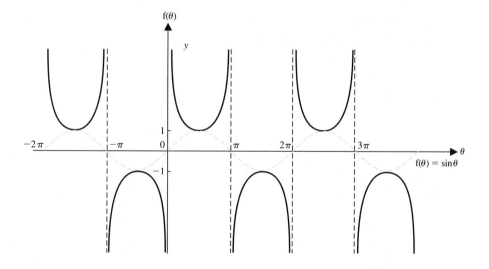

From this graph we can see that

the cosec function is not continuous, being undefined when θ is any integral multiple of π (we would expect this because these are values of θ where $\sin \theta = 0$ and the reciprocal of 0 is $\pm\infty$). It is however periodic, with a pattern that repeats after every interval of 2π.

The pattern of the graph of $f(\theta) = \sec \theta$ is similar to that of the cosec graph, as would be expected.

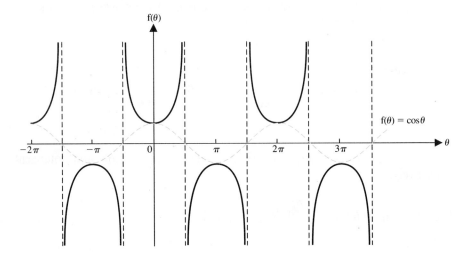

The sec function also is periodic, with period 2π. It is not continuous, being undefined when θ is any odd multiple of $\frac{1}{2}\pi$, i.e. when $\theta = \frac{1}{2}(2n - 1)\pi$ for $n = \ldots -2, -1, 0, 1, 2, \ldots$

The graph of $f(\theta) = \cot \theta$ is given below.

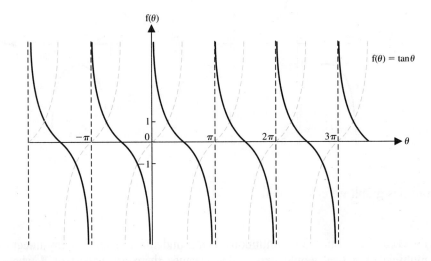

Hence the cotangent function is periodic, with period π (i.e. half that of the cosec and sec functions), and it is undefined when θ is any multiple of π.

Example 6a

For $0 \leqslant \theta \leqslant 360$, find the values of θ for which $\operatorname{cosec} \theta = -8$

$$\operatorname{cosec} \theta = \frac{1}{\sin \theta}$$

$$\therefore \quad \operatorname{cosec} \theta = -8 \quad \Rightarrow \quad \frac{1}{\sin \theta} = -8$$

$$\Rightarrow \quad \sin \theta = -\tfrac{1}{8}$$

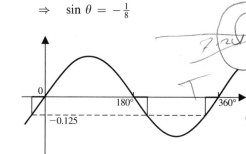

$\therefore$ from a calculator $\theta = -7.2°$

From the sketch, the required values of θ are 187.2° and 352.8°

SOLUTION OF EQUATIONS INVOLVING RECIPROCAL TRIG RATIOS

One way to deal with a trig equation that involves a reciprocal trig ratio is to convert this ratio to a standard one and then use one of the methods given in *Module A*.

EXERCISE 6a

1. Find, for values of θ in the range $0 \leqslant \theta \leqslant 360°$, the values of θ for which

   (a) $\sec \theta = 2$ (b) $\cot \theta = 0.6$ (c) $\operatorname{cosec} \theta = 1.5$

2. Within the range $-180° \leqslant \theta \leqslant 180°$ find the values of θ for which

   (a) $\cot \theta = 1.2$ (b) $\sec \theta = -1.5$ (c) $\operatorname{cosec} \theta = -2$

3. Given that $\tan \theta = \dfrac{\sin \theta}{\cos \theta}$, write $\cot \theta$ in terms of $\sin \theta$ and $\cos \theta$. Hence show that $\cot \theta - \cos \theta = 0$ can be written in the form $\cos \theta (1 - \sin \theta) = 0$, provided that $\sin \theta \neq 0$.
   Thus find the values in the range $-\pi \leqslant \theta \leqslant \pi$ for which $\cot \theta - \cos \theta = 0$.

4. Find, in surd form, the values of

   (a) $\cot \frac{1}{4}\pi$ (b) $\sec \frac{5}{4}\pi$ (c) $\operatorname{cosec} \frac{11}{6}\pi$

5. Sketch the graph of $f(\theta) = \sec(\theta - \frac{1}{4}\pi)$ for $0 \leqslant \theta \leqslant 2\pi$ and give the values of θ for which $f(\theta) = 1$.

6. Sketch the graph of $f(\theta) = \cot(\theta + \frac{1}{3}\pi)$ for $-\pi \leqslant \theta \leqslant \pi$. Hence give the values of θ in this range for which $f(\theta) = 1$.

7. Use this diagram to show that the cotangent of an angle is equal to the tangent of its complement.

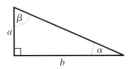

8. Sketch the graph of $f(\theta) = \tan(\frac{1}{2}\pi - \theta)$ for $0 \leqslant \theta \leqslant \pi$ and compare your sketch with the graph of $f(\theta) = \cot \theta$.
   What can you deduce about the relationship between $\tan(\frac{1}{2}\pi - \theta)$ and $\cot \theta$?

IDENTITIES

At this stage it is important to remember the difference between identities and equations.

This is an equation: $(x - 1)^2 = 4$
The equality is true only when $x = 3$ or when $x = -1$
In any equation, the equality is valid only for a restricted set of values.

This is an identity: $(x - 1)^2 = x^2 - 2x + 1$
The RHS is just a different way of expressing the LHS, and the equality is true for all values of x.

In an identity, the equality is true for *any* value of the variable.
The symbol $\equiv$ means 'is identical to', so, strictly, we should write

$$(x - 1)^2 \equiv x^2 - 2x + 1$$

In practice however the symbol $=$ is often used in an algebraic identity.

In this chapter we concentrate on some trigonometric identities and some of their uses.
One such identity, introduced in *Module A*, is

$$\tan \theta \equiv \frac{\sin \theta}{\cos \theta}$$

THE PYTHAGOREAN IDENTITIES

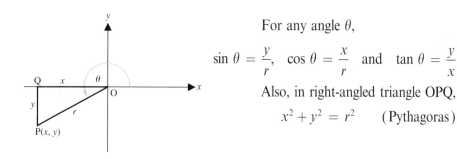

For any angle θ,

$$\sin\theta = \frac{y}{r}, \quad \cos\theta = \frac{x}{r} \quad \text{and} \quad \tan\theta = \frac{y}{x}$$

Also, in right-angled triangle OPQ,

$$x^2 + y^2 = r^2 \qquad (\text{Pythagoras})$$

Therefore, $\quad (\cos\theta)^2 + (\sin\theta)^2 = \left(\dfrac{x}{r}\right)^2 + \left(\dfrac{y}{r}\right)^2 = \dfrac{x^2+y^2}{r^2} = 1$

Using the notation $\cos^2\theta$ to mean $(\cos\theta)^2$, etc., we have

$$\cos^2\theta + \sin^2\theta \equiv 1 \qquad\qquad [1]$$

Using the identity $\tan\theta \equiv \dfrac{\sin\theta}{\cos\theta}$ we can write [1] in two other forms.

$$[1] \div \cos^2\theta \quad\Rightarrow\quad 1 + \frac{\sin^2\theta}{\cos^2\theta} \equiv \frac{1}{\cos^2\theta}$$

$$\Rightarrow \qquad\qquad 1 + \tan^2\theta \equiv \sec^2\theta$$

$$[1] \div \sin^2\theta \quad\Rightarrow\quad \frac{\cos^2\theta}{\sin^2\theta} + 1 \equiv \frac{1}{\sin^2\theta}$$

$$\Rightarrow \qquad\qquad \cot^2 + 1 \equiv \operatorname{cosec}^2\theta$$

These identities can be used to

    simplify trig expressions,
    eliminate trig terms from pairs of equations,
    derive a variety of further trig relationships,
    calculate other trig ratios of any angle for which one trig ratio is known.

These identities are also very useful in the solution of certain types of trig equation and we will look at this application later in this chapter.

Examples 6b

1. Simplify $\dfrac{\sin \theta}{1 + \cot^2 \theta}$

$$\frac{\sin \theta}{1 + \cot^2 \theta} \equiv \frac{\sin \theta}{\cosec^2 \theta} \equiv \sin^3 \theta$$

Using $\quad 1 + \cot^2 \theta \equiv \cosec^2 \theta \quad$ and $\quad \cosec \theta \equiv \dfrac{1}{\sin \theta}$

2. Eliminate θ from the equations $\quad x = 2 \cos \theta \quad$ and $\quad y = 3 \sin \theta$

$$\cos \theta = \tfrac{1}{2}x \quad \text{and} \quad \sin \theta = \tfrac{1}{3}y$$

Using $\quad \cos^2 \theta + \sin^2 \theta \equiv 1 \quad$ gives

$$\left(\frac{x}{2}\right)^2 + \left(\frac{y}{3}\right)^2 = 1$$

$\Rightarrow \qquad 9x^2 + 4y^2 = 36$

In Example 2, both x and y initially depend on θ, a variable angle. Used in this way, θ is called a *parameter*, and is a type of variable that plays an important part in the analysis of curves and functions.

3. If $\sin A = -\tfrac{1}{3}$ and A is in the third quadrant, find $\cos A$ without using a calculator.

There are two ways of doing this problem. The first method involves drawing a quadrant diagram and working out the remaining side of the triangle, using Pythagoras theorem.

From the diagram, $x = -2\sqrt{2}$

$\therefore \qquad \cos A = \dfrac{x}{r} = -\dfrac{2\sqrt{2}}{3}$

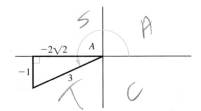

The second method uses the identity $\cos^2 A + \sin^2 A \equiv 1$ giving

$$\cos^2 A + \left(\frac{1}{3}\right)^2 = 1 \quad \Rightarrow \quad \cos A = \pm\sqrt{\frac{8}{9}} = \pm\frac{2\sqrt{2}}{3}$$

As A is between π and $\tfrac{3}{2}\pi$, $\cos A$ is negative, i.e.

$$\cos A = -\frac{2\sqrt{2}}{3}$$

4. Prove that $(1 - \cos A)(1 + \sec A) \equiv \sin A \tan A$

Because the relationship has yet to be proved, we must not assume its truth by using the complete identity in our working. The left and right hand sides must be isolated throughout the proof, preferably by working on only one of these sides.

Consider the LHS:

$$(1 - \cos A)(1 + \sec A) \equiv 1 + \sec A - \cos A - \cos A \sec A$$

$$\equiv 1 + \sec A - \cos A - \cos A \left(\frac{1}{\cos A}\right)$$

$$\equiv \sec A - \cos A$$

$$\equiv \frac{1}{\cos A} - \cos A \equiv \frac{1 - \cos^2 A}{\cos A}$$

$$\equiv \frac{\sin^2 A}{\cos A} \qquad (\cos^2 A + \sin^2 A \equiv 1)$$

$$\equiv \sin A \left[\frac{\sin A}{\cos A}\right]$$

$$\equiv \sin A \tan A \equiv \text{RHS}$$

EXERCISE 6b

1. Without using a calculator, complete the following table.

	$\sin \theta$	$\cos \theta$	$\tan \theta$	type of angle
(a)		$-\frac{5}{13}$		reflex
(b)	$\frac{3}{5}$			obtuse
(c)			$\frac{7}{24}$	acute
(d)				straight line

Simplify the following expressions.

2. $\dfrac{1 - \sec^2 A}{1 - \operatorname{cosec}^2 A}$

3. $\dfrac{\sin \theta}{\sqrt{(1 - \cos^2 \theta)}}$

4. $\dfrac{\sin \theta}{\cos \theta} + \dfrac{\cos \theta}{\sin \theta}$

5. $\dfrac{\sqrt{(1 + \tan^2 \theta)}}{\sqrt{(1 - \sin^2 \theta)}}$

6. $\dfrac{1}{\cos \theta \sqrt{(1 + \cot^2 \theta)}}$

7. $\dfrac{\sin \theta}{1 + \cot^2 \theta}$

Eliminate θ from the following pairs of equations.

8. $x = 4 \sec \theta$
   $y = 4 \tan \theta$

9. $x = a \operatorname{cosec} \theta$
   $y = b \cot \theta$

10. $x = 2 \tan \theta$
    $y = 3 \cos \theta$

11. $x = 1 - \sin \theta$
    $y = 1 + \cos \theta$

12. $x = 2 + \tan \theta$
    $y = 2 \cos \theta$

13. $x = a \sec \theta$
    $y = b \sin \theta$

Prove the following identities.

14. $\cot \theta + \tan \theta \equiv \sec \theta \operatorname{cosec} \theta$

15. $\dfrac{\cos A}{1 - \tan A} + \dfrac{\sin A}{1 - \cot A} \equiv \sin A + \cos A$

16. $\tan^2 \theta + \cot^2 \theta \equiv \sec^2 \theta + \operatorname{cosec}^2 \theta - 2$

17. $\dfrac{\sin A}{1 + \cos A} \equiv \dfrac{1 - \cos A}{\sin A}$ (Hint. Multiply top and bottom of LHS by $(1 - \cos A)$.)

18. $\dfrac{\sin A}{1 + \cos A} + \dfrac{1 + \cos A}{\sin A} \equiv \dfrac{2}{\sin A}$

SOLVING EQUATIONS

We have already solved some simple trig equations in *Module A*. We can now solve a greater variety of equations using the Pythagorean identities.

Examples 6c

1. Solve the equation $2 \cos^2 \theta - \sin \theta = 1$ for values of θ in the range 0 to 2π.

   The given equation is quadratic, but it involves the sine and the cosine of θ, so we use $\cos^2 \theta + \sin^2 \theta \equiv 1$ to express the equation in terms of $\sin \theta$ only.

   $$2 \cos^2 \theta - \sin \theta = 1$$

   $\Rightarrow \qquad 2(1 - \sin^2 \theta) - \sin \theta = 1$

   $\Rightarrow \qquad 2 \sin^2 \theta + \sin \theta - 1 = 0$

   $\Rightarrow \qquad (2 \sin \theta - 1)(\sin \theta + 1) = 0 \quad \Rightarrow \quad \sin \theta = \tfrac{1}{2} \quad \text{or} \quad -1$

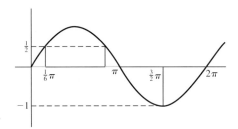

If $\sin \theta = \frac{1}{2}$, $\theta = \frac{1}{6}\pi$, $\frac{5}{6}\pi$

If $\sin \theta = -1$, $\theta = \frac{3}{2}\pi$

Therefore the solution of the equation is $\theta = \frac{1}{6}\pi$, $\frac{5}{6}\pi$, $\frac{3}{2}\pi$

2. Solve the equation $\cot x = \sin x$ for values of x from 0 to 360

Using $\cot x \equiv \dfrac{\cos x}{\sin x}$ gives $\dfrac{\cos x}{\sin x} = \sin x$

Both sides of an equation can be multiplied by any number *except* zero. We can now multiply the equation by $\sin x = 0$ provided that $\sin x = 0$ is excluded from the solution set.

Multiplying both sides of $\dfrac{\cos x}{\sin x} = \sin x$ by $\sin x$ gives

$$\cos x = \sin^2 x \quad (\sin x \neq 0)$$

$\Rightarrow$ $\qquad\qquad \cos^2 x + \cos x - 1 = 0$

This equation does not factorise, so we use the formula for solving a quadratic equation, giving

$$\cos x = \tfrac{1}{2}(-1 \pm \sqrt{5})$$

$\therefore$ $\qquad \cos x = -1.618$ and there is no value of x for which this is true.

or $\qquad \cos x = 0.618$

$\Rightarrow$ $\qquad x = 51.8°$ or $308.2°$

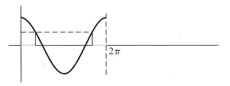

Solve the following equations for angles in the range $0 \leqslant \theta \leqslant 360°$

1. $\sin^2 \theta = \frac{1}{4}$

2. $9 \cos^2 \theta = 1$

3. $25 \sin^2 \theta = 9$

4. $\sec^2 \theta = 2 \tan \theta$

5. $\cos^2 \theta + 3 \sin \theta = 3$

6. $\sin^2 \theta = \cos^2 \theta - 1$

7. $\sec^2 \theta + \tan^2 \theta = 6$

8. $4 \cos^2 \theta + 5 \sin \theta = 3$

9. $\cot^2 \theta = \operatorname{cosec} \theta$

10. $\tan \theta + \cot \theta = 2 \sec \theta$

11. $\tan \theta + 3 \cot \theta = 5 \sec \theta$

12. $\sec \theta = 1 - 2 \tan^2 \theta$

Solve the following equations for angles in the range $-\pi \leqslant \theta \leqslant \pi$

13. $5 \cos \theta - 4 \sin^2 \theta = 2$

14. $4 \cot^2 \theta + 12 \operatorname{cosec} \theta + 1 = 0$

15. $4 \sec^2 \theta - 3 \tan \theta = 5$

16. $2 \cos \theta - 4 \sin^2 \theta + 2 = 0$

17. $2 \sin \theta \cos \theta + \sin \theta = 0$

18. $2 \cos \theta = \cos \theta \operatorname{cosec} \theta$

19. $\sqrt{3} \tan \theta = 2 \sin \theta$

20. $\cot \theta = \cos \theta$

EQUATIONS INVOLVING MULTIPLE ANGLES

Many trig equations involve ratios of a multiple of θ, for example

$$\cos 2\theta = \tfrac{1}{2} \qquad \tan 3\theta = -2$$

Simple equations of this type can be solved by finding first the values of the multiple angle and then, by division, the corresponding values of θ.

However it must be remembered that if values of θ are required in the range $\alpha \leqslant \theta \leqslant \beta$ then values of 2θ will be needed in double that range, i.e. $2\alpha \leqslant 2\theta \leqslant 2\beta$ and similarly for other multiples of θ, e.g. if values of θ are needed for $0 \leqslant \theta \leqslant 360°$, then 2θ must be found in the range $0 \leqslant 2\theta \leqslant 720°$, $\frac{1}{2}\theta$ should be found in the range $0 \leqslant \frac{1}{2}\theta \leqslant 180°$, and so on.

Examples 6d

1. Find the angles in the interval $\left[-\frac{1}{2}\pi, \frac{1}{2}\pi\right]$ which satisfy the equation $\cos 2\theta = \frac{1}{2}$

As values of θ in the interval $\left[-\frac{1}{2}\pi, \frac{1}{2}\pi\right]$ are required, we need to find values of 2θ in the interval $[-\pi, \pi]$.

In the interval $[-\pi, \pi]$,

the solutions of $\cos 2\theta = \frac{1}{2}$ are $2\theta = \pm\frac{1}{3}\pi$

Hence $\theta = \pm\frac{1}{6}\pi$

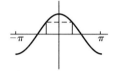

EQUATIONS INVOLVING COMPOUND ANGLES

When a compound angle appears in a trig equation such as

$$\cos\left(\theta - \tfrac{1}{4}\pi\right) = \tfrac{1}{2}$$

the equation can be solved by first finding values of the compound angle. The values of θ can then be found from a simple linear equation. If values of θ are required in the range $0 \leqslant \theta \leqslant \pi$ say, then we must find the values of $\theta + \alpha$ in the interval $[0 + \alpha, \pi + \alpha]$.

Examples 6d (continued)

2. Solve the equation $\cos(\theta - 20°) = -\frac{1}{2}$ for values of θ in the range $-180° \leqslant \theta \leqslant 180°$

As values of θ are required in the interval $[-180°, 180°]$, we need values of $(\theta - 20°)$ in the interval $[-200°, 160°]$.

In the interval $[-200°, 160°]$, the solutions of $\cos(\theta - 20°) = -\frac{1}{2}$ are $\theta - 20° = \pm120°$

Hence

$\theta = 120° + 20°$ or $-120° + 20°$

$= 140°$ or $-100°$

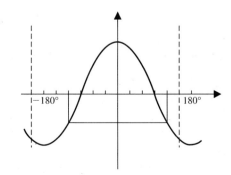

EXERCISE 6d

Find the values of θ in the range $0 \leqslant \theta \leqslant 360°$ which satisfy the following equations.

1. $\tan 2\theta = 1$

2. $\cos 3\theta = -0.5$

3. $\sin \frac{1}{2}\theta = -\frac{\sqrt{2}}{2}$

4. $\sec 5\theta = 2$

5. $\cot \frac{1}{3}\theta = -4$

6. $\cos 2\theta = 0.63$

7. $\cos (\theta - 45°) = 0$

8. $\sin (\theta + 30°) = -1$

9. $\tan (\theta - 60°) = 0$

10. $\cos (\theta + 60°) = \frac{1}{2}$

Solve the following equations for values of θ in the range $-\pi \leqslant \theta \leqslant \pi$

11. $\cos (\theta + \frac{1}{4}\pi) = \frac{1}{2}$

12. $\tan (\theta - \frac{1}{3}\pi) = -1$

13. $\sin (\theta + \frac{1}{6}\pi) = \frac{1}{2}$

14. $\cos (\theta - \frac{1}{3}\pi) = -\frac{1}{2}$

Solve these equations for values of θ in the range $-180° \leqslant \theta \leqslant 180°$

15. $\tan 2\theta = 1.8$

16. $\sin 3\theta = 0.7$

17. $\cos \frac{1}{2}\theta = 0.85$

Solve these equations for values of θ in the range $0 \leqslant \theta \leqslant 2\pi$

18. $\tan 4\theta = -\sqrt{3}$

19. $\sec 5\theta = 2$

20. $\cot \frac{1}{2}\theta = -1$

MIXED EXERCISE 6

1. Eliminate α from the equations $x = \cos \alpha, \ y = \operatorname{cosec} \alpha$

2. If $\cos \beta = 0.5$, find possible values for $\sin \beta$ and $\tan \beta$, giving your answers in exact form.

3. Simplify the expression $\dfrac{1}{1 + \cos \theta} + \dfrac{1}{1 - \cos \theta}$. Hence solve the equation
   $\dfrac{1}{1 + \cos \theta} + \dfrac{1}{1 - \cos \theta} = 4$ for values of θ in the range $0 \leqslant \theta \leqslant 2\pi$

4. Solve the equation $\sec \theta + \tan^2 \theta = 5$
   Give the answer in degrees, in the range $0 \leqslant \theta \leqslant 180°$

5. Sketch the graph, in the range $-2\pi \leqslant \theta \leqslant 2\pi$, of
   (a) $y = \operatorname{cosec} 2\theta$
   (b) $y = \operatorname{cosec} (2\theta - \frac{1}{2}\pi)$

6. Find the value of a for which the point $(\pi, 0)$ lies on the graph of $y = a + \sec \theta$

7. Prove that $(\cot \theta + \operatorname{cosec} \theta)^2 \equiv \dfrac{1 + \cos \theta}{1 - \cos \theta}$

8. Find the values of θ for which $\tan(3\theta - \frac{1}{3}\pi) = 1$ in the interval $[-\pi, \pi]$

9. Eliminate θ from the equations
   (a) $x - 2 = \sin \theta$, $y + 1 = \cos \theta$
   (b) $x = \sec \theta - 3$, $y = 2 - \tan \theta$

10. Solve the equation $\tan 2\alpha = \cot 2\alpha$ for $0 \leqslant \alpha \leqslant \pi$

11. Prove that $(\cos A + \sin A)^2 + (\cos A - \sin A)^2 \equiv 2$

12. Simplify $(1 + \cos A)(1 - \cos A)$

13. Find in degrees the solutions of the equation $\tan \theta = 3 \sin \theta$ for which $0 \leqslant \theta \leqslant 360°$

14. Simplify $\sec^4 \theta - \sec^2 \theta$

CHAPTER 7

COMPOUND ANGLE IDENTITIES

COMPOUND ANGLES

It is often useful to be able to express a trig ratio of an angle A + B in terms of trig ratios of A and of B.

It is dangerously easy to think, for instance, that $\sin(A+B)$ is $\sin A + \sin B$. However, this is *false* as can be seen by considering

$$\sin(45° + 45°) = \sin 90° \qquad = 1$$

whereas $\qquad \sin 45° + \sin 45° = \tfrac{1}{2}\sqrt{2} + \tfrac{1}{2}\sqrt{2} \neq 1$

Thus the sine function is *not distributive* and neither are the other trig functions.

The correct identify is $\sin(A+B) = \sin A \cos B + \cos A \sin B$.

This is proved geometrically when A and B are both acute, from the diagram below.

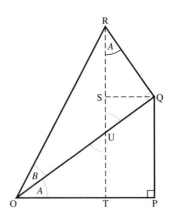

The right-angled triangles OPQ and OQR contain angles A and B as shown. From the diagram, $\angle URQ = A$

$$\sin(A+B) = \frac{TR}{OR} = \frac{TS+SR}{OR} = \frac{PQ+SR}{OR}$$

$$= \frac{PQ}{OQ} \times \frac{OQ}{OR} + \frac{SR}{QR} \times \frac{QR}{OR}$$

$$\therefore \quad \sin(A+B) \equiv \sin A \cos B + \cos A \sin B$$

This identity is in fact valid for all angles and it can be adapted to give the full set of compound angle formulae. The reader is left to do this in the following exercise.

EXERCISE 7a

1. In the identity $\sin(A+B) \equiv \sin A \cos B + \cos A \sin B$, replace B by $-B$ to show that $\sin(A-B) \equiv \sin A \cos B - \cos A \sin B$

2. In the identity $\sin(A-B) \equiv \sin A \cos B - \cos A \sin B$, replace A by $(\frac{1}{2}\pi - A)$ to show that $\cos(A+B) \equiv \cos A \cos B - \sin A \sin B$

3. In the identity $\cos(A+B) \equiv \cos A \cos B - \sin A \sin B$, replace B by $-B$ to show that $\cos(A-B) \equiv \cos A \cos B + \sin A \sin B$

4. Use $\dfrac{\sin(A+B)}{\cos(A+B)}$ to show that $\tan(A+B) \equiv \dfrac{\tan A + \tan B}{1 - \tan A \tan B}$

5. Replace B by $-B$ in the formula for $\tan(A+B)$ to show that $\tan(A-B) \equiv \dfrac{\tan A - \tan B}{1 + \tan A \tan B}$

Collecting these results we have:

$$\sin(A+B) \equiv \sin A \cos B + \cos A \sin B$$

$$\sin(A-B) \equiv \sin A \cos B - \cos A \sin B$$

$$\cos(A+B) \equiv \cos A \cos B - \sin A \sin B$$

$$\cos(A-B) \equiv \cos A \cos B + \sin A \sin B$$

$$\tan(A+B) \equiv \frac{\tan A + \tan B}{1 - \tan A \tan B}$$

$$\tan(A-B) \equiv \frac{\tan A - \tan B}{1 + \tan A \tan B}$$

1. Find exact values for   (a) $\sin 75$ (b) $\cos 105$

To find exact values, we need to express the given angle in terms of angles whose trig ratios are known as exact values, e.g. $30°$, $60°$, $45°$, $90°$, $120°$, ...
Now $75° = 45° + 30°$ (or $120° - 45°$ or other alternative compound angles).

(a) $\sin 75° = \sin(45° + 30°) = \sin 45° \cos 30° + \cos 45° \sin 30°$

$$= \left(\frac{\sqrt{2}}{2}\right)\left(\frac{\sqrt{3}}{2}\right) + \left(\frac{\sqrt{2}}{2}\right)\left(\frac{1}{2}\right)$$

$$= \frac{\sqrt{2}}{4}(\sqrt{3} + 1)$$

(b) $\cos 105° = \cos(60° + 45°) = \cos 60° \cos 45° - \sin 60° \sin 45°$

$$= \left(\frac{1}{2}\right)\left(\frac{\sqrt{2}}{2}\right) - \left(\frac{\sqrt{3}}{2}\right)\left(\frac{\sqrt{2}}{2}\right)$$

$$= \frac{\sqrt{2}}{4}(1 - \sqrt{3})$$

2. A is obtuse and $\sin A = \frac{3}{5}$. B is acute and $\sin B = \frac{12}{13}$. Find the exact value of $\cos(A + B)$.

$$\cos(A + B) \equiv \cos A \cos B - \sin A \sin B$$

In order to use this formula, we need values for $\cos A$ and $\cos B$. These can be found using Pythagoras' theorem in the appropriate right-angled triangle.

$$\cos A = -\tfrac{4}{5} \qquad\qquad \cos B = \tfrac{5}{13}$$

$\therefore$ $\cos(A + B) = \left(-\frac{4}{5}\right)\left(\frac{12}{13}\right) - \left(\frac{3}{5}\right)\left(\frac{12}{13}\right) = -\frac{56}{65}$

3. Simplify $\sin\theta\cos\frac{1}{3}\pi - \cos\theta\sin\frac{1}{3}\pi$ and hence find the smallest positive value of θ for which the expression has a minimum value.

$\sin\theta\cos\frac{1}{3}\pi - \cos\theta\sin\frac{1}{3}\pi$ is the expansion of $\sin(A - B)$ with $A = \theta$ and $\beta = \frac{1}{3}\pi$

$$\sin\theta\cos\tfrac{1}{3}\pi - \cos\theta\sin\tfrac{1}{3}\pi = \sin(\theta - \tfrac{1}{3}\pi) = f(\theta) \text{ say}$$

Now the graph $f(\theta)$ is a sine wave, but translated $\frac{1}{3}\pi$ in the direction of the positive θ-axis.

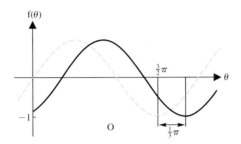

Therefore $f(\theta)$ has a minimum value of -1 and the smallest +ve value of θ at which this occurs is $\frac{3}{2}\pi + \frac{1}{3}\pi = \frac{11}{6}\pi$

4. Prove that $\dfrac{\sin (A - B)}{\cos A \cos B} \equiv \tan A - \tan B$

Expanding the numerator, the LHS becomes

$$\frac{\sin A \cos B - \cos A \sin B}{\cos A \cos B} \equiv \frac{\sin A \cos B}{\cos A \cos B} - \frac{\cos A \sin B}{\cos A \cos B}$$

$$\equiv \tan A - \tan B \equiv \text{RHS}$$

5. Find, in the range $0 \leqslant \theta \leqslant 2\pi$, the solution of the equation $2 \cos \theta = \sin (\theta + \frac{1}{6}\pi)$

$$2 \cos \theta = \sin (\theta + \tfrac{1}{6}\pi)$$

$$= \sin \theta \cos \tfrac{1}{6}\pi + \cos \theta \sin \tfrac{1}{6}\pi = \tfrac{\sqrt{3}}{2} \sin \theta + \tfrac{1}{2} \cos \theta$$

$\therefore$ $\quad \tfrac{3}{2} \cos \theta = \tfrac{\sqrt{3}}{2} \sin \theta$

$\Rightarrow \quad \dfrac{3}{\sqrt{3}} = \dfrac{\sin \theta}{\cos \theta} \quad \Rightarrow \quad \tan \theta = \sqrt{3}$

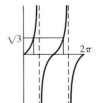

Now $\quad \tan \frac{1}{3}\pi = \sqrt{3}$

Therefore the solution is $\theta = \frac{1}{3}\pi, \frac{4}{3}\pi$

Find the exact value of each expression, leaving your answer in surd form where necessary.

1. $\cos 40° \cos 50° - \sin 40° \sin 50°$ 2. $\sin 37° \cos 7° - \cos 37° \sin 7°$

3. $\cos 75°$ 4. $\tan 105°$

5. $\sin 165°$ 6. $\cos 15°$

Simplify each of the following expressions.

7. $\sin \theta \cos 2\theta + \cos \theta \sin 2\theta$ 8. $\cos \alpha \cos (90° - \alpha) - \sin \alpha \sin (90° - \alpha)$

9. $\dfrac{\tan A + \tan 2A}{1 - \tan A \tan 2A}$ 10. $\dfrac{\tan 3\beta - \tan 2\beta}{1 + \tan 3\beta \tan 2\beta}$

11. A is acute and $\sin A = \frac{7}{25}$, B is obtuse and $\sin B = \frac{4}{5}$. Find an exact expression for

   (a) $\sin (A + B)$ (b) $\cos (A + B)$ (c) $\tan (A + B)$

12. Find the greatest value of each expression and the value of θ between 0 and 360° at which it occurs.

   (a) $\sin \theta \cos 25° - \cos \theta \sin 25°$ (b) $\sin \theta \sin 30° + \cos \theta \cos 30°$
   (c) $\cos \theta \cos 50° - \sin \theta \sin 50°$ (d) $\sin 60° \cos \theta - \cos 60° \sin \theta$

Prove the following identities

13. $\cot (A + B) \equiv \dfrac{\cot A \cot B - 1}{\cot A + \cot B}$

14. $(\sin A + \cos A)(\sin B + \cos B) \equiv \sin (A + B) + \cos (A - B)$

15. $\sin (A + B) + \sin (A - B) \equiv 2 \sin A \cos B$

16. $\cos (A + B) + \cos (A - B) \equiv 2 \cos A \cos B$

17. $\dfrac{\sin (A + B)}{\cos A \cos B} \equiv \tan A + \tan B$

Solve the following equations for values of θ in the range $0 \leqslant \theta \leqslant 360°$

18. $\cos (45° - \theta) = \sin \theta$

19. $3 \sin \theta = \cos (\theta + 60°)$

20. $\tan (A - \theta) = \frac{2}{3}$ and $\tan A = 3$

21. $\sin (\theta + 60°) = \cos \theta$

THE DOUBLE ANGLE IDENTITIES

The compound angle formulae deal with any two angles A and B and can therefore be used for two equal angles, i.e. when $B = A$.

Replacing B by A in the trig identities for $(A + B)$ gives the following set of double angle identities.

$$\sin 2A \equiv 2 \sin A \cos A$$

$$\cos 2A \equiv \cos^2 A - \sin^2 A$$

$$\tan 2A \equiv \frac{2 \tan A}{1 - \tan^2 A}$$

The second of these identities can be expressed in several forms because

$$\cos^2 A - \sin^2 A \equiv \begin{cases} (1 - \sin^2 A) - \sin^2 A \equiv 1 - 2 \sin^2 A \\ \cos^2 A - (1 - \cos^2 A) \equiv 2 \cos^2 A - 1 \end{cases}$$

i.e.

$$\cos 2A \equiv \begin{cases} \cos^2 A - \sin^2 A \\ 1 - 2 \sin^2 A \\ 2 \cos^2 A - 1 \end{cases}$$

There is one other group of identities that can be useful, and these are derived from two of the forms of the cosine double angle formulae.

Starting with $\cos 2A \equiv 2 \cos^2 A - 1$ we have

$$\cos^2 A \equiv \tfrac{1}{2}(1 + \cos 2A)$$

Similarly starting with $\cos 2A \equiv 1 - 2 \sin^2 A$, we get

$$\sin^2 A \equiv \tfrac{1}{2}(1 - \cos 2A)$$

1. If $\tan \theta = \frac{3}{4}$, find the values of $\tan 2\theta$ and $\tan 4\theta$

Using $\tan 2A \equiv \dfrac{2 \tan A}{1 - \tan^2 A}$ with $A = \theta$ and $\tan \theta = \frac{3}{4}$ gives

$$\tan 2\theta = \frac{2\left(\frac{3}{4}\right)}{1 - \left(\frac{3}{4}\right)^2} = \frac{24}{7}$$

Using the identity for $\tan 2A$ again, but this time with $A = 2\theta$, gives

$$\tan 4\theta = \frac{2 \tan 2\theta}{1 - \tan^2 2\theta} = \frac{2\left(\frac{24}{7}\right)}{1 - \left(\frac{24}{7}\right)^2} = -\frac{336}{527}$$

2. Eliminate θ from the equations $x = \cos 2\theta$, $y = \sec \theta$

Using $\cos 2\theta \equiv 2 \cos^2 \theta - 1$ gives

$$x = 2 \cos^2 \theta - 1 \quad \text{and} \quad y = \frac{1}{\cos \theta}$$

$$\therefore \qquad x = 2\left(\frac{1}{y}\right)^2 - 1$$

$$\Rightarrow \qquad (x + 1)y^2 = 2$$

Note that this is a Cartesian equation which has been obtained by *eliminating the parameter θ* from a *pair of parametric equations*.

3. Prove that $\sin 3A \equiv 3 \sin A - 4 \sin^3 A$

$$\sin 3A \equiv \sin (2A + A)$$
$$\equiv \sin 2A \cos A + \cos 2A \sin A$$
$$\equiv (2 \sin A \cos A) \cos A + (1 - 2 \sin^2 A) \sin A$$
$$\equiv 2 \sin A \cos^2 A + \sin A - 2 \sin^3 A$$
$$\equiv 2 \sin A (1 - \sin^2 A) + \sin A - 2 \sin^3 A$$
$$\equiv 3 \sin A - 4 \sin^3 A$$

4. Find the solution of the equation $\cos 2x + 3 \sin x = 2$ giving values of θ in the interval $[-\pi, \pi]$

When a trig equation involves different multiples of an angle, it is usually sensible to express the equation in a form where the trig ratios are all of the same angle and, when possible, only one trig ratio is included.

Using $\cos 2x \equiv 1 - 2\sin^2 x$ gives

$$1 - 2\sin^2 x + 3\sin x = 2$$

$$\Rightarrow \qquad 2\sin^2 x - 3\sin x + 1 = 0$$

$$\Rightarrow \qquad (2\sin x - 1)(\sin x - 1) = 0$$

$$\therefore \qquad \sin x = \tfrac{1}{2} \quad \text{or} \quad \sin x = 1$$

When $\quad \sin x = \tfrac{1}{2}, \quad x = \tfrac{1}{6}\pi, \tfrac{5}{6}\pi$

When $\quad \sin x = 1,$

$$x = \tfrac{1}{2}\pi$$

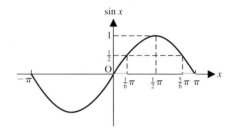

Therefore the solution is $\quad x = \tfrac{1}{6}\pi, \tfrac{5}{6}\pi, \tfrac{1}{2}\pi$

5. Express $4\cos^2 x + 1$ in terms of the angle $2x$

Using $\cos^2 x = \tfrac{1}{2}(1 + \cos 2x)$ gives

$$4\cos^2 x + 1 = 4 \times \tfrac{1}{2}(1 + \cos 2x) + 1$$
$$= 2(1 + \cos 2x) + 1$$
$$= 3 + 2\cos 2x$$

6. (a) Use the formula $\tan 2\theta = \dfrac{2\tan\theta}{1 - \tan^2\theta}$ to mark two sides of a right-angled triangle in terms of $\tan\theta$ where one of the angles in the triangle is 2θ. Use Pythagoras' theorem to find the length of the third side in terms of $\tan\theta$.

(b) Hence write down a formula for $\sin 2\theta$ in terms of $\tan\theta$.

(c) Use the formula from part (b) to find the smallest positive value of θ for which $2\tan\theta = \operatorname{cosec} 2\theta$.

(a)

```
       ╱|
   2θ ╱ |
     ╱  | x
1-tan²θ ╱   |
   ╱____|
   2 tan θ
```

Using Pythagoras's theorem,

$$x^2 = (2\tan\theta)^2 + (1 - \tan^2\theta)^2$$
$$= 4\tan^2\theta + 1 - 2\tan^2\theta + 4\tan^4\theta$$
$$= 1 + 2\tan^2\theta + \tan^4\theta$$
$$= (1 + \tan^2\theta)^2$$
$$\Rightarrow \qquad x = 1 + \tan^2\theta$$

(b) $\sin 2\theta = \dfrac{\text{opp}}{\text{hyp}} = \dfrac{2\tan\theta}{1+\tan^2\theta}$

(c) $2\tan\theta = \csc 2\theta = \dfrac{1}{\sin 2\theta}$

Using the formula from (b) gives $2\tan\theta = \dfrac{1+\tan^2\theta}{2\tan\theta}$

$\Rightarrow \qquad 4\tan^2\theta = 1+\tan^2\theta$

i.e. $\qquad 3\tan^2\theta = 1 \quad \Rightarrow \quad \tan\theta = \dfrac{1}{\sqrt{3}}$

The smallest positive value of θ for which $\tan\theta = \dfrac{1}{\sqrt{3}}$ is $\dfrac{\pi}{6}$

EXERCISE 7c

Simplify, giving an exact value where this is possible.

1. $2\sin 15° \cos 15°$

2. $\cos^2\frac{1}{8}\pi - \sin^2\frac{1}{8}\pi$

3. $\sin\theta\cos\theta$

4. $1 - 2\sin^2 4\theta$

5. $\dfrac{2\tan 75°}{1-\tan^2 75°}$

6. $\dfrac{2\tan 3\theta}{1-\tan^2 3\theta}$

7. $2\cos^2\frac{3}{8}\pi - 1$

8. $1 - 2\sin^2\frac{1}{8}\pi$

9. Find the value of $\cos 2\theta$ and $\sin 2\theta$ when θ is acute and when

(a) $\cos\theta = \frac{3}{5}$ 　　　　(b) $\sin\theta = \frac{7}{25}$ 　　　　(c) $\tan\theta = \frac{12}{5}$

10. If $\tan\theta = -\frac{7}{24}$ and θ is obtuse, find

(a) $\tan 2\theta$ 　　(b) $\cos 2\theta$ 　　(c) $\sin 2\theta$ 　　(d) $\cos 4\theta$

11. Eliminate θ from the following pairs of equations.

(a) $x = \tan 2\theta, \quad y = \tan\theta$ 　　　　(b) $x = \cos 2\theta, \quad y = \cos\theta$

(c) $x = \cos 2\theta, \quad y = \csc\theta$ 　　　　(d) $x = \sin 2\theta, \quad y = \sec 4\theta$

12. Express in terms of $\cos 2x$

(a) $2\sin^2 x - 1$ 　　　　　　　　(b) $4 - 2\cos^2 x$

(c) $2\cos^2 x + \sin^2 x$ 　　　　　　(d) $2\cos^2 x (1 + \cos^2 x)$

(e) $\cos^4 x$　(Hint: $\cos^4 x \equiv (\cos^2 x)^2$) 　　(f) $\sin^4 x$

13. Prove the following identities.

(a) $\dfrac{1 - \cos 2A}{\sin 2A} \equiv \tan A$

(b) $\sec 2A + \tan 2A \equiv \dfrac{\cos A + \sin A}{\cos A - \sin A}$

(c) $\cos 4A \equiv 8 \cos^4 A - 8 \cos^2 A + 1$

(d) $\sin 2\theta \equiv \dfrac{2 \tan \theta}{1 + \tan^2 \theta}$

(e) $\cos 2\theta \equiv \dfrac{1 - \tan^2 \theta}{1 + \tan^2 \theta}$

14. Find the solutions of the following equations in the interval $[0, 2\pi]$.

(a) $\cos 2x = \sin x$ (b) $\sin 2x + \cos x = 0$

(c) $\cos 2x = \cos x$ (d) $\sin 2x = \cos x$

(e) $4 - 5 \cos \theta = 2 \sin^2 \theta$ (f) $\sin 2\theta - 1 = \cos 2\theta$

15. (a) Use the identities in question 13 parts (d) and (e) to express in terms of $\tan \theta$

(i) $\cos 2\theta - \sin 2\theta$ (ii) $\dfrac{\sin 2\theta}{1 - \cos 2\theta}$

(b) Hence find, in degrees, the smallest value of θ greater than 0 for which

(i) $\cos 2\theta - \sin 2\theta = 1$ (ii) $\dfrac{\sin 2\theta}{1 - \cos 2\theta} = 1$

MIXED EXERCISE 7

1. Eliminate θ from the equations $x = \sin \theta$ and $y = \cos 2\theta$

2. Prove the identity $\dfrac{\sin 2\theta}{1 + \cos 2\theta} \equiv \tan \theta$

3. Prove that $\tan \left(\theta + \frac{1}{4}\pi\right) \tan \left(\frac{1}{4}\pi - \theta\right) \equiv -1$

4. If $\cos A = \frac{4}{5}$ and $\cos B = \frac{5}{13}$ find the possible values of $\cos (A + B)$

5. Eliminate θ from the equations $x = \cos 2\theta$ and $y = \cos^2 \theta$

6. Solve the equation $8 \sin \theta \cos \theta = 3$ for values of θ from $-180°$ to $180°$

7. Find in the interval $[-\pi, \pi]$ the solution of the equation $\cos^2 \theta - \sin^2 \theta = 1$

8. Prove the identity $\cos^4 \theta - \sin^4 \theta \equiv \cos 2\theta$

9. Simplify the expression $\dfrac{1 + \cos 2x}{1 - \cos 2x}$

10. Find the values of A between 0 and 360° for which
$\sin(60° - A) + \sin(120° - A) = 0$

11. (a) Express $2\sin^2\theta + 1$ in terms of $\cos 2\theta$

    (b) Express $4\cos^2 2A$ in terms of $\cos 4A$ (Hint: Use $2A = x$)

*12. You will need a graphics calculator or graph drawing software on a computer for this question.

    (a) Plot the graph of $y = 3\sin x + 4\cos x$ for $0 \leqslant x \leqslant 4\pi$

    (b) Your graph should suggest another form for the equation. What do you think this is?

    (c) Test your answer to (b) by plotting your suggested form.

    (d) Repeat parts (a) to (c) for $y = 5\sin x + 13\cos x$.

    (e) Try drawing graphs of $y = a\cos x + b\sin x$ for a variety of values of a and b. Hence suggest a way in which $a\cos x + b\sin x$ can be expressed as a single trig ratio.

CONSOLIDATION B

Logarithms

$$\log_a b = c \quad \Rightarrow \quad a^c = b$$

$$\log ab = \log a + \log b$$

$$\log a/b = \log a - \log b$$

$$\log a^n = n \log a$$

$$\log_a b = \frac{\log_c b}{\log_c a}$$

Trigonometric Identities

$$\tan \theta \equiv \frac{\sin \theta}{\cos \theta}$$

$$\cos^2 \theta + \sin^2 \theta \equiv 1$$

$$1 + \tan^2 \theta \equiv 1$$

Compound Angle Identities

$$\sin (A \pm B) \equiv \sin A \cos B \pm \cos A \sin B$$

$$\cos (A \pm B) = \cos A \cos B \mp \sin A \sin B$$

$$\tan (A \pm B) \equiv \frac{\tan A \pm \tan B}{1 \mp \tan A \tan B}$$

Note that using the upper sign throughout gives one formula and the lower sign gives another. No other combination of signs can be used.

Double Angle Identities

$$\sin 2A \equiv 2 \sin A \cos A$$

$$\cos 2A \equiv \begin{cases} \cos^2 A - \sin^2 A \\ 2 \cos^2 A - 1 \\ 1 - 2 \sin^2 A \end{cases} \quad \text{and} \quad \begin{cases} \cos^2 A = \frac{1}{2}(1 + \cos 2A) \\ \sin^2 A = \frac{1}{2}(1 - \cos 2A) \end{cases}$$

$$\tan 2A \equiv \frac{2 \tan A}{1 - \tan^2 A}$$

MULTIPLE CHOICE EXERCISE B

TYPE I

1. $\log 5 - 2 \log 2 + \frac{3}{2} \log 16$ is equal to

   A $\log 80$ **B** 10 **C** 0

   D $2 \log 12$ **E** 1

2. Given that $x = \cos^2 \theta$ and $y = \sin^2 \theta$,

   A $x^2 + y^2 = 1$ **B** $x + y = 0$ **C** $0 \leqslant y \leqslant 1$

   D $y = x - \frac{1}{2}\pi$ **E** $x + y = 1$

3. The value of $\log_5 0.04$ is

   A 4 **B** 5 **C** $\frac{1}{2}$ **D** -2 **E** 0.25

4. One of these is not an identity. Which one is it?

   A $\cos^2 \theta = 1 - \sin^2 \theta$ **B** $\cos 2\theta = 2 \cos^2 \theta + 1$

   C $\cos^2 \theta = \frac{1}{2}(\cos 2\theta + 1)$ **D** $1 + \tan^2 \theta = \sec^2 \theta$

   E $\tan 2\theta = \dfrac{2 \tan \theta}{1 - \tan^2 \theta}$

5. If $\log_x y = 2$ then

   A $x = 2y$ **B** $x = y^2$ **C** $x^2 = y$

   D $y = 2x$ **E** $y = \sqrt{x}$

6. Given that $e^{2x} + e^x = 2$,

   A $2x + x = \ln 2$ **B** $e^{3x} = 2$

   C $e^x = 1$ satisfies the equation   **D** $x^3 = 1$

7. $\ln x + 3 \ln y = 4$

   A $\ln xy^3 = 4$ **B** $3xy = e^4$

   C $y = 1$ when $x = 1$ **D** $x + y^3 = 4$

8. One of the following expressions is not identical to any of the others. Which one is it?

   A $\dfrac{2 \tan \theta}{1 + \tan^2 \theta}$ **B** $2 \cos^2 \tfrac{1}{2}\theta$ **C** $1 - \sin^2 \theta$

   D $\dfrac{1}{1 + \tan^2 \theta}$ **E** $\sin 2\theta$

TYPE II

9. The function $f(\theta) = \sec \theta$ is undefined when $\theta = \tfrac{1}{2}\pi$

10. $\tfrac{1}{2} \log 16 - 1$ can be expressed as a single logarithm.

11. $\dfrac{\operatorname{cosec} x}{\sec x} \equiv \tan x$

12. $3 \log x + 1 = \log 10x^3$ is an equation.

13. $-1 \leqslant \sec x \leqslant 1$

14. The equation $\cot \theta = 3$ has two roots in the interval $-\pi \leqslant \theta \leqslant \pi$

MISCELLANEOUS EXERCISE B

1. Find, correct to the nearest degree, all the values of θ between $0°$ and $360°$ satisfying the equation $8 \cos^2 \theta + 2 \sin \theta = 7$.   (WJEC)

2. Solve the equation $3^{2x} = 4^{2-x}$, giving your answer to three significant figures.   (UCLES)

3. Find all values of θ, such that $0° \leqslant \theta \leqslant 180°$, which satisfy the equation $2 \sin 2\theta = \tan \theta$ (UCLES)

4. Solve the equation $x^{1/2} + x^{-1/2} = 2(x^{1/2} - x^{-1/2})$.   (UCLES)

5. Solve the equation $9 \cos^2 x - 6 \cos x - 0.21 = 0$, $0° \leqslant x \leqslant 360°$, giving each answer in degrees to 1 decimal place.   (ULEAC)

6. Solve the inequality $(0.8)^x > 5$.   (UCLES)$_s$

7. On a single diagram, sketch the graphs of $y = \ln(10x)$ and $y = \dfrac{6}{x}$, and explain

how you can deduce that the equation $\ln(10x) = \dfrac{6}{x}$ has exactly one real root.

Given that the root is close to 2, use the iteration

$$x_{n+1} = \frac{6}{\ln(10x_n)}$$

to evaluate the root correct to three decimal places.   (UCLES)$_{sp}$

8. Show that $\tan\theta + \cot\theta = \dfrac{2}{\sin 2\theta}$.

Hence, or otherwise, solve the equation $\tan\theta + \cot\theta = 4$, giving all the values
of θ between $0°$ and $360°$.   (UCLES)$_s$

9. Solve the equation $4\tan^2 x + 12\sec x + 1 = 0$, giving all the solutions in degrees,
to the nearest degree, in the interval $-180° < x < 180°$.   (AEB)$_s$

10. (a) Find the values of $\cos x$ for which

$$6\sin^2 x = 5 + \cos x.$$

(b) Find all the values of x in the interval $180° < x < 540°$ for which

$$6\sin^2 x = 5 + \cos x.$$ (ULEAC)$_s$

11. (i)

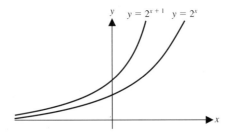

The diagram shows the graphs of $y = 2^x$ and $y = 2^{x+1}$.
Describe two different geometrical transformations which map the graph
of $y = 2^x$ onto the graph of $y = 2^{x+1}$.

(ii) Given that $3^x = 2^{ax}$ for all values of x, find a, giving your answer in terms
of natural logarithms.
Describe a geometrical transformation which maps the graph of $y = 2^x$
onto the graph of $y = 3^x$.   (UCLES)$_s$

12. Given that $-90 < x < 90$ find the values of x for which

(a) $4\sin^2 x° = 3$,

(b) $\sec(2x - 15)° = 2$.   (ULEAC)

13. (a)   Given that $(3x + 2)$ is a factor of $3x^3 + Ax^2 - 4x - 4$, show that $A = 5$.

    (b)   Factorise $3x^3 + 5x^2 - 4x - 4$ completely.

    (c)   Given that $0° \leqslant t \leqslant 360°$, find the values of t, to the nearest degree, for which

$$3 \sin^3 t + 5 \sin^2 t - 4 \sin t - 4 = 0. \qquad \text{(ULEAC)}$$

14. Solve the equation

$$2 \log_3 x = 1 + \log_3 (18 - x) \qquad \text{(AEB)}$$

15. Given that $\log(2x - 4) + \log 3 = 3 \log y$ find an expression for x in terms of y.

CHAPTER 8

DIFFERENTIATION

In *Module A* we established that, for a curve $y = f(x)$,

the gradient of the curve at A

$\quad$ = gradient of the tangent to the curve at A

$\quad = \displaystyle\lim_{\delta x \to 0} \frac{\delta y}{\delta x}$

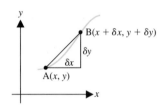

The gradient is denoted by $\dfrac{dy}{dx}$ or $f'(x)$,

i.e. $\quad \dfrac{dy}{dx} = \displaystyle\lim_{\delta x \to 0} \frac{\delta y}{\delta x} \quad$ and $\quad f'(x) = \displaystyle\lim_{\delta x \to 0} \frac{f(x + \delta x) - f(x)}{\delta x}$

The following particular results were also established.

y	$\dfrac{dy}{dx}$
a	0
ax	a
x^n	nx^{n-1}
e^x	e^x
e^{ax}	ae^{ax}
$\ln x$	$1/x$

GRADIENTS OF TANGENTS AND NORMALS

If the equation of a curve is known, and the gradient function can be found, then the gradient, m say, at a particular point A on that curve can be calculated. This is also the gradient of the tangent to the curve at A.

The normal at A is perpendicular to the tangent at A, therefore its gradient is $-1/m$.

Examples 8a

1. The equation of a curve is $s = 6 - 3t - 4t^2 - t^3$. Find the gradient of the tangent and of the normal to the curve at the point $(-2, 4)$.

$$s = 6 - 3t - 4t^2 - t^3 \quad \Rightarrow \quad \frac{ds}{dt} = 0 - 3 - 8t - 3t^2$$

At the point $(-2, 4)$, $\dfrac{ds}{dt} = -3 - 8(-2) - 3(-2)^2 = 1$

Therefore the gradient of the tangent at $(-2, 4)$ is 1 and the gradient of the normal is $-1/1$, i.e. -1

2. Find the coordinates of the points on the curve $y = 2x^3 - 3x^2 - 8x + 7$ where the gradient is 4.

$$y = 2x^3 - 3x^2 - 8x + 7 \quad \Rightarrow \quad \frac{dy}{dx} = 6x^2 - 6x - 8$$

If the gradient is 4 then $\dfrac{dy}{dx} = 4$

i.e. $\quad 6x^2 - 6x - 8 = 4 \quad \Rightarrow \quad 6x^2 - 6x - 12 = 0$

$$\Rightarrow \quad x^2 - x - 2 = 0$$

$\therefore \qquad (x - 2)(x + 1) = 0 \quad \Rightarrow \quad x = 2 \quad \text{or} \quad -1$

When $\quad x = 2, \quad y = 16 - 12 - 16 + 7 = -5$

when $\quad x = -1, \quad y = -2 - 3 + 8 + 7 = 10$

Therefore the gradient is 4 at the points $(2, -5)$ and $(-1, 10)$

EXERCISE 8a

Find the gradient of the tangent and the gradient of the normal at the given point on the given curve.

1. $y = x^2 + 4$ where $x = 1$

2. $y = 3/x$ where $x = -3$

3. $y = \sqrt{z}$ where $z = 4$

4. $s = 2t^3$ where $t = -1$

5. $v = 2 - \dfrac{1}{u}$ where $u = 1$

6. $y = (x + 3)(x - 4)$ where $x = 3$

7. $y = z^3 - z$ where $z = 2$

8. $s = t + 3t^2$ where $t = -2$

9. $z = x^2 - \dfrac{2}{x}$ where $x = 1$

10. $y = \sqrt{x} + \dfrac{1}{\sqrt{x}}$ where $x = 9$

11. $s = \sqrt{t}(1 + \sqrt{t})$ where $t = 4$

12. $y = \dfrac{x^2 - 4}{x}$ where $x = -2$

Find the coordinates of the point(s) on the given curve where the gradient has the value specified.

13. $y = 3 - \dfrac{2}{x}; \quad \dfrac{1}{2}$ 14. $z = x^2 - x^3; \quad -1$

15. $s = t^3 - 12t + 9; \quad 15$ 16. $v = u + \dfrac{1}{u}; \quad 0$

17. $s = (t + 3)(t - 5); \quad 0$ 18. $y = \dfrac{1}{x^2}; \quad \dfrac{1}{4}$

19. $y = (2x - 5)(x + 1); \quad -3$ 20. $y = z^3 - 3z; \quad 0$

DIFFERENTIATING A FUNCTION OF A FUNCTION

Suppose that we want to differentiate $(2x-1)^3$. We could expand the bracket and differentiate term by term, but this is tedious and, for powers higher than three, very long and not easy. We obviously need a more direct method for differentiating an expression of this kind.

Now $(2x-1)^3$ is a cubic function of the linear function $(2x-1)$, i.e. it is a *function of a function*.

A function of this type is of the form $gf(x)$.

Consider any equation of the form $y = gf(x)$.

If we make the substitution $u = f(x)$ then $y = gf(x)$ can be expressed in two simple parts,

i.e. $\quad y = g(u) \quad$ where $\quad u = f(x)$

A small increase of δx in the value of x causes a corresponding small increase of δu in the value of u.

Then if $\delta x \to 0$, it follows that $\delta u \to 0$

Hence
$$\frac{dy}{dx} = \lim_{\delta x \to 0} \left(\frac{\delta y}{\delta x} \right) = \lim_{\delta x \to 0} \left(\frac{\delta y}{\delta u} \right) \left(\frac{\delta u}{\delta x} \right)$$

$\Rightarrow$
$$\frac{dy}{dx} = \left(\lim_{\delta u \to 0} \frac{\delta y}{\delta u} \right) \times \left(\lim_{\delta x \to 0} \frac{\delta u}{\delta x} \right)$$

i.e.
$$\frac{dy}{dx} = \frac{dy}{du} \times \frac{du}{dx}$$

This is known as *the chain rule*.

Example 8b

Find $\dfrac{dy}{dx}$ if $y = (2x-4)^4$

If $\quad u = 2x - 4 \quad$ then $\quad y = u^4$

Then $\quad \dfrac{dy}{dx} = \dfrac{dy}{du} \times \dfrac{du}{dx} \quad$ gives $\quad \dfrac{dy}{dx} = (4u^3)(2) = 8u^3$

But $\quad u = 2x - 4$

$\therefore \quad \dfrac{dy}{dx} = 8(2x-4)^3$

This example is a particular case of the equation $y = (ax+b)^n$.

Similar working shows that, in general,

$$\text{if} \quad y = (ax + b)^n \quad \text{then} \quad \frac{dy}{dx} = an(ax + b)^{n-1}$$

This fact is needed very often and is quotable.

Use the quotable result above to differentiate each function with respect to x.

1. $(3x + 1)^2$
2. $(x - 3)^4$
3. $(4x + 5)^5$

4. $(2 + 3x)^7$
5. $(6x - 2)^3$
6. $(4 - 2x)^5$

7. $(1 - 5x)^2$
8. $(3 - 2x)^3$
9. $(4 - 3x)^4$

10. $(3x + 1)^{-1}$
11. $(2x - 5)^{-4}$
12. $(1 - 2x)^{-5}$

13. $(2x + 3)^{1/2}$
14. $(8 - 3x)^{1/3}$
15. $\sqrt{(1 - 4x)}$

16. $\dfrac{1}{2 - 7x}$
17. $\dfrac{1}{\sqrt{(3 - x)}}$
18. $\dfrac{1}{(1 - 5x)^2}$

To Differentiate e^u where $u = f(x)$

If $\quad y = e^u \quad$ then $\quad \dfrac{dy}{dx} = \dfrac{dy}{du} \times \dfrac{du}{dx} \quad$ gives $\quad \dfrac{dy}{dx} = e^u \times \dfrac{du}{dx}$

This can also be expressed in the form

$$\frac{d}{dx} e^{f(x)} = e^{f(x)} f'(x)$$

i.e. $$\frac{d}{dx} e^{f(x)} = f'(x) e^{f(x)}$$

e.g. $\quad$ if $\quad y = e^{(x^2 + 1)} \quad$ then $\quad \dfrac{dy}{dx} = 2x e^{(x^2 + 1)}$

The case when u is a linear function of x is particularly useful,

i.e. $$y = e^{(ax + b)} \quad \Rightarrow \quad \frac{dy}{dx} = a e^{(ax + b)}$$

To Differentiate ln u where $u = f(x)$

If $\quad y = \ln u \quad$ then $\quad \dfrac{dy}{dx} = \dfrac{dy}{du} \times \dfrac{du}{dx} \quad$ gives $\quad \dfrac{dy}{dx} = \dfrac{1}{u} \times \dfrac{du}{dx}$

This can also be expressed in the form

$$\frac{d}{dx}\{\ln f(x)\} = \frac{1}{f(x)} \times f'(x) = \frac{f'(x)}{f(x)}$$

e.g. $\quad$ if $\quad y = \ln(2 + x^3) \quad$ then $\quad \dfrac{dy}{dx} = \dfrac{3x^2}{2 + x^3}$

Again the case when u is $ax + b$ occurs frequently and is worth noting,

i.e. $\qquad\qquad y = \ln(ax + b) \qquad \dfrac{dy}{dx} = \dfrac{a}{ax + b}$

Example 8c

Find $\dfrac{dy}{dx}$ when $y = \ln 5x^{-1/2}$

We could use the standard result to differentiate this equation but the algebraic manipulation required is not straightforward. The differentiation is simpler if we use the laws of logs to simplify $\ln 5x^{-1/2}$ *before* it is differentiated.
In this case we use $\log ab \equiv \log a + \log b$ and $\log a^n \equiv n \log a$

$$\ln 5x^{-1/2} = \ln 5 + \ln x^{-1/2}$$

$$= \ln 5 - \tfrac{1}{2}\ln x$$

$$\therefore \qquad \frac{dy}{dx} = 0 - \left(\frac{1}{2}\right)\left(\frac{1}{x}\right) = \frac{1}{2x}$$

EXERCISE 8c

Use the quotable results above to differentiate each function with respect to x.

1. $\ln(x^3)$

2. $e^{(x^2 + 1)}$

3. $\ln 3x$

4. $\sqrt{\{e^{(x+1)}\}}$

5. $\ln x^{-2}$

6. $\ln(3 - x^2)$

7. $\ln(3x - 4)$

8. e^{-2x}

9. $\ln(2\sqrt{x})$

10. $\ln(2x-3)^3$ 11. $\ln \dfrac{x}{x+1}$ 12. $\ln \sqrt{(x-1)}$

13. If f and g are the functions defined by $f : x \rightarrow x^2$ and $g : x \rightarrow e^x$
    (a) write down the functions $fg(x)$ and $gf(x)$ with respect to x.
    (b) differentiate $fg(x)$ and $gf(x)$ with respect to x.

Using the Second Derivative to distinguish between Stationary Points

This section starts with a brief reminder of the work on stationary values in *Module A*.

At a stationary point on a curve, y is momentarily neither increasing nor decreasing, so $\dfrac{dy}{dx} = 0$, i.e. the tangent to the curve is horizontal.

The three types of stationary point are shown in this diagram.

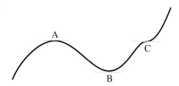

A is a maximum point $\left.\begin{array}{l} \\ \end{array}\right\}$ A and B are turning points
B is a minimum point

C is a point of inflexion.

The nature of a stationary point, P, can be identified by considering two points close to, and on either side of, P and using either y values on each side of the stationary point, or the sign of $\dfrac{dy}{dx}$ at each side of stationary point.

There is a third method for distinguishing between the various types of stationary point, which will now be considered.

The way in which $\dfrac{dy}{dx}$ changes in the region of a stationary point varies with the type of stationary point. This behaviour pattern provides another way to identify the nature of that point.

Now the rate at which $\dfrac{dy}{dx}$ increases with respect to x can be written

$\dfrac{d}{dx}\left(\dfrac{dy}{dx}\right)$ which is condensed to $\dfrac{d^2y}{dx^2}$

(we say 'd 2 y by d x squared')

If, for example, $\dfrac{dy}{dx}$ is *increasing* as x increases, we can say that $\dfrac{d^2y}{dx^2}$ is positive.

$\dfrac{d^2y}{dx^2}$ is called the *second derivative* with respect to x of y and, if $y = f(x)$, the second derivative can also be denoted by $f''(x)$.

Now we can examine the behaviour of $\dfrac{dy}{dx}$ at each type of stationary point.

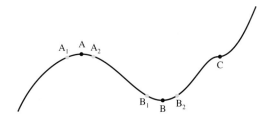

For the maximum point A,

at A_1, $\dfrac{dy}{dx}$ is +ve and at A_2, $\dfrac{dy}{dx}$ is −ve

so, passing through A, $\dfrac{dy}{dx}$ goes from + to −, i.e. $\dfrac{dy}{dx}$ decreases

$\Rightarrow$ at A, $\dfrac{d^2y}{dx^2}$ is negative.

For the minimum point B,

at B_1, $\dfrac{dy}{dx}$ is −ve and at B_2, $\dfrac{dy}{dx}$ is +ve

so, passing through B, $\dfrac{dy}{dx}$ goes from − to +, i.e. $\dfrac{dy}{dx}$ increases

$\Rightarrow$ at B, $\dfrac{d^2y}{dx^2}$ is positive.

At the point of inflexion, C, although $\dfrac{dy}{dx}$ can become zero, it does not change

sign so $\dfrac{d^2y}{dx^2}$ is zero.

Unfortunately it is also possible for $\dfrac{d^2y}{dx^2}$ to be zero at a turning point, so

finding that $\dfrac{d^2y}{dx^2} = 0$ does not provide a definite conclusion.

Summing up we have:

	Maximum	Minimum
Sign of $\dfrac{d^2y}{dx^2}$	Negative (or zero)	Positive (or zero)

Note that, if $\dfrac{d^2y}{dx^2}$ is zero, one of the other two methods must be used to
determine the nature of a stationary point.

Examples 8d

1. Locate the stationary points on the curve $y = x^3 - 6x^2 + 9x + 5$ and determine the
    nature of each one. Sketch the curve, marking the coordinates of the stationary
    points.

$$y = x^3 - 6x^2 + 9x + 5 \quad \Rightarrow \quad \dfrac{dy}{dx} = 3x^2 - 12x + 9$$

At stationary points $\dfrac{dy}{dx} = 0$

i.e. $3x^2 - 12x + 9 = 0 \quad \Rightarrow \quad 3(x-1)(x-3) = 0$

When $x = 1$, $y = 9$ and when $x = 3$, $y = 5$
i.e. the stationary points are $(1, 9)$ and $(3, 5)$.

Differentiating $\dfrac{dy}{dx}$ w.r.t. x gives $\dfrac{d^2y}{dx^2} = 6x - 12$

When $x = 1$, $\dfrac{\mathrm{d}^2 y}{\mathrm{d}x^2}$ is negative, so $(1, 9)$ is a maximum point

When $x = 3$, $\dfrac{\mathrm{d}^2 y}{\mathrm{d}x^2}$ is positive, so $(3, 5)$ is a minimum point

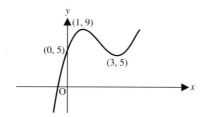

2. Find the stationary values of the function $f : x \to x^5$ and investigate their nature.

$$f(x) = x^5 \quad \Rightarrow \quad f'(x) = 5x^4$$

At stationary values $f'(x) = 0$, i.e. $5x^4 = 0 \quad \Rightarrow \quad x = 0$

i.e. there is just one stationary point, where $x = 0$

When $x = 0$, $f(x) = 0$

$\therefore$ the stationary value of $f(x)$ is 0.

Differentiating $f'(x)$ w.r.t. x gives $f''(x) = 20x^3$

when $x = 0$, $f''(x) = 0$

This is inconclusive so we will look at the signs of $f'(x)$ on either side of $x = 0$

x	-1	0	1
$f'(x)$	$+$	0	$+$
Gradient	$/$	$-$	$/$

From this table we see that the stationary value at $x = 0$, is a point of inflexion.

EXERCISE 8d

Find the stationary point(s) on the following curves and distinguish between them.

1. $y = x^3$ 2. $y = 12x - 4x^3$ 3. $y = x(x^2 - 3)$

4. $y = x - x^2$ 5. $y = x + \dfrac{4}{x}$ 6. $y = \dfrac{1}{x^2} + x^2$

7. $y = 2x^5 - 5x^3$ 8. $y = x^4$ 9. $y = \dfrac{x^2}{2} + \dfrac{1}{x}$

10. This is a sketch of the curve $y = x^4$

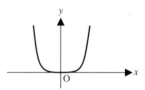

  (a) Describe the behaviour of y as x increases from negative to positive values.

  (b) Describe the behaviour of $\dfrac{dy}{dx}$ as x goes from negative to positive values.

  (c) Find $\dfrac{dy}{dx}$ as a function of x and sketch the graph of $\dfrac{dy}{dx}$ against values of x.

  (d) From your sketch $\left(\text{i.e. do not find } \dfrac{d^2y}{dx^2} \right)$ deduce the value of $\dfrac{d^2y}{dx^2}$ when $x = 0$ and give a reason for your answer.

*11. Use a graphics calculator, or computer, to draw, on the same set of axes

  (a) $y = x^2$, $y = x^4$, $y = x^6$ (b) $y = x^3$, $y = x^5$, $y = x^7$

  Describe briefly the effect on the shape of the graph of raising the power of x.

DIFFERENTIATING A PRODUCT

In *Module A*, products were differentiated by multiplying them out and differentiating term by term.
Clearly this is suitable only for fairly simple products and a more general method is needed.

Suppose that $y = uv$ where u and v are both functions of x, e.g. $y = x^2(x^4 - 1)$.

It is dangerously tempting to think that $\dfrac{dy}{dx}$ is given by $\left(\dfrac{du}{dx} \right)\left(\dfrac{dv}{dx} \right)$. But this is *not so* as is clearly shown by a simple example such as $y = (x^2)(x^3)$ where, because $y = x^5$, we know that $\dfrac{dy}{dx} = 5x^4$ which is *not* equal to $(2x)(3x^2)$.

i.e.   differentiation is *not* distributive across a product.

Returning to $y = uv$ where $u = f(x)$ and $v = g(x)$, we see that if x increases by a small amount δx then there are corresponding small increases of δu, δv and δy in the values of u, v and y.

$$\therefore \qquad y + \delta y = (u + \delta u)(v + \delta v)$$

$$= uv + u\delta v + v\delta u + \delta u \delta v$$

But $\quad y = uv \quad$ so $\quad \delta y = u\delta v + v\delta u + \delta u \delta v$

$$\Rightarrow \qquad \frac{\delta y}{\delta x} = u\frac{\delta v}{\delta x} + v\frac{\delta u}{\delta x} + \delta u\frac{\delta v}{\delta x}$$

Now as $\delta x \to 0$, $\quad \dfrac{\delta v}{\delta x} \to \dfrac{dv}{dx}, \quad \dfrac{\delta u}{\delta x} \to \dfrac{du}{dx} \quad$ and $\quad \delta u \to 0$

Therefore $\quad \dfrac{dy}{dx} = \lim\limits_{\delta x \to 0} \dfrac{\delta y}{\delta x} = u\dfrac{dy}{dx} + v\dfrac{du}{dx} + 0$

i.e. $$\frac{d}{dx}(uv) = v\frac{du}{dx} + u\frac{dv}{dx}$$

This formula is verified by the simple example we considered above, i.e. $y = (x^2)(x^3)$.

Using $u = x^2$ and $v = x^3$ gives $\dfrac{dy}{dx} = (x^3)(2x) + (x^2)(3x^2) = 5x^4$ which is correct.

Example 8e

Differentiate with respect to x

(a) $x^2 e^x$ (b) $(x+1)^3(2x-5)^2$ (c) $\dfrac{(x-1)^2}{(x+2)}$

(a) $u = x^2 \quad$ so $\quad \dfrac{du}{dx} = 2x$

$\quad v = e^x \quad$ so $\quad \dfrac{dv}{dx} = e^x$

$\dfrac{d}{dx}(uv) = v\dfrac{du}{dx} + u\dfrac{dv}{dx} \quad$ gives $\quad \dfrac{d}{dx}(x^2 e^x) = 2xe^x + x^2 e^x$

(b) If $u = (x+1)^3$, then $\dfrac{du}{dx} = 3(x+1)^2$

and if $v = (2x-5)^2$, then $\dfrac{dv}{dx} = \{2(2)(2x-5)\}$

$\dfrac{d}{dx}(uv) = v\dfrac{du}{dx} + u\dfrac{dv}{dx}$ gives

$\dfrac{d}{dx}(x+1)^3(2x-5)^2 = \{(2x-5)^2\}\{3(x+1)^2\}$

$\qquad\qquad\qquad + \{(x+1)^3\}\{2(2)(2x-5)\}$

$\qquad\qquad\quad = (2x-5)(x+1)^2\{3(2x-5)+4(x+1)\}$

$\qquad\qquad\quad = (2x-5)(x+1)^2(10x-11)$

(c) If we write $\dfrac{(x-1)^2}{(x+2)}$ as $(x-1)^2(x+2)^{-1}$

then $u = (x-1)^2$ gives $\dfrac{du}{dx} = 2(x-1)$

and $v = (x+2)^{-1}$ gives $\dfrac{dv}{dx} = -(x+2)^{-2}$

Using $\dfrac{d}{dx}(uv) = v\dfrac{du}{dx} + u\dfrac{dv}{dx}$ we have

$\dfrac{d}{dx}\left[\dfrac{(x-1)^2}{(x+2)}\right] = (x+2)^{-1}\{2(x-1)\} + (x-1)^2\{-(x+2)^{-2}\}$

$\qquad\qquad\qquad = \dfrac{(x-1)}{(x+2)^2}\{2(x+2)-(x-1)\}$

$\qquad\qquad\qquad = \dfrac{(x-1)(x+5)}{(x+2)^2}$

EXERCISE 8e

Differentiate each function with respect to x.

1. $x(x-3)^2$

2. $(x-6)\sqrt{x}$

3. $(x+2)(x-2)^5$

4. $x(2x+3)^3$

5. $(x+1)(x-1)^4$

6. $\sqrt{x}(x-3)^3$

7. $\dfrac{(x+5)^4}{(x-3)}$

8. $\dfrac{x}{(3x+2)}$

9. $\dfrac{(2x-7)^{3/2}}{x}$

10. $x^3\sqrt{(x-1)}$

11. $x(x+3)^{-1}$

12. $x^2(2x-3)^2$

13. $(x-1)e^x$

14. $x^2 e^{2x}$

15. $x\ln x$

16. $x\ln(2x-1)$

17. xe^{3x-1}

18. $(x^2-1)\ln\sqrt{x}$

DIFFERENTIATING A QUOTIENT

To differentiate a function of the form u/v, where u and v are both functions of x, it is sometimes convenient to rewrite the function as uv^{-1} and differentiate it as a product. This method was used in part (c) of the previous worked example but it is not always the neatest way to differentiate a quotient. The alternative is to apply the formula derived below.

When a function is of the form u/v, where u and v are both functions of x, a small increase of δx in the value of x causes corresponding small increases of δu and δv in the values of u and v. Then, as $\delta x \rightarrow 0$, δu and δv also tend to zero.

$$\text{If} \qquad y = \frac{u}{v} \quad \text{then} \quad y + \delta y = \frac{(u + \delta u)}{(v + \delta v)}$$

$$\therefore \qquad \delta y = \frac{u + \delta u}{v + \delta v} - \frac{u}{v} = \frac{v\delta u - u\delta v}{v(v + \delta v)}$$

$$\therefore \qquad \frac{\delta y}{\delta x} = \left(v\frac{\delta u}{\delta x} - u\frac{\delta v}{\delta x} \right) \Big/ v(v + \delta v)$$

$$\Rightarrow \qquad \frac{dy}{dx} = \lim_{\delta x \to 0} \frac{\delta y}{\delta x} = \left(v\frac{du}{dx} - u\frac{dv}{dx} \right) \Big/ v^2$$

$$\text{i.e.} \qquad \frac{dy}{dx} = \frac{v\dfrac{du}{dx} - u\dfrac{dv}{dx}}{v^2}$$

Example 8f

If $y = \dfrac{(4x - 3)^6}{(x + 2)}$ find $\dfrac{dy}{dx}$

Using $u = (4x - 3)^6$ gives $\dfrac{du}{dx} = 24(4x - 3)^5$

and $v = x + 2$ gives $\dfrac{dv}{dx} = 1$

Then $\dfrac{dy}{dx} = \left(v\dfrac{du}{dx} - u\dfrac{dv}{dx} \right) \Big/ v^2$

$$= \frac{(x + 2)\{24(4x - 3)^5\} - (4x - 3)^6}{(x + 2)^2}$$

$$= \frac{(4x - 3)^5(20x + 51)}{(x + 2)^2}$$

Use the quotient formula to differentiate each of the following functions with respect to x

1. $\dfrac{(x-3)^2}{x}$

2. $\dfrac{x^2}{(x+3)}$

3. $\dfrac{(4-x)}{x^2}$

4. $\dfrac{(x+1)^2}{x^3}$

5. $\dfrac{4x}{(1-x)^3}$

6. $\dfrac{2x^2}{(x-2)}$

7. $\dfrac{x^{5/3}}{(3x-2)}$

8. $\dfrac{(1-2x)^3}{x^3}$

9. $\dfrac{\sqrt{(x+1)^5}}{x}$

10. $\dfrac{e^x}{e^x-1}$

11. $\dfrac{\ln x}{2x+1}$

12. $\dfrac{\ln x}{\ln(x+1)}$

IDENTIFYING THE CATEGORY OF A FUNCTION

Before any of the techniques explained earlier can be used to differentiate a given function, it is important to recognise the category to which the function belongs, i.e. is it a product or a function of a function or, if it is a fraction, is it one which would better be expressed as a product.

A product comprises two parts, each of which is an *independent* function of x, whereas *if one operation is carried out on another function of x we have a* function of a function.

Example 8g

Differentiate $\ln\{x\sqrt{(x^2-4)}\}$, w.r.t. x

This is the log of a function of x, i.e. it is a function of a function.
However we can simplify the expression by using the laws of logarithms to change it to a sum.

$$\ln\{x\sqrt{(x^2-4)}\} = \ln x + \ln\sqrt{(x^2-4)} = \ln x + \tfrac{1}{2}\ln(x^2-4)$$

$$\therefore \quad \frac{d}{dx}\ln\{x\sqrt{(x^2-4)}\} = \frac{d}{dx}\ln x + \frac{d}{dx}\{\tfrac{1}{2}\ln(x^2-4)\}$$

$$= \frac{1}{x} + \frac{1}{2}\left(\frac{2x}{x^2-4}\right)$$

$$= \frac{1}{x} + \frac{x}{x^2-4}$$

Simplifying the given function at the start, made the differentiation in this problem much easier. *Before differentiating any function, all possible simplification should be done*, particularly when complicated log expressions are involved.

This exercise contains a mixture of compound functions. In each case first identify the type of function and then use the appropriate method to find its derivative.

1. $x\sqrt{(x+1)}$

2. $(x^2 - 8)^3$

3. $x/(x^2 + 1)$

4. $\sqrt[3]{(2 - x^4)}$

5. $(x^2 + 1)/(x^2 + 2)$

6. $x^2(\sqrt{x} - 2)$

7. $(x^2 - 2)^3$

8. $\sqrt{(x - x^2)}$

9. $x/(\sqrt{x} + 1)$

10. $x^2\sqrt{(x - 2)}$

11. $\sqrt{(x + 1)}/x^2$

12. $(x^4 + x^2)^3$

13. $\sqrt{(x^2 - 8)}$

14. $x^3(x^2 - 6)$

15. $(x^2 - 6)^3$

16. $x/(x^2 - 6)$

17. $(x^4 + 3)^{-2}$

18. $\sqrt{x}(2 - x)^3$

19. $\sqrt{x}/(2 - x)^3$

20. $(x - 1)(x - 2)^2$

21. $(2x^3 + 4)^5$

22. xe^x

23. $x^2 \ln x$

24. $e^x(x^3 - 2)$

25. $x^2 \ln (x - 2)^6$

26. $(x - 1)e^x$

27. $(x^2 + 4) \ln \sqrt{x}$

28. $x\sqrt{(2 + x)}$

29. $x \ln \sqrt{(x - 5)}$

30. $(x^2 - 2)e^x$

31. $\dfrac{x}{e^x}$

32. $\dfrac{e^x}{x^2}$

33. $\dfrac{(\ln x)}{x^3}$

34. $\dfrac{\sqrt{(x + 1)}}{\ln x}$

35. $\dfrac{e^x}{x^2 - 1}$

36. $\dfrac{e^x}{e^x - e^{-x}}$

37. e^{4x}

38. $\ln (x^2 - 1)$

39. e^{x^2}

40. $6e^{(1-x)}$

41. $e^{(x^2 + 1)}$

42. $\ln \sqrt{(x + 2)}$

43. $(\ln x)^2$

44. $1/(\ln x)$

45. $\sqrt{(e^x)}$

46. $x \ln x$

47. $(4x - 1)^{2/3}$

48. $\dfrac{e^x}{x - 1}$

49. 10^{3x}

50. $\dfrac{(1 + 2x^2)}{1 + x^2}$

51. $e^{-2/x}$

52. $\ln (1 - e^x)$

53. $e^{3x} x^3$

54. $\dfrac{e^{x/2}}{x^5}$

55. $\ln 4x^3 (x+3)^2$ 56. $(\ln x)^4$ 57. $\dfrac{(x+3)^3}{x^2+2}$ 58. $4 \ln (x^2+1)$

Find and simplify $\dfrac{dy}{dx}$ and hence find $\dfrac{d^2y}{dx^2}$ if

59. $y = \ln \dfrac{x}{x+1}$ 60. $y = \dfrac{e^x}{e^x - 4}$

*61. The equation of a curve is $y = \dfrac{x^2 - 3}{e^x}$

(a) Find the stationary points on the curve.

(b) Determine the nature of these stationary points

(c) What happens to the value of y as (i) $x \to \infty$ (ii) $x \to -\infty$

(d) Use the information from parts (a) to (c) to *sketch* the curve. Use a graphics calculator *as a check*.

(e) How many roots are there of the equation $\dfrac{x^2 - 3}{e^x} = 0$? Justify your answer.

(f) Find these roots.

*62. The graph shows the number of stick insects in a vivarium at various times.

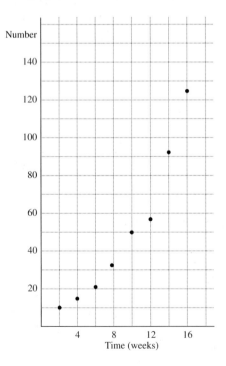

(a) Find the growth factor between the numbers of stick insects in consecutive time intervals of two weeks. Hence find a relationship between x (time in days) and y (number of insects) that can be used as an approximate model for the data given in the graph.

(b) Use the model to estimate
   (i) the number of insects 45 days after the first recorded entry on the graph
   (ii) how many days after the first recorded entry the number of insects had grown to 100.
   Explain why these results can only be estimates.

(c) Find the rate at which the numbers are growing 10 weeks after the first entry and comment on accuracy of the result.

CHAPTER 9

THE BINOMIAL EXPANSION

POWER SERIES

A series such as $x + x^2 + x^3 + \ldots$ is called a power series because the terms involve powers of a variable quantity. A series, such as $1 + 2 + 4 + 8 + \ldots$, each of whose terms has a fixed numerical value, is called a number series.

THE BINOMIAL THEOREM

When an expression such as $(1 + x)^4$ is expanded, the coefficients of the terms in the expansion can be obtained from Pascal's Triangle. Now $(1 + x)^{20}$ could be expanded in the same way but as the construction of the triangular array would be tedious, we need a more general method to expand powers of $(1 + x)$.

This general method uses the *binomial theorem* which states that

when *n* is a positive integer

$$(1 + x)^n = 1 + nx + \frac{n(n-1)}{2}x^2 + \frac{n(n-1)(n-2)}{(2)(3)}x^3 + \frac{n(n-1)(n-2)(n-3)}{(2)(3)(4)}x^4 + \ldots + x$$

The coefficients of the powers of x are called *binomial coefficients*. For powers of x greater than 3, the coefficients are lengthy to write out in full so we use the notation $\binom{n}{r}$ to represent the coefficient of x^r.

The pattern shown in the coefficients given above shows that

$$\binom{n}{r} = \frac{n(n-1)(n-2)\ldots(n-r+1)}{(2)(3)(4)\ldots(r)}$$

Notice that the expansion of $(1 + x)^n$ is a finite series which has $(n + 1)$ terms and that the term in x^2 is the third term, the term in x^3 is the fourth term and the *term in x^r is the $(r + 1)$th term*.

Factorial Notation

The coefficients of powers of x involve expressions such as $1 \times 2 \times 3 \times 4$; these are tedious to write out so we introduce a special notation for such products. The product $1 \times 2 \times 3 \times 4 \times 5 \times 6 \times 7 \times 8$ is written as $8!$ and called 'eight factorial'. Similarly $5!$ means $1 \times 2 \times 3 \times 4 \times 5$.

The product of all the whole numbers from 1 up to any positive integer n is called n factorial and is written as $n!$

The binomial coefficient represented by $\binom{n}{r}$ can be expressed concisely using the factorial notation;

$$\frac{n(n-1)(n-2)\ldots(n-r+1)}{2 \times 3 \times 4 \times \ldots \times r}$$

$$= \frac{n(n-1)(n-2)\ldots(n-r+1)(n-r)(n-r-1)\ldots(3)(2)(1)}{(2 \times 3 \times 4 \times \ldots \times r)(n-r)(n-r-1)\ldots(3)(2)(1)}$$

$$= \frac{n!}{r!(n-r)!}$$

i.e.
$$\binom{n}{r} = \frac{n!}{r!(n-r)!}$$

Now $\binom{n}{0} = 1$ as it is the coefficient of x^0, i.e. the first term in the expansion of $(1+x)^n$.

Using the factorial notation, $\binom{n}{0} = \frac{n!}{0!n!}$

and $\dfrac{n!}{0!n!} = 1$ only if we define $0!$ as 1.

i.e.
$$0! = 1$$

Using the Σ notation, the binomial theorem may be written as

$$(1+x)^n = \sum_{r=0}^{n} \binom{n}{r} x^r \quad \text{for} \quad n = 0, 1, 2, \ldots$$

A proof of the binomial theorem follows.

From our knowledge of multiplying brackets, we know that, when n is a positive integer, $(1 + x)^n$ can be expanded as a series of ascending powers of x, starting with 1 and ending with x^n,

i.e. $\quad (1 + x)^n \equiv 1 + a_1 x + a_2 x^2 + a_3 x^3 + \ldots + a_r x^r + \ldots + x^n \qquad [1]$

where $a_1, a_2, \ldots$ are constants.

Differentiating both sides of this identity w.r.t. x gives

$$n(1 + x)^{n-1} \equiv a_1 + 2a_2 x + 3a_3 x^2 + \ldots + ra_r x^{r-1} + \ldots + nx^{n-1} \qquad [2]$$

As [2] is also an identity it is true for all values of x, so when $x = 0$ we have

$$n(1)^{n-1} = a_1 \quad \Rightarrow \quad a_1 = n$$

Differentiating both sides of [2] w.r.t. x gives

$$n(n-1)(1+x)^{n-2} \equiv 2a_2 + (2)(3)a_3 x + \ldots + r(r-1)a_r x^{r-2}$$

$$+ \ldots + n(n-1)x^{n-2} \qquad [3]$$

When $x = 0$, [3] gives

$$n(n-1)(1)^{n-2} = 2a_2 \quad \Rightarrow \quad a_2 = \frac{n(n-1)}{2} = \frac{n!}{2!(n-2)!}$$

Differentiating both sides of [3] w.r.t. x and then using $x = 0$ gives

$$n(n-1)(n-2)(1)^{n-3} = (2)(3)a_3 \quad \Rightarrow \quad a_3 = \frac{n(n-1)(n-2)}{(2)(3)}$$

$$= \frac{n!}{3!(n-3)!}$$

This process can be repeated to give all the coefficients but as the pattern is now clear we can deduce that

$$a_r = \frac{n(n-1)(n-2)\ldots(n-r+1)}{(2)(3)(4)\ldots(r)} = \frac{n!}{r!(n-r)!}$$

where a_r is the coefficient of x^r.

1. Write down the first three terms in the expansion in ascending powers of x of

(a) $\left(1 - \dfrac{x}{2}\right)^{10}$ (b) $(3 - 6x)^8$

(a) Using the binomial theorem and replacing x by $-\dfrac{x}{2}$ and n by 10 we have

$$\left(1 - \frac{x}{2}\right)^{10} = 1 + (10)\left(-\frac{x}{2}\right) + \frac{10 \times 9}{2}\left(-\frac{x}{2}\right)^2 + \ldots$$

$$= 1 - 5x + \tfrac{45}{4}x^2 + \ldots$$

(b) Writing $(3 - 6x)^8$ as $3^8(1 - 2x)^8$ we can use the binomial theorem replacing x by $-2x$ and n by 8

$$(3 - 6x)^8 = 3^8(1 - 2x)^8 = 3^8\left[1 + (8)(-2x) + \frac{(8)(7)}{2}(-2x)^2 + \ldots\right]$$

Therefore the first three terms of the expansion are

$$3^8 + (3^8)(8)(-2x) + \frac{(3^8)(8)(7)}{2}(4x^2)$$

i.e. $3^8 - (3^8)(16)x + (3^8)(112)x^2$

2. Expand $(2x - 1)^5$

Writing $(2x - 1)^5$ as $(-1)^5(1 - 2x)^5$ we can use the binomial theorem replacing x by $-2x$ and n by 5

$$(2x - 1)^5 = (-1)^5(1 - 2x)^5$$

$$= (-1)\left[1 + (5)(-2x) + \frac{(5)(4)}{2}(-2x)^2 + \frac{(5)(4)(3)}{(2)(3)}(-2x)^3\right.$$

$$\left. + \frac{(5)(4)(3)(2)}{(2)(3)(4)}(-2x)^4 + (-2x)^5\right]$$

$$= -1 + 10x - 40x^2 + 80x^3 - 80x^4 + 32x^5$$

3. Write down the first three terms in the binomial expansion of

$$(1 - 2x)(1 + \tfrac{1}{2}x)^{10}$$

The third term in the binomial expansion is the term containing x^2, so we start by expanding $(1 + \tfrac{1}{2}x)^{10}$ as far as the term in x^2

$$(1 + \tfrac{1}{2}x)^{10} = 1 + (10)(\tfrac{1}{2}x) + \frac{(10)(9)}{2}(\tfrac{1}{2}x)^2 + \dots$$

$$= 1 + 5x + \tfrac{45}{4}x^2 + \dots$$

$\therefore \quad (1 - 2x)(1 + \tfrac{1}{2}x)^{10} = (1 - 2x)(1 + 5x + \tfrac{45}{4}x^2 + \dots)$

$$= 1 + 5x + \tfrac{45}{4}x^2 + \dots -2x - 10x^2 + \dots$$

$$= 1 + 3x + \tfrac{5}{4}x^2 + \dots$$

Notice that we do not write down the product of $-2x$ and $\tfrac{45}{4}x^2$, as terms in x^3 are not required.

4. Find the sixth term in the expansion of $(a + b)^{20}$ as a series of ascending powers of b.

$$(a + b)^{20} = a^{20}\left(1 + \frac{b}{a}\right)^{20}$$

The sixth term in the binomial expansion of $(1 + x)^n$ is the term in x^5, i.e. $\binom{n}{6}x^5$ Replacing x by $\dfrac{b}{a}$ and n by 20 gives

the sixth term in the expansion of $(a + b)^{20}$ is

$$(a)^{20}\frac{(20)(19)(18)(17)(16)}{(2)(3)(4)(5)}\left(\frac{b}{a}\right)^5 = 15\,504\,a^{15}b^5$$

5. If the first two terms in the expansion of $(2 - ax)^6$ are $b + 12x$, find the values of a and b.

$$(2 - ax)^6 = 2^6\left(1 - \frac{a}{2}x\right)^6$$

$$= 2^6\left[1 + (6)\left(\frac{a}{2}x\right) + \dots\right]$$

$\therefore \quad 2^6 + (2^6)(6)\left(\frac{a}{2}x\right) \equiv b + 12x$

$\Rightarrow \quad b = 2^6 \quad$ and $\quad (2^6)(6)\left(\frac{a}{2}\right) = 12$

i.e. $\quad b = 64 \quad$ and $\quad a = \tfrac{1}{16}$

1. Write down the first four terms in the binomial expansion of

(a) $(1+3x)^{12}$ (b) $(1-2x)^9$ (c) $(1+5x)^7$

(d) $\left(1-\dfrac{x}{3}\right)^{20}$ (e) $(1-\tfrac{2}{3}x)^6$ (f) $(1+\tfrac{3}{5}x)^{20}$

2. Write down the first three terms in the binomial expansion of

(a) $(2+x)^{10}$ (b) $\left(2-\tfrac{3}{2}x\right)^7$ (c) $\left(\tfrac{3}{2}+2x\right)^9$

3. Write down the term indicated in the binomial expansion of each of the following functions.

(a) $(1-4x)^7$, 3rd term   (b) $\left(1-\dfrac{x}{2}\right)^{20}$, 2nd term

(c) $(1+2x)^{12}$, 4th term   (d) $(1-\tfrac{1}{2}x)^9$, 3rd term

(e) $(2-x)^{15}$, 4th term   (f) $(1-2x)^{12}$, the term in x^4

(g) $\left(2+\dfrac{x}{2}\right)^9$, the term in x^5 (h) $(p-2q)^{10}$, 5th term

(i) $(a+b)^8$, the term in a^3 (j) $(a+2b)^8$, 2nd term

4. Write down the binomial expansion of each function as a series of ascending powers of x as far as, and including, the term in x^2.

(a) $(1+x)(1-x)^9$ (b) $(1-x)(1+2x)^{10}$

(c) $(2+x)\left(1-\dfrac{x}{2}\right)^{20}$ (d) $(1+x)^2(1-5x)^{14}$

5. Write down the full expansion of each of the following expressions.

(a) $(2x-1)^6$ (b) $(3x-2)^5$ (c) $(x-4)^9$

6. Expand $(1-3x)^2(1+2x)^5$ as a series of ascending powers of x as far as and including the term in x^3.

7. Factorise $5-4x-x^2$. Hence expand $(5-4x-x^2)^7$ as a series of ascending powers of x as far as and including the term in x^2.

8. Use the binomial theorem to find the first three terms in the expansion of $(1+2x-x^2)^{20}$ as a series of ascending powers of x; first write $(1+2x-x^2)$ in the form $(1+X)$ where $X=f(x)$ and expand $(1+X)^{20}$.

USING SERIES TO FIND APPROXIMATIONS

Consider $(1+x)^{20}$ and its binomial expansion,

$$(1+x)^{20} = 1 + 20x + \frac{(20)(19)}{2}x^2 + \frac{(20)(19)(18)}{(2)(3)}x^3 + \ldots + x^{20}$$

This is valid for all values of x so if, for example, $x = 0.01$ we have

$$(1.01)^{20} = 1 + 20(0.01) + \frac{(20)(19)}{2}(0.01)^2$$

$$+ \frac{(20)(19)(18)}{(2)(3)}(0.01)^3 + \ldots + (0.01)^{20}$$

i.e. $\quad (1.01)^{20} = 1 + 0.2 + 0.019 + 0.001\,14 + 0.000\,048\,45 + \ldots + 10^{-40}$

Because the value of x (i.e. 0.01) is small, we see that adding successive terms of the series makes progressively smaller contributions to the accuracy of $(1.01)^{20}$. In fact, taking only the first two terms gives $(1.01)^{20} \approx 1.2$

This approximation is correct to two significant figures as the third and succeeding terms do not add anything to alter this.

In general, if x is small so that successive powers of x quickly become negligible in value, then the sum of the first few terms in the expansion of $(1+x)^n$ gives an approximation for $(1+x)^n$.

The worked examples illustrate how a series expansion enables us to find a simple function which can be used as an approximation to a given function when x has values that are close to zero.

Examples 9b

1. If x is so small that x^2 and higher powers can be neglected show that

$$(1-x)^5 \left(2 + \frac{x}{2}\right)^{10} \approx 2^9(2 - 5x)$$

Using the binomial expansion of $(1-x)^5$ and neglecting terms containing x^2 and higher powers of x we have

$$(1-x)^5 \approx 1 - 5x$$

Similarly
$$\left(2+\frac{x}{2}\right)^{10} \equiv 2^{10}\left(1+\frac{x}{4}\right)^{10}$$

$$\approx 2^{10}\left[1+10\left(\frac{x}{4}\right)\right]$$

Therefore $(1-x)^5\left(2+\frac{x}{2}\right)^{10} \approx 2^{10}(1-5x)\left(1+\frac{5x}{2}\right)$

$$= 2^9(1-5x)(2+5x)$$

$$\approx 2^9(2-5x)$$

again neglecting the term in x^2

The graphical significance of the approximation in the last example is interesting.

If $y=(1-x)^5\left(2+\frac{x}{2}\right)^{10}$

then, for values of x close to zero, $y \approx 2^9(2-5x)$ which is the equation of a straight line,

i.e. $y=2^9(2-5x)$ is the tangent to $y=(1-x)^5\left(2+\frac{x}{2}\right)^{10}$ at the point where $x=0$

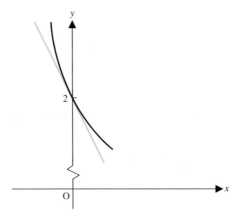

Note that the function $2^9(2-5x)$ is called a *linear approximation* for the function $(1-x)^5\left(2+\frac{x}{2}\right)^{10}$ in the region where $x \approx 0$

2. The shape of the curve $y = (2x - 1)^9$ near the origin can be found by using $y \approx ax^2 + bx + c$. Find the values of a, b and c and hence find, correct to 2 sf, the error involved in taking y as $ax^2 + bx + c$ when (a) $x = 0.01$ (b) $x = 0.1$

Using the binomial theorem gives

$$(2x - 1)^9 = (-1)^9 (1 - 2x)^9 = (-1)(1 - 2x)^9$$

$$= -\left[1 + 9(-2x) + \frac{9 \times 8}{2}(-2x)^2 + \ldots \right]$$

$$= -1 + 18x - 144x^2 + \ldots$$

Near the origin, x is small so ignoring powers of x greater than 2 gives

$$(2x - 1)^9 \approx -144x^2 + 18x - 1$$

and $(2x - 1)^9 \approx ax^2 + bx + c$

i.e. $a = -144$, $b = 18$ and $c = -1$

(a) When $x = 0.01$, $y = (2x - 1)^9 = -0.833\,74\ldots$

$$y \approx -144x^2 + 18x - 1 = -0.8344$$

The error involved is -6.5×10^{-4}

(b) When $x = 0.1$, $y = (2x - 1)^9 = -0.134\,21\ldots$

$$y \approx -144x^2 + 18x - 1 = -0.64$$

The error involved is -0.51

This error is huge compared to the value of y, so we deduce that the approximation is reasonable only very close to the origin.

1. By neglecting x^2 and higher powers of x find linear approximations for the following functions for values of x close to zero.

(a) $(1 - 2x)^{10}$ (b) $\left(1 + \frac{x}{5}\right)^6$ (c) $(2 + x)^7$

2. Show that, if x is small enough for x^2 and higher powers of x to be neglected, the function $(x - 2)(1 + 3x)^8$ has a linear approximation of $-2 - 47x$.

3. If x is so small that x^3 and higher powers of x are negligible, show that $(2x + 3)(1 - 2x)^{10} \approx 3 - 58x + 500x^2$.

4. By neglecting x^3 and higher powers of x find quadratic approximations for the following functions in the immediate neighbourhood of $x = 0$

(a) $(1 - 5x)^{10}$ (b) $(2 - x)^8$ (c) $(1 + x)(1 - x)^{20}$

5. For values of x near the origin the curve $y = ax^2 + bx + c$ can be used as an approximation for the curve $(1 - 2x)^2(1 + x)^{20}$. Find the values of a, b and c.

6. Find the equation of the tangent to the curve $y = (1 - \frac{1}{2}x)^8$ at the point where the curve crosses the y-axis by

(a) differentiation

(b) using the binomial theorem to find a linear function that approximates to $(1 - \frac{1}{2}x)^8$ when x is small.

*7. For small values of x, the value of y on the curve $y = (2x + 3)(1 - x)^{10}$ can be found approximately by using $y \approx ax^2 + bx + c$.

(a) Find the values of a, b and c.

(b) Find the error involved in using $y \approx ax^2 + bx + c$ when $x = 0.05$

(c) By trying different values of x, find correct to 2 sf a range of values of x for which the error in using the quadratic approximation is less than 5%.

*8. (a) Write down the first six terms in the expansion of $(1 + x)^{100}$. Do not simplify the coefficients.

(b) Deduce a recurrence relation between successive terms of the series.

CHAPTER 10

VARIATION

When two quantities, x and y, are related, e.g. $y = 2x - 3$, then as the value of x varies, the way in which y varies depends on the relationship between x and y.

In Module A we considered exponential growth and decay which is one particular form of variation.

In this section we look at two more forms of variation that occur frequently in everyday situations.

Direct Proportion

When two variable quantities are always in the same ratio, they are said to be *directly proportional*.

For example, if 1 kg of apples costs 98 p then 2 kg costs 98×2 p and x kg costs $98 \times x$ p. Hence the ratio of cost to weight, $98x : x$, is constant and is $98 : 1$.

So if y pence is the cost of x kg, $y : x = 98 : 1$

$$\Rightarrow \qquad y = 98x$$

and we say that 'y is directly proportional to x'
or that 'y varies directly as x varies'.

The symbol $\propto$ is used to mean 'is proportional to'
so we can write $y \propto x$.

In general, as y varies directly with x, the relationship between them is of the form $y = kx$ where k is a constant.

Similarly, if y varies directly with the square of x, then $y = kx^2$.

The constant can be evaluated when a pair of corresponding values of x and y are known.

Inverse Proportion

When the product of two variable quantities is constant they are said to be *inversely proportional.*

The average speed of a train travelling between two stations, for example, can vary, in which case the time taken to cover that constant distance will also vary in accordance with the rule 'speed $\times$ time = distance'.

If the speed is y km/h and the time taken to cover k km is x hours, then

$$y \times x = k \quad \Rightarrow \quad y = \frac{k}{x}$$

We say that 'y is inversely proportional to x' or 'y varies inversely with x' and write this in symbols as $y \propto \dfrac{1}{x}$.

In general, if y *is inversely proportional to x*, the relationship between them is of the form $y = \dfrac{k}{x}$ where k is a constant.

Examples 10a

1. Two quantities x and y vary so that y is directly proportional to x^3. Given that $y = 4$ when $x = 2$, find y when $x = 4$.

$$y \propto x^3 \quad \Rightarrow \quad y = kx^3$$

$$y = 4 \quad \text{when} \quad x = 2 \quad \Rightarrow \quad 4 = k(2)^3,$$

$$\therefore \qquad k = 0.5$$

i.e. $\qquad y = (0.5)x^3$

When $\quad x = 4, \ y = (0.5)(4)^3 = 32$

2. The resistance, R ohms, in a wire varies inversely as the square of the current, I amps flowing through it. It is known that when the current is 5 amps, the resistance is 3 ohms. What is the resistance when the current is increased to 15 amps?

$$R \propto \frac{1}{I^2} \quad \Rightarrow \quad R = \frac{k}{I^2}$$

When $R = 3$, $I = 5$, i.e. $3 = \dfrac{k}{25} \Rightarrow k = 75$

$\therefore \quad R = \dfrac{75}{I^2}$

When $I = 15$, $R = \dfrac{75}{15^2} = \dfrac{1}{3}$

The resistance is $\frac{1}{3}$ ohm when the current is 15 amps.

1. If y is directly proportional to x and, when $x = 3$, $y = 8$, find y when $x = 5$.

2. Give that y varies inversely as x varies and that $y = 2$ when $x = 6$, find y when $x = 9$.

3. Two quantities, p and q, vary such that p is inversely proportional to q. If $p = 0.7$ when $q = 1.3$, find p when $q = 0.2$.

4. If y varies directly as the square of x, find y when $x = 2$, given that when $x = 5$, $y = 30$.

5. The mass, m grams of mould on a culture dish is found to vary directly with the time, t hours, for which the dish is in an incubator for t in the range $1 \leqslant t \leqslant 24$. If $m = 10$ when $t = 2$, find m when $t = 12$. Can a value for m be found when $t = 48$?

6. In question 5, find m when $t = 24$. It is found that if the dish is left out of the incubator after 24 hours, then m varies inversely as t varies. Find m when $t = 36$.

7. It is known that the area, A, of a circle varies directly as the square of its radius r. What is the value of the constant of proportion?

8. Express, in terms of proportion, the relationship between the volume of a cube and the length of one of its edges.

9. A ball falls through a liquid in a tank in such a way that the time for which it has been falling, t s, is inversely proportional to its distance h m from the bottom of the tank. If, when the ball has been falling for 6 seconds it is 0.5 m above the bottom of the tank, where is it when it has been falling for 10 seconds? State with reasons whether this relationship is likely to be valid when the ball is very close to the bottom of the tank.

REDUCTION OF A RELATIONSHIP TO A LINEAR LAW

In this part of the chapter we look at a practical application of the equation $y = mx + c$.

Linear Relationships

If it is thought that a certain relationship exists between two variable quantities, this hypothesis can be tested by experiment, i.e. by giving one variable certain values and measuring the corresponding values of the other variable. The experimental data collected can then be displayed graphically. If the graph shows points that lie approximately on a straight line (allowing for experimental error) then a linear relationship between the variables (i.e. a relationship of the form $Y = mX + c$) is indicated. Further, the gradient of the line (m) and the vertical axis intercept (c) provide the values of the constants.

Examples 10b

1. An elastic string is fixed at one end and a variable weight is hung on the other end. It is believed that the length of the string is related to the weight by a linear law. Use the following experimental data to confirm this belief and find the particular relationship between the length of the string and the weight.

Weight (W) in newtons	1	2	3	4	5	6	7	8
Length (l) in metres	0.33	0.37	0.4	0.45	0.5	0.53	0.56	0.6

If l and W are related by a linear law then, allowing for experimental error, we expect that the points will lie on a straight line. Plotting l against W gives the following graph.

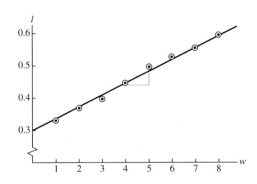

These points do lie fairly close to a straight line.
From the graph, l and W are connected by a linear relationship,

i.e.   a relationship of the form $l = aW + b$

Now we draw the line of 'closest fit'. This is the line that has the points distributed above and below it as evenly as possible; it is not necessarily the line which goes through the most points.

By measurement from the graph

the gradient $= 0.04$

the intercept on the vertical axis $= 0.3$

So comparing $\left.\begin{array}{c} l = aW + b \\ Y = mX + c \end{array}\right\}$ we have $a = 0.04, \quad b = 0.3$
with

i.e.   within the limits of experimental accuracy

$l = 0.04W + 0.3$

Relationship Reducible to a Linear Form

If the relationship between two variables is not linear, the points on the corresponding graph lie on a curve. Because it is very difficult to identify the equation of a curve from a section of it, the form of a non-linear relationship can rarely be verified in this way. Such relationships, however, can often be reduced to a linear form.
When attempting to reduce a relationship between two variables to a form from which a straight line graph can be drawn, the given equation must be expressed in the form

$$Y = mX + c$$

where X and Y are variable terms, values for which must be calculable from the given data, i.e. X and Y must not contain unknown constants. On the other hand, m and c must be constants and may be unknown.

For example, if $y = ax^2 + b,$ using $y = Y$ and $x^2 = X$ gives

$$Y = mX + c$$

Then if values of Y are plotted against values of X, i.e. if values of y are plotted against values of x^2, a straight line will result. The gradient, m, of this line gives the value of a and the intercept, c, on the y-axis gives the value of b.

Now, in general, X and Y may be functions of one or both variables; the following worked example illustrates a case where both variables appear in the Y term.

When trying to reduce a non-linear relationship to a linear one, we aim to

● express it in a form containing three terms

● make one of those terms constant.

Examples 10b (continued)

2. In an experiment, the mass, y grams, of a substance is measured at various times, x seconds. The results are shown in the table below.

x	5	8	10	15	17	25
y	5	13	21	46	60	100

It is thought that x and y are related by a law of the form $y = ax^2 + bx$.

(a) Confirm this graphically, showing that one result does not conform to this law.

(b) Find approximate values for a and b.

(c) Explain, with reasons, whether it is sensible to use these results to predict the mass when $x = 30$.

(a) One of the terms must be made into a constant.

If $y = ax^2 + bx$, dividing both sides by x gives

$$\frac{y}{x} = ax + b$$

which is of the form $\qquad Y = mX + c$

where $\qquad Y = \frac{y}{x}$, $X = x$, $m = a$ and $c = b$,

so $\frac{y}{x}$ and x have a linear relationship.

To confirm this graphically, we need first to tabulate corresponding values of x and y/x.

x	5	8	10	15	17	25
$\frac{y}{x}$	1	1.6	2.1	3.1	3.5	4

Plotting values of $\dfrac{y}{x}$ against values of x gives this graph.

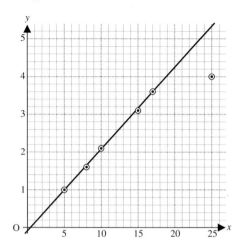

The straight line shows that, except where $x = 25$, there is a linear relationship between $\dfrac{y}{x}$ and x, confirming that $y = ax^2 + bx$

(b) From the graph, the gradient is 0.21,

$$\therefore \qquad a \approx 0.2$$

and the intercept on the y-axis is -0.1,

$$\therefore \qquad b \approx -0.1$$

(c) From the evidence given, the relationship between x and y applies for $5 \leqslant x \leqslant 17$. It clearly does not apply when $x = 25$; this may be because this result is an error or because the conditions for the relationship to exist no longer apply so it is not sensible to use the relationship to predict the mass for a value of x above 17.

3. Two variables, s and t, are related by the law $\dfrac{1}{s} + \dfrac{1}{t} = \dfrac{1}{a}$.

State how this law can be reduced to a linear form so that a straight line graph can be drawn from experimental data.

Rearranging $\dfrac{1}{s} + \dfrac{1}{t} = \dfrac{1}{a}$ in the form $\dfrac{1}{s} = -\dfrac{1}{t} + \dfrac{1}{a}$

and comparing with $\qquad\qquad Y = mX + c$

it can be seen that plotting values of $\dfrac{1}{s}$ against values of $\dfrac{1}{t}$, will give a straight line whose gradient is -1 and whose intercept on the vertical axis is $\dfrac{1}{a}$.

1. Reduce each of the given relationships to the form $Y = mX + c$. In each case give the functions equivalent to X and Y and the constants equivalent to m and c.

(a) $\dfrac{1}{y} = ax + b$

(b) $y = \dfrac{x + a}{x}$

(c) $y^2 = ax + b$

(d) $xy + by = ax^2$

In Questions 2 to 4, the table gives sets of values for the related variables and the law which relates the variables. By drawing a straight line graph find approximate values for a and b.

2. $y = ax + ab$

x	3	5	7	10
y	-2	2	6	12

3. $r^2 = a\theta - b$

θ	1	4	10	25	40
r	1.6	2	2.6	3.8	4.7

4. $\dfrac{a}{V} + \dfrac{b}{L} = 1$

V	2.5	3	5.5	7	12
L	2.5	1.5	0.79	0.7	0.6

Relationships of the Form $y = ax^n$

A relationship of the form $y = ax^n$ where a is a constant can be reduced to a linear relationship by taking logarithms, since

$$y = ax^n \iff \ln y = n \ln x + \ln a$$

(Although any base can be used, it is sensible to use either e or 10 as these are built into most calculators.)

Comparing $\quad \ln y = n \ln x + \ln a$

with $\qquad\qquad Y = mX + c$

we see that plotting values of $\ln y$ against values of $\ln x$ gives a straight line whose gradient is n and whose intercept on the vertical axis is $\ln a$.

1. The following data, collected from an experiment is believed to obey a law of the form $p = aq^n$. Verify this graphically and find the values of a and n.

q	1	2	3	4	5	6
p	0.5	0.63	0.72	0.8	0.85	0.9

If the relationship $p = aq^n$ is correct, then $\ln p = n \ln q + \ln a$

comparing with $\qquad\qquad\qquad\qquad y = mx + c$

we see that $\ln p$ and $\ln q$ are related by a linear law.

First a table of values of $\ln p$ and $\ln q$ is needed.

$\ln q$	0	0.69	1.10	1.39	1.61	1.79
$\ln p$	-0.69	-0.47	-0.33	-0.22	-0.16	-0.11

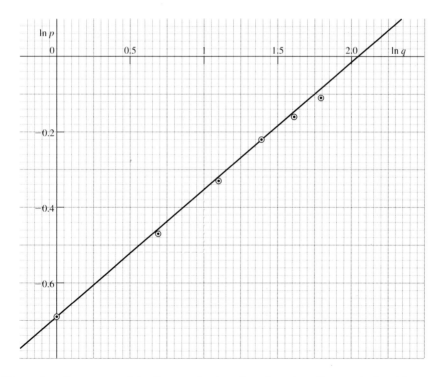

The points lie on a straight line confirming that there is a linear relationship between $\ln q$ and $\ln p$.

From the graph, the gradient of the line is 0.33, $\Rightarrow$ $n = 0.33$ and the intercept on the vertical axis is -0.69, so

$$\ln a = -0.69 \quad \Rightarrow \quad a = 0.5$$

Therefore the data does obey a law of the form $p = aq^n$, where $a \approx 0.5$ and $n \approx 0.33$

(Using the tabulated values of $\ln q$ and $\ln p$ and a computer programme, gives $n = 0.327$ and $\ln a = -0.687$).

Relationships of the Form $y = ab^x$

A relationship of the form $y = ab^x$ where a and b are constant can be reduced to a linear relationship by taking logs, since

$$y = ab^x \iff \log y = x \log b + \log a$$

Comparing $\qquad \log y = x \log b + \log a$

with $\qquad\qquad Y = mX + c$

we see that plotting values of $\log y$ against corresponding values of x gives a straight line whose gradient is $\log b$ and whose intercept on the vertical axis is $\log a$.

Note that in all graphical work the scales should be chosen to give the greatest possible accuracy, i.e. the range of values given in the table should have as much spread as possible.

This sometimes means that the horizontal scale does not include zero and the value of c cannot then be read from the graph.

In these circumstances which arise in the next example, we find c by using the equation $Y = mX + c$ together with the measured value of m and the coordinates of any point P on the graph (*not* a pair of values from the table).

Examples 10c (continued)

2. In an experiment, the mass, y grams, of a substance is measured at various times, x seconds.

   The results are shown in the table below. It is believed that x and y are related by a law of the form $2y = ab^{(x-3)}$.

x	10	12	15	20	21
y	37.5	90	320	2440	3700

   (a) Confirm this graphically.

   (b) Find approximate values of a and b and interpret the meaning of b.

(a)   If $2y = ab^{(x-3)}$,   taking logs of both sides gives

$$\log 2y = (x-3)\log b + \log a$$

which is of the form $Y = mX + c$

where $Y = \log 2y$, $X = x-3$ and $m = \log b$, $c = \log a$
i.e. $\log 2y$ and $x - 3$ obey a linear law.

So we need to tabulate corresponding values of $(x-3)$ and $\log 2y$ from the given values of x and y

$x-3$	7	9	12	17	18
$\log 2y$	1.9	2.3	2.8	3.7	3.9

Then plotting $\log 2y$ against $x - 3$ gives the graph below.

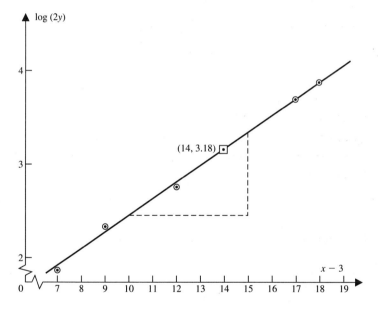

The straight line shows that there is a linear relationship between $\log 2y$
and $x - 3$, confirming that $2y = ab^{x-3}$.

(b)   From the graph, the gradient is 0.175

$\therefore$ $\log b \approx 0.175$ $\Rightarrow$ $b \approx 1.49$

Using the point $P(14, 3.18)$ and $m = 0.175$ then $y = mX + c$ gives

$$3.18 = (0.175)(14) + c   \Rightarrow   c = 0.73$$

i.e. $\log a \approx 0.73$ $\Rightarrow$ $a \approx 5.37$

The relationship $2y = ab^{(x-3)}$ represents a form of exponential growth
where b is the growth factor.

1. Reduce each of the given relationships to the form $Y = mX + c$. In each case give the functions equivalent to X and Y and the constants equivalent to m and c

   (a) $ae^x = y(y - b)$ (b) $y = ax^{n+2}$ (c) $y^a = e^{x+\lambda}$

   In Questions 2 and 3, the table gives sets of values for the related variables and the law which relates the variables. By drawing a straight line graph find approximate values for a and b.

2. $s = ab^{-t}$

t	1	2	3	4
s	1.5	0.4	0.1	0.02

3. $ay = b^x$

x	5	6	7	8
y	1.07	2.13	4.27	8.53

4. The variables x and y are believed to satisfy a relationship of the form $y = k(x+1)^n$. Show that the experimental values shown in the table do satisfy the relationship. Find approximate values for k and n.

x	4	8	15	19	24
y	4.45	4.60	4.80	4.89	5.00

5. Two variables s and t are related by a law of the form $s = ke^{-nt}$. The values in the table were obtained from an experiment. Show graphically that these values do verify the relationship and use the graph to find approximate values of k and n.

t	1	1.5	2	2.5	3
s	1230	590	260	140	60

*6. It is thought that the value of a second-hand car reduces exponentially with age and, if A is the initial value of a car, its value £y after time x years can be modelled by the relationship

$$y - 2000 = ab^{-x}$$

   (a) If the initial value of a car was £12 000 and its value 2 years later was £8 500, find the values of a and b.

   (b) Reduce the relationship to a linear law and represent it graphically for $0 \leqslant x \leqslant 10$.

(c)   The table gives the values of another car for various values of x.

x	0	2	4	6	8	10
y	12 000	8800	7300	6100	5100	4300

Plot these points on your graph. Explain why the given model does not predict the changing values of this car.

(d)   Suggest a relationship between x and y that does reflect the values given in the table.

*7.   When doing an emergency stop, it is though that the distance, s metres, travelled by a car from the time the brakes are applied until it stops can be modelled by the equation

$$s = ku^2$$

where u km/h is the initial speed of the car and k is a constant.

(a)   A car initially moving at 60 km/h covers 120 m before it stops. Assuming the model is valid, find the value of k.

(b)   Reduce the relationship to a linear law, and using the value of k found in (a), represent this graphically for $0 \leqslant u \leqslant 200$.

In the same car on another day and under different conditions, the following information was recorded.

u	50	100	130	150	200
s	100	210	310	780	620

(c)   Show that these values do not satisfy a relationship of the form $s = ku^2$, even allowing for error.

(d)   Given that there is one rogue value, identify it and suggest a relationship that could be used to model the information in the table.

CHAPTER 11

DIFFERENTIATION OF
TRIGONOMETRIC FUNCTIONS

INVESTIGATING THE GRADIENT FUNCTION WHEN
$y = \sin x$

The diagram shows the graph of $y = \sin x$ for values of x between $-\pi$ and $\frac{5}{2}\pi$.

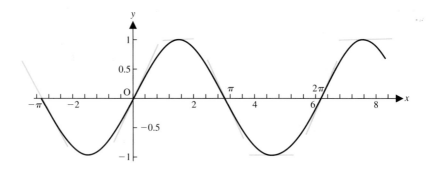

The tangents to the curve are drawn by eye and their gradients used to complete the following table.

x	$-\pi$	$-\dfrac{\pi}{2}$	0	$\dfrac{\pi}{2}$	π	$\dfrac{3\pi}{2}$	2π	$\dfrac{5\pi}{2}$
approximate gradient of $\sin x$, i.e. $\dfrac{\mathrm{d}(\sin x)}{\mathrm{d}x}$	-1	0	1	0	-1	0	1	0

Plotting these approximate values of $\dfrac{\mathrm{d}(\sin x)}{\mathrm{d}x}$ against x, and drawing a smooth curve through them, gives the curve opposite which looks remarkably like the cosine wave.

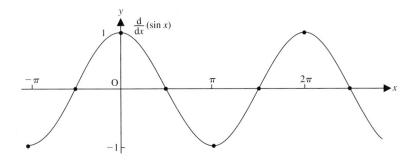

The next exercise explores this relationship further.

For this exercise you will need a large plot of the graph of $y = \sin x$ for values of x in the range $0 \leqslant x \leqslant 2\pi$ drawn with equal scales on the axes.

There is a copy of this on p. 380, but if you have access to the appropriate hardware and software, a landscape print of this plot on A4 paper is easier to use.

1. Draw tangents to the curve $y = \sin x$ at each value of x (in radians) given in the table and then find the gradient of each tangent to complete the second row of the table.

x	0	1	2	3	4	5	6
approximate gradient							
$\cos x$							

Now use your calculator to complete the third row of the table. Comment on the values in the last two rows of the completed table.

2. Repeat question 1 using a plot of $y = \sin x°$ for values of x from 0 to 360 at intervals of 60 in the table.

Is $\dfrac{d}{dx}(\sin x°)$ approximately equal to $\cos x°$?

3. (a) Use a *sketch* graph of $y = \cos x$ for $-\pi \leqslant x \leqslant 3\pi$ to fill in the second row of the table.

x	$-\pi$	$-\dfrac{\pi}{2}$	0	$\dfrac{\pi}{2}$	π	$\dfrac{3\pi}{2}$	2π	$\dfrac{5\pi}{2}$	3π
gradient of $\cos x$									

(b) Plot these values of the gradient of $\cos x$ against x and draw a curve through the points.

(c) Deduce the equation of the curve.

(d) State, with a reason, whether the result for (c) is valid when x is measured in degrees.

THE DERIVATIVES OF SIN x AND COS x

The results from the last exercise demonstrate that, when x is measured in radians the gradient function of $\sin x$ is $\cos x$, and the gradient function of $\cos x$ is $-\sin x$.

i.e.   **if $y = \sin x$ then $\dfrac{dy}{dx} = \cos x$**

and   **if $y = \cos x$ then $\dfrac{dy}{dx} = -\sin x$**

These two results can be quoted whenever they are needed.

It is important to realise that they are valid only when x is measured in radians.

These results can be obtained by differentiating from first principles, but the trigonometric knowledge needed to simplify the expressions is beyond the scope of this book.

Throughout all subsequent work on the calculus of trig functions, the angle is measured in radians unless it is stated otherwise.

Examples 11b

1. Find the smallest positive value of x for which there is a stationary value of the function $x + 2 \cos x$

$$f(x) = x + 2 \cos x \quad \Rightarrow \quad f'(x) = 1 - 2 \sin x$$

where $f'(x) = \dfrac{d}{dx} f(x)$

For stationary values $f'(x) = 0$

i.e. $1 - 2 \sin x = 0 \quad \Rightarrow \quad \sin x = \tfrac{1}{2}$

The smallest positive angle with a sine of $\tfrac{1}{2}$ is $\tfrac{1}{6}\pi$

NOTE that the answer *must* be given in radians because the rule used to differentiate $\cos x$ is valid only for an angle in radians.

2. Find the smallest positive value of θ for which the curve $y = 2\theta - 3 \sin \theta$ has a gradient of $\frac{1}{2}$

$$y = 2\theta - 3 \sin \theta \quad \text{gives} \quad \frac{dy}{d\theta} = 2 - 3 \cos \theta$$

when $\dfrac{dy}{d\theta} = \frac{1}{2}$, $\quad 2 - 3 \cos \theta = \frac{1}{2}$

$$3 \cos \theta = \tfrac{3}{2}$$

$$\cos \theta = \tfrac{1}{2}$$

The smallest positive value of θ for which $\cos \theta = \frac{1}{2}$, is $\frac{1}{3}\pi$

EXERCISE 11b

1. Find $\dfrac{dy}{dx}$ when

(a) $y = 3 \sin x$

(b) $y = x - \cos x$

(c) $y = 2 \cos x + \sin x$

(d) $y = 4x^2 - 3 \sin x$

2. Write down the derivative of each of the following expressions.

(a) $\sin x - \cos x$

(b) $\sin \theta + 4$

(c) $3 \cos \theta$

(d) $5 \sin \theta - 6$

(e) $2 \cos \theta + 3 \sin \theta$

(f) $4 \sin x - 5 - 6 \cos x$

3. Find the gradient of each curve at the point whose x-coordinate is given.

(a) $y = \cos x; \quad \frac{1}{2}\pi$

(b) $y = \sin x; \quad 0$

(c) $y = \cos x + \sin x; \quad \pi$

(d) $y = x - \sin x; \quad \frac{1}{2}\pi$

(e) $y = 2 \sin x - x^2; \quad -\pi$

(f) $y = -4 \cos x; \quad \frac{1}{2}\pi$

4. For each of the following curves find the smallest positive value of θ at which the gradient of the curve has the given value.

(a) $y = 2 \cos \theta; \quad -1$

(b) $y = \theta + \cos \theta; \quad \frac{1}{2}$

(c) $y = \sin \theta + \cos \theta; \quad 0$

(d) $y = \sin \theta + 2\theta; \quad 1$

5. Considering only positive values of x, locate the first two turning points on each of the following curves and determine whether they are maximum or minimum points.

(a) $2 \sin x - x$

(b) $x + 2 \cos x$

In each case illustrate your solution by a sketch.

6. Find the equation of the tangent to the curve $y = \cos \theta + 3 \sin \theta$ at the point where $\theta = \frac{1}{2}\pi$

7. Find the equation of the normal to the curve $y = x^2 + \cos x$ at the point where $x = \pi$

8. Find the coordinates of a point on the curve $y = \sin x + \cos x$ at which the tangent is parallel to the line $y = x$

COMPOUND FUNCTIONS

The variety of functions which can be handled when they occur in products, quotients and functions of a function, now includes the sine and cosine ratios.

Differentiation of $\sin f(x)$

If $y = \sin f(x)$ then using $u = f(x)$ gives $y = \sin u$

Then $\quad \dfrac{dy}{dx} = \dfrac{dy}{du} \times \dfrac{du}{dx} \quad \Rightarrow \quad \dfrac{dy}{dx} = \cos u \left(\dfrac{du}{dx}\right)$

i.e. $\qquad\qquad \dfrac{d}{dx}\{\sin f(x)\} = f'(x)\cos f(x)$

Similarly $\qquad\qquad \dfrac{d}{dx}\{\cos f(x)\} = -f'(x)\sin f(x)$

e.g. $\quad \dfrac{d}{dx}\sin e^x = e^x \cos e^x \quad$ and $\quad \dfrac{d}{dx}\cos \ln x = -\dfrac{1}{x}\sin \ln x$

In particular $\qquad\qquad \dfrac{d}{dx}(\sin ax) = a\cos ax$

and $\qquad\qquad \dfrac{d}{dx}(\cos ax) = -a\sin ax$

These results are quotable.

1. Differentiate $\cos (\frac{1}{6}\pi - 3x)$ with respect to x

$$\frac{d}{dx} \{\cos (\tfrac{1}{6}\pi - 3x)\} = -(-3) \sin (\tfrac{1}{6}\pi - 3x)$$

$$= 3 \sin (\tfrac{1}{6}\pi - 3x)$$

2. Find the derivative of $\dfrac{e^x}{\sin x}$

$$y = \frac{e^x}{\sin x} = \frac{u}{v}$$

where $u = e^x$ and $v = \sin x$

$\Rightarrow \qquad \dfrac{du}{dx} = e^x$ and $\dfrac{dv}{dx} = \cos x$

$$\frac{dy}{dx} = \left(v \frac{du}{dx} - u \frac{dv}{dx} \right) \div v^2 = \frac{e^x \sin x - e^x \cos x}{\sin^2 x}$$

$\therefore \qquad \dfrac{d}{dx} \left(\dfrac{e^x}{\sin x} \right) = \dfrac{e^x}{\sin^2 x} (\sin x - \cos x)$

3. Find $\dfrac{dy}{d\theta}$ if $y = \cos^3\theta$

$$y = \cos^3\theta = [\cos \theta]^3$$

$\therefore \qquad y = u^3$ where $u = \cos \theta$

$$\frac{dy}{d\theta} = \frac{dy}{du}\frac{du}{d\theta} = (3u^2)(-\sin \theta) = 3(\cos \theta)^2(-\sin \theta)$$

$\therefore \qquad y = \cos^3\theta \quad \Rightarrow \quad \dfrac{dy}{d\theta} = -3 \cos^2\theta \sin \theta$

This is one example of a general rule, i.e.

$$\text{if} \quad y = \cos^n x \quad \text{then} \quad \frac{dy}{dx} = -n \cos^{n-1}x \sin x$$

and

$$\text{if} \quad y = \sin^n x \quad \text{then} \quad \frac{dy}{dx} = n \sin^{n-1}x \cos x$$

Differentiate each of the following functions with respect to x

1. $\sin 4x$

2. $\cos(\pi - 2x)$

3. $\sin(\frac{1}{2}x + \pi)$

4. $\dfrac{\sin x}{x}$

5. $\dfrac{\cos x}{e^x}$

6. $\sqrt{\sin x}$

7. $\sin^2 x$

8. $\sin x \cos x$

9. $e^{\sin x}$

10. $\ln(\cos x)$

11. $e^x \cos x$

12. $x^2 \sin x$

13. $\sin x^2$

14. $e^{\cos x}$

15. $\ln \sin^3 x$

16. $\sec x$, i.e. $\dfrac{1}{\cos x}$

17. $\tan x$, i.e. $\dfrac{\sin x}{\cos x}$

18. $\csc x$

19. $\cot x$

Using the answers to Questions 16 to 19, we can now make a complete list of the derivatives of the basic trig functions:

function	derivative
$\sin x$	$\cos x$
$\cos x$	$-\sin x$
$\tan x$	$\sec^2 x$
$\cot x$	$-\csc^2 x$
$\sec x$	$\sec x \tan x$
$\csc x$	$-\csc x \cot x$

MIXED EXERCISE 11 *Even numbers.*

This exercise contains a variety of functions of all the types. Consider carefully what method to use in each case and do not forget to check first whether a given function has a standard derivative.

Find the derivative of each function in Questions 1 to 19.

1. (a) $-\sin 4\theta$

   (b) $\theta - \cos \theta$

   (c) $\sin^3 \theta + \sin 3\theta$

2. (a) $x^3 + e^x$

   (b) $e^{(2x+3)}$

   (c) $e^x \sin x$

3. (a) $\ln \frac{1}{3} x^{-3}$

   (b) $\ln \dfrac{2}{x^2}$

   (c) $\ln \dfrac{\sqrt{x}}{4}$

4. (a) $3 \sin x - e^{-x}$ (b) $\ln x^{1/2} - \frac{1}{2} \cos x$

  (c) $x^4 + 4e^x - \ln 4x$ (d) $\frac{1}{2}e^{-x} + x^{-1/2} - \ln \frac{1}{2}x$

5. $(x+1) \ln x$ 6. $\sin^2 3x$ 7. $(4x-1)^{2/3}$

8. $(3\sqrt{x} - 2x)^2$ 9. $\dfrac{(x^4-1)}{(x+1)^3}$ 10. $\dfrac{\ln x}{\ln (x-1)}$

11. $\ln \cot x$ 12. $x^2 \sin x$ 13. $\dfrac{e^x}{x-1}$

14. $\dfrac{1+\sin x}{1-\sin x}$ 15. $x^2\sqrt{(x-1)}$ 16. $(1-x^2)(1-x)^2$

17. $\ln \sqrt{\dfrac{(x+3)^3}{(x^2+2)}}$ 18. $\sin x \cos^3 x$ 19. $e^{\cos^2 x}$

20. Find the value(s) of x for which the following functions have stationary values.

  (a) $3x - e^x$ (b) $x^2 - 2 \ln x$ (c) $\ln \dfrac{1}{x} + 4x$

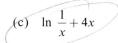

In each Question from 21 to 24, find

  (a) the gradient of the curve at the given point,

  (b) the equation of the tangent to the curve at that point,

  (c) the equation of the normal to the curve at that point.

21. $y = \sin x - \cos x$; $x = \frac{1}{2}\pi$ 22. $y = x + e^x$; $x = 1$

23. $y = 1 + x + \sin x$; $x = 0$ 24. $y = 3 - x^2 + \ln x$; $x = 1$

25. Considering only positive values of x, locate the first two turning points, if there are two, on each of the following curves and determine whether they are maximum or minimum points.

  (a) $y = 1 - \sin x$ (b) $y = \frac{1}{2}x + \cos x$ (c) $y = e^x - 3x$

26. Find the coordinates of a point on the curve where the tangent is parallel to the given line.

  (a) $y = 3x - 2 \cos x$; $y = 4x$ (b) $y = 2 \ln x - x$; $y = x$

*27. Using $f'(x) = \lim\limits_{\delta x \to 0} \dfrac{f(x + \delta x) - f(x)}{\delta x}$ when $f(x) = \sin x$ gives

$$\frac{d}{dx}(\sin x) = \lim\limits_{\delta x \to 0} \frac{\sin(x + \delta x) - \sin x}{\delta x}$$

The diagram shows part of a circle whose centre is 0 and whose radius is 1.

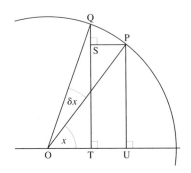

(a) Use trigonometry in right-angled triangles to show that

$$QS = \sin(x + \delta x) - \sin x$$

(b) Treating QP as an arc, express $\dfrac{QS}{QP}$ in terms of x and δx.

(c) Treating QP as a straight line, show that $\angle PQS \approx x$ if δx is small. Hence express $\dfrac{QS}{QP}$ in terms of x (not δx).

(d) What is the condition for the chord QP to be approximately equal to the length of the arc QP?

(e) Complete the argument to show that $\dfrac{d}{dx}(\sin x) = \cos x$.

CHAPTER 12

DIFFERENTIATING IMPLICIT AND PARAMETRIC FUNCTIONS

IMPLICIT FUNCTIONS

All the differentiation carried out so far has involved equations that can be expressed in the form $y = f(x)$.

However the equations of some curves, for example $x^2 - y^2 + y = 1$, cannot easily be written in this way, as it is too difficult to isolate y. A relationship of this type, where y is not given explicitly as a function of x, is called an *implicit function*, i.e. it is *implied* in the equation that $y = f(x)$.

TO DIFFERENTIATE AN IMPLICIT FUNCTION

The method we use is to differentiate, term by term, with respect to x, but first we need to know how to differentiate terms like y^2 with respect to x.

If $\quad g(y) = y^2 \quad$ and $\quad y = f(x)$

then $\quad g(y) = \{f(x)\}^2 \quad$ which is a function of a function.

Using the mental substitution $u = f(x)$ we have

$$\frac{d}{dx}\{f(x)\}^2 = 2\{f(x)\}\left(\frac{d}{dx}f(x)\right) = 2y\left(\frac{dy}{dx}\right) = \left(\frac{d}{dy}g(y)\right)\left(\frac{dy}{dx}\right)$$

In general, $\qquad \dfrac{\mathbf{d}}{\mathbf{dx}}\,\mathbf{g(y)} = \left(\dfrac{\mathbf{d}}{\mathbf{dy}}\,\mathbf{g(y)}\right)\left(\dfrac{\mathbf{dy}}{\mathbf{dx}}\right)$

e.g. $\quad \dfrac{d}{dx}y^3 = 3y^2\,\dfrac{dy}{dx} \quad$ and $\quad \dfrac{d}{dx}e^y = e^y\,\dfrac{dy}{dx}$

161

We can now differentiate, term by term with respect to x, the example considered above, i.e.

if $\qquad x^2 - y^2 + y = 1$

then $\qquad \dfrac{d}{dx}(x^2) - \dfrac{d}{dx}(y^2) + \dfrac{dy}{dx} = \dfrac{d}{dx}(1)$

$\Rightarrow \qquad\qquad 2x - 2y\,\dfrac{dy}{dx} + \dfrac{dy}{dx} = 0$

Hence $\quad 2x = \dfrac{dy}{dx}(2y - 1) \quad \Rightarrow \quad \dfrac{dy}{dx} = \dfrac{2x}{(2y - 1)}$

Examples 12a

1. Differentiate each equation with respect to x and hence find $\dfrac{dy}{dx}$ in terms of x and y.

   (a) $x^3 + xy^2 - y^3 = 5$ $\qquad\qquad\qquad$ (b) $y = xe^y$

(a) If $x^3 + xy^2 - y^3 = 5$ then, differentiating term by term,

$$\frac{d}{dx}(x^3) + \frac{d}{dx}(xy^2) - \frac{d}{dx}(y^3) = \frac{d}{dx}(5)$$

The term xy^2 is a product so we differentiate it using the product rule, i.e.

$$\frac{d}{dx}(xy^2) = y^2\,\frac{d}{dx}(x) + x\frac{d}{dx}(y^2) = y^2 + (x)(2y)\frac{dy}{dx}$$

$$\therefore \qquad 3x^2 + y^2 + 2xy\,\frac{dy}{dx} - 3y^2\,\frac{dy}{dx} = 0$$

Hence $\quad \dfrac{dy}{dx} = \dfrac{(3x^2 + y^2)}{y(3y - 2x)}$

(b) If $y = xe^y$ then $\dfrac{dy}{dx} = \dfrac{d}{dx}(xe^y)$

$$= e^y\,\frac{d}{dx}(x) + x\,\frac{d}{dx}(e^y)$$

$\Rightarrow \qquad\qquad\qquad \dfrac{dy}{dx} = e^y + xe^y\,\dfrac{dy}{dx}$

Hence $\qquad\qquad \dfrac{dy}{dx} = \dfrac{e^y}{1 - xe^y}$

2. If $e^x y = \sin x$ show that $\dfrac{d^2 y}{dx^2} + 2\dfrac{dy}{dx} + 2y = 0$

In a problem of this type it is tempting to express $e^x y = \sin x$ in the form $y = e^{-x} \sin x$, find $\dfrac{dy}{dx}$ and $\dfrac{d^2 y}{dx^2}$ and show that they satisfy the given equation, which is called a *differential equation*. However it is much more direct to differentiate the implicit equation as given.

Differentiating $e^x y = \sin x$ w.r.t. x gives

$$e^x y + e^x \frac{dy}{dx} = \cos x$$

Differentiating again w.r.t. x gives

$$\left(e^x y + e^x \frac{dy}{dx} \right) + \left(e^x \frac{dy}{dx} + e^x \frac{d^2 y}{dx^2} \right) = -\sin x = -e^x y$$

Hence $\quad e^x \dfrac{d^2 y}{dx^2} + 2e^x \dfrac{dy}{dx} + 2e^x y = 0$

There is no finite value of x for which $e^x = 0$ so we can divide the equation by e^x.

i.e. $\quad \dfrac{d^2 y}{dx^2} + 2\dfrac{dy}{dx} + 2y = 0$

3. Find the equation of the tangent at the point (x_1, y_1) to the curve with equation $x^2 - 2y^2 - 6y = 0$

To find the equation of the tangent we need the gradient of the curve and in this case it must be found by implicit differentiation.

$$x^2 - 2y^2 - 6y = 0 \quad \Rightarrow \quad 2x - 2\left(2y\frac{dy}{dx} \right) - 6\frac{dy}{dx} = 0$$

$$\Rightarrow \quad \frac{dy}{dx} = \frac{x}{(3 + 2y)}$$

$\therefore$ the gradient of the tangent at the point (x_1, y_1) is $\dfrac{x_1}{(3 + 2y_1)}$

$\therefore$ the equation of the tangent is $y - y_1 = \dfrac{x_1}{(3 + 2y_1)}(x - x_1)$

which simplifies to $xx_1 - 2yy_1 - 3(y + y_1) = x_1^2 - 2y_1^2 - 6y_1$

Now (x_1, y_1) is on the given curve, $x_1^2 - 2y_1^2 - 6y_1 = 0$, so the equation of the tangent becomes

$$xx_1 - 2yy_1 - 3(y + y_1) = 0$$

Note that in the last example the equation of the curve can be converted into the equation of the tangent by changing x^2 into xx_1, y^2 into yy_1 and y into $\frac{1}{2}(y + y_1)$.

In fact, for any curve whose equation is of degree two, the equation of the tangent at (x_1, y_1) can be written down directly by making the replacements listed above, together with two more, i.e.

$$x \to \tfrac{1}{2}(x + x_1) \quad \text{and} \quad xy \to \tfrac{1}{2}(xy_1 + x_1 y)$$

This property can be applied to advantage when the numerical values of the coordinates of the point of contact are known,

e.g. the equation of the tangent at the point $(1, -1)$ to the curve

$$3x^2 - 7y^2 + 4xy - 8x = 0$$

can be written down as

$$3x(1) - 7y(-1) + 2\{x(-1) + (1)y\} - 4(x + 1) = 0$$

i.e. $9y - 3x = 4$

Question 18 in the following exercise gives the reader the opportunity to justify using these mechanical replacements in the equation of a curve, to give the equation of a tangent.

Note that, although this method allows the equation of a tangent to be *written down*, its use is not suitable when the *derivation* of the equation is required.

EXERCISE 12a

Differentiate the following equations with respect to x

1. $x^2 + y^2 = 4$

2. $x^2 + xy + y^2 = 0$

3. $x(x + y) = y^2$

4. $\dfrac{1}{x} + \dfrac{1}{y} = e^y$

5. $\dfrac{1}{x^2} + \dfrac{1}{y^2} = \dfrac{1}{4}$

6. $\dfrac{x^2}{4} - \dfrac{y^2}{9} = 1$

7. $\sin x + \sin y = 1$

8. $\sin x \cos y = 2$

9. $x e^y = x + 1$

10. $(1 + y)(1 + x) = x$

11. Find $\dfrac{dy}{dx}$ as a function of x if $y^2 = 2x + 1$

12. Find $\dfrac{d^2y}{dx^2}$ as a function of x if $\sin y + \cos y = x$

13. Find the gradient of $x^2 + y^2 = 9$ at the points where $x = 1$

14. If $y \cos x = e^x$ show that $\dfrac{d^2y}{dx^2} - 2\tan x\,\dfrac{dy}{dx} - 2y = 0$

15. Write down the equation of the tangent to

    (a) $x^2 - 3y^2 = 4y$ (b) $x^2 + xy + y^2 = 3$

    at the point (x_1, y_1).

16. Show that the equation of the tangent to $x^2 + xy + y = 0$ at the point (x_1, y_1) is

    $$x(2x_1 + y_1) + y(x_1 + 1) + y_1 = 0$$

*17. Write down the equation of the tangent at $(1, \frac{1}{3})$ to the curve whose equation is

    $$2x^2 + 3y^2 - 3x + 2y = 0$$

*18. Show that the equation of the tangent at (x_1, y_1) to the curve $ax^2 + by^2 + cxy + dx = 0$ is

    $$axx_1 + byy_1 + \tfrac{1}{2}c(xy_1 + yx_1) + \tfrac{1}{2}d(x + x_1) = 0$$

*19. Given that $\sin y = 2\sin x$ show that $\left(\dfrac{dy}{dx}\right)^2 = 1 + 3\sec^2 y$. By differentiating this equation with respect to x show that

    $$\dfrac{d^2y}{dx^2} = 3\sec^2 y \tan y$$

    and hence that $\cot y\,\dfrac{d^2y}{dx^2} - \left(\dfrac{dy}{dx}\right)^2 + 1 = 0$

LOGARITHMIC DIFFERENTIATION

The advantage of simplifying a logarithmic expression before attempting to differentiate it has already been noted.

We are now going to examine some equations which are awkward to differentiate as they stand, but which are much easier to deal with if we first take logs of both sides of the equation. One type of equation where this method is useful is one in which the variable is contained in an index.

1. Differentiate x^x with respect to x

$$y = x^x$$

$$\ln y = x \ln x$$

thus
$$\frac{1}{y}\frac{dy}{dx} = x\frac{1}{x} + \ln x$$

$\Rightarrow$
$$\frac{dy}{dx} = y(1 + \ln x)$$

Therefore
$$\frac{d}{dx}(x^x) = x^x(1 + \ln x)$$

2. Differentiate the equation $x = y^x$ with respect to x

$$x = y^x \quad \Rightarrow \quad \ln x = x \ln y$$

$\therefore$
$$\frac{1}{x} = \ln y + (x)\left(\frac{1}{y}\frac{dy}{dx}\right)$$

i.e.
$$x^2 \frac{dy}{dx} + xy \ln y = y$$

Note that in the second example it is not easy to express $\dfrac{dy}{dx}$ as a function of x, because it is difficult in the first place to find y in terms of x. So although the usual practice is to give a derived function in terms of x it is not always possible, or sensible, to do so.

Differentiation of a^x where a is a Constant

The basic rule for differentiating an exponential function applies when the base is e but not for any other base. So for a^x, where we need another approach, we use logarithmic differentiation.

Using $\qquad y = a^x \quad$ gives $\quad \ln y = x \ln a$

Differentiating w.r.t. x gives $\qquad \dfrac{1}{y}\dfrac{dy}{dx} = \ln a$

Hence $\dfrac{dy}{dx} = y \ln a = a^x \ln a$

i.e. $\dfrac{d}{dx} a^x = a^x \ln a$

This result is quotable.

In each question find $\dfrac{dy}{dx}$.

1. $y = 3^x$ 2. $y = 2(1.5)^x$ 3. $y = 3^{2x}$

4. $y = x3^x$ 5. $y = 3^{-x}$ 6. $y = 5a^x$

Differentiate each equation with respect to x

7. $x^y = e^x$ 8. $x^y = (y+1)$ 9. $y = x^{2x}$

10. Find the gradient of the curve $y = x2^x$ where $x = 2$.

11. The equation of a curve is $y = 2^x(x-1)$

   (a) Find the x coordinate of the stationary point.

   (b) Give the coordinates of the point where the curve crosses the x-axis.

   (c) Sketch the curve.

PARAMETRIC EQUATIONS

Sometimes a direct relationship between x and y is awkward to analyse and in such cases it may be easier to express x and y each in terms of a third variable, called a *parameter*.

Consider, for example, the equations

$$x = t^2$$
$$y = t - 1$$

By giving t any value we choose, we get a pair of corresponding values of x and y. For example, when $t = 3$, $x = 9$ and $y = 2$, therefore $(9, 2)$ is a point of the curve that represents the relationship.

The direct relationship between x and y can be found by eliminating t from these two *parametric equations*. In this case it is $(y+1)^2 = x$.

While the Cartesian equation can be used to find the gradient and general shape of a curve, as well as the equations of tangents and normals, it is often easier to derive them from the parametric equations.

Sketching a Curve from Parametric Equations

Consider the curve whose parametric equations are

$$x = t^2 \quad \text{and} \quad y = t - 1$$

To get an idea of the shape of this curve, we can plot some points by finding the values of x and y that correspond to some chosen values of t,

t	-2	-1	0	1	2
x	4	1	0	1	4
y	-3	-2	-1	0	1

Also we can observe what happens to x and to y as t varies:

$x \geqslant 0$ for all values of t,

as $t \to \infty$, $x \to \infty$ and $y \to \infty$

as $t \to -\infty$, $x \to \infty$ and $y \to -\infty$

there are no values of t for which either x or y is undefined so it is reasonable to assume that the curve is continuous.

Based on this information, a sketch of the curve can now be made.

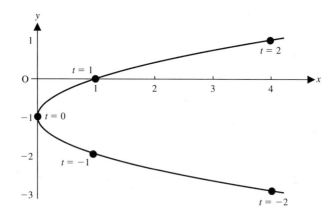

FINDING THE GRADIENT FUNCTION USING PARAMETRIC EQUATIONS

If both x and y are given as functions of t, then a small increase of δt in the value of t results in corresponding small increases of δx and δy in the values of x and y.

As $\delta t \to 0$, δx and δy also approach zero, therefore

$$\frac{dy}{dx} = \frac{dy}{dt} \times \frac{dt}{dx}$$

But $\qquad \dfrac{dt}{dx} = 1 \Big/ \dfrac{dx}{dt}$

Therefore $\qquad \dfrac{dy}{dx} = \dfrac{dy}{dt} \Big/ \dfrac{dx}{dt}$

Hence, for the parametric equations considered above, i.e. $x = t^2$ and $y = t - 1$, we have

$$\frac{dy}{dx} = \frac{1}{2t}$$

Each point on the curve is defined by a value of t which also gives the value of $\dfrac{dy}{dx}$ at that point. For example, when $t = 1$, $\dfrac{dy}{dx} = \dfrac{1}{2}$.

Conversely, the values of t where $\dfrac{dy}{dx}$ has a special value lead to the coordinates of the relevant points on the curve.

For example, when $\dfrac{dy}{dx} = 2$, $\dfrac{1}{2t} = 2 \Rightarrow t = \frac{1}{4}$

and when $t = \frac{1}{4}$, $x = \frac{1}{16}$ and $y = -\frac{3}{4}$;

i.e. the gradient of the curve is 2 at the point $\left(\frac{1}{16}, -\frac{3}{4}\right)$.

Note that there are no values of t for which $\dfrac{dy}{dx} = 0$, so there are no stationary points on this curve.

Examples 12c

1. Find the Cartesian equation of the curve whose parametric equations are

(a) $x = t^2$ (b) $x = \cos \theta$ (c) $x = 2t$

  $y = 2t$ $y = \sin \theta$ $y = \dfrac{2}{t}$

(a) $y = 2t \Rightarrow t = \frac{1}{2}y$

  $\therefore \qquad x = t^2 \Rightarrow x = \left(\frac{1}{2}y\right)^2 = \frac{1}{4}y^2 \Rightarrow y^2 = 4x$

(b) Using $\cos^2\theta + \sin^2\theta = 1$ where $\cos \theta = x$ and $\sin \theta = y$ gives

  $x^2 + y^2 = 1$

(c) $y = \dfrac{2}{t} \Rightarrow t = \dfrac{2}{y}$

  $\therefore \qquad x = 2t \Rightarrow x = \dfrac{4}{y} \Rightarrow xy = 4$

2. Find the stationary point on the curve whose parametric equations are $x = t^3$,
   $y = (t + 1)^2$ and determine its nature.

  $$\frac{dy}{dx} = \frac{dy}{dt} \bigg/ \frac{dx}{dt} = \frac{2(t+1)}{3t^2}$$

At stationary points $\dfrac{dy}{dx} = 0$ i.e. $t = -1$

When $t = -1$, $x = -1$ and $y = 0$

Therefore the stationary point is $(-1, 0)$.

To determine the nature of the stationary point we examine the sign of $\dfrac{dy}{dx}$ in the neighbourhood of the point by first choosing appropriate values for x and then finding the corresponding values of t.

The equations $x = t^3$ and $y = (t + 1)^2$ show that there is no finite value of t for which either x or y is not defined, so the curve is continuous. Also there is no other stationary point.

Value of x	-2	-1	0
Value of t	$-\sqrt[3]{2}$	-1	0
Sign of $\dfrac{dy}{dx}$	$-$	0	$+$
	$\searrow$	$-$	$\nearrow$

Hence $(-1, 0)$ is a minimum point.

3. Find the equation of the normal to the curve $x = t^2$, $y = t + 2/t$, at the point where $t = 1$. Show, without sketching the curve, that this normal does not cross the curve again.

$$x = t^2 \quad \text{and} \quad y = t + \frac{2}{t} \quad \text{give} \quad \frac{dy}{dt} = 1 - \frac{2}{t^2} \quad \text{and} \quad \frac{dx}{dt} = 2t$$

$$\therefore \qquad \frac{dy}{dx} = \frac{dy}{dt} \div \frac{dx}{dt} = \frac{1 - 2/t^2}{2t} = \frac{t^2 - 2}{2t^3}$$

When $t = 1$; $x = 1$, $y = 3$ and $\dfrac{dy}{dx} = -\dfrac{1}{2}$

Therefore the gradient of the normal at $P(1, 3)$ is $\dfrac{-1}{-\frac{1}{2}} = 2$

The equation of this normal is $\qquad\qquad y - 3 = 2(x - 1)$

i.e. $\qquad\qquad\qquad\qquad\qquad\qquad y = 2x + 1$

All points for which $x = t^2$ and $y = t + 2/t$ are on the given curve, therefore, for any point that is on both the curve and the normal, these coordinates also satisfy the equation of the normal, i.e.,

at points common to the curve and the normal,

$$t + \frac{2}{t} = 2t^2 + 1 \quad \Rightarrow \quad 2t^3 - t^2 + t - 2 = 0 \qquad\qquad [1]$$

If a cubic equation can be factorised, each factor equated to zero gives a root of the equation, just as in the case of a quadratic equation.

Now we know that $t = 1$ at one point where the curve and normal meet so $t = 1$ is a root of $[1]$ and $(t - 1)$ is a factor of the LHS,

i.e. $\qquad (t - 1)(2t^2 + t + 2) = 0$

Therefore, at any other point where the normal meets the curve, the value of t is a root of the equation $2t^2 + t + 2 = 0$

Checking the value of $b^2 - 4ac$ shows that this equation has no real roots so there are no more points where the normal meets the curve.

EXERCISE 12c

1. Find the gradient function of each of the following curves in terms of the parameter.

(a) $x = 2t^2$, $y = t$ \qquad\qquad (b) $x = \sin \theta$, $y = \cos \theta$

(c) $x = t$, $y = \dfrac{4}{t}$

2. Sketch each curve in question 1.

3. If $x = \dfrac{t}{1-t}$ and $y = \dfrac{t^2}{1-t}$, find $\dfrac{dy}{dx}$ in terms of t. What is the value of $\dfrac{dy}{dx}$ at the point where $x = 1$?

4. (a) If $x = t^2$ and $y = t^3$, find $\dfrac{dy}{dx}$ in terms of t

   (b) If $y = x^{3/2}$, find $\dfrac{dy}{dx}$

   (c) Explain the connection between these two results.

5. Find the Cartesian equation of each of the curves given in Question 1 and hence find $\dfrac{dy}{dx}$. Show in each case that $\dfrac{dy}{dx}$ agrees with the gradient function found in Question 1.

6. Find the turning points of the curve whose parametric equations are $x = t$, $y = t^3 - t$, and distinguish between them.

7. A curve has parametric equations $x = \theta - \cos \theta$, $y = \sin \theta$. Find the smallest positive value of θ at which the gradient of this curve is zero.

8. Find the equation of the tangent to the curve $x = t^2$, $y = 4t$ at the point where $t = -1$

9. Find the equation of the normal to the curve $x = \cos \theta$, $y = \sin \theta$ at the point where $\theta = \tfrac{1}{4}\pi$. Find the coordinates of the point where this normal cuts the curve again.

MIXED EXERCISE 12   صفحة 1 ← 5

1. Differentiate with respect to x

   (a) y^4 (b) xy^2 (c) $1/y$ (d) $x \ln y$

   (e) $\sin y$ (f) e^y (g) $y \cos x$ (h) $y \cos y$

   In each Question from 2 to 13, find $\dfrac{dy}{dx}$

2. $x^2 - 2y^2 = 4$ 3. $\dfrac{1}{x} + \dfrac{1}{y} = 2$

4. $x^2 y^3 = 9$ 5. $y = 3(1.1)^x$

6. $y = \dfrac{(x-1)^4}{(x+3)}$ 7. $x^2 y^2 = \dfrac{(y+1)}{(x+1)}$

8. $x = t^2, \ y = t^3$ 9. $x = (t+1)^2, \ y = t^2 - 1$

10. $x = \sin^2\theta, \ y = \cos^3\theta$ 11. $x = 4t, \ y = \dfrac{4}{t}$

12. $y^2 - 2xy + 3y = 7x$ 13. $x = \dfrac{t}{1-t}, \ y = \dfrac{t^2}{1-t}$

14. If $x = \sin t$ and $y = \cos 2t$, find $\dfrac{dy}{dx}$ in terms of x and prove that

$$\frac{d^2y}{dx^2} + 4 = 0$$

15. If $x = e^t - t$ and $y = e^{2t} - 2t$, show that $\dfrac{dy}{dx} = 2(e^t + 1)$

*16. You will need a computer with graph drawing software or a calculator capable of plotting parametric equations for this question.
Many interesting shapes can be obtained by using quite simple parametric equations; these are a few suggestions.

(a) Plot (i) $x = \cos \theta, \ y = \sin \theta$ for $0 \leqslant \theta \leqslant 2\pi$
          (ii) $x = \cos \theta, \ y = \sin 2\theta$
          (iii) $x = \cos \theta, \ y = \sin 3\theta$

(b) Plot $x = \cos \theta, \ y = \sin^n \theta$ for various values of n.

(c) Plot $x = t^2, \ y = t^3$ for $-6 \leqslant t \leqslant 6$

   (i) Try different powers of t
   (ii) Plot $x = (t+1)^2, \ y = t^3$
   (iii) Try other functions of t

(d) Plot $x = t^2, \ y = t \sin t$

(e) Plot $x = t^2, \ y = t^n \sin t$ for various values of n.

The Binomial Theorem

If n is a positive integer then $(1 + x)^n$ can be expanded as a finite series,

where $(1 + x)^n = 1 + nx + \binom{n}{2} x^2 + \binom{n}{3} x^3 + \ldots + x^n$

and where $\binom{n}{r} = \dfrac{n!}{r!(n-r)!} = \dfrac{n(n-1)(n-2)\ldots(n-r+1)}{r!}$

Variation

When a quantity p varies directly as another quantity q, the relationship between them is of the form $p = kq$ where k is a constant.

When a quantity p varies inversely as another quantity q, the relationship between them is of the form $p = \dfrac{k}{q}$ where k is a constant.

Differentiation

STANDARD RESULTS

$f(x)$	$\dfrac{d}{dx} f(x)$
x^n	nx^{n-1}
$\sin x$	$\cos x$
$\cos x$	$-\sin x$
$\tan x$	$\sec^2 x$
e^x	e^x
$\ln x$	$1/x$
a^x	$a^x \ln a$
$(ax + b)^n$	$na(ax + b)^{n-1}$
$\sin(ax + b)$	$a\cos(ax + b)$
e^{ax+b}	ae^{ax+b}
$\ln(ax + b)$	$a/(ax + b)$

COMPOUND FUNCTIONS

If u and v are both functions of x, then

$$y = uv \quad \Rightarrow \quad \frac{dy}{dx} = v\frac{du}{dx} + u\frac{dv}{dx}$$

$$y = \frac{u}{v} \quad \Rightarrow \quad \frac{dy}{dx} = \left(v\frac{du}{dx} - u\frac{dv}{dx} \right) \Big/ v^2$$

$$y = f(u) \quad \text{where} \quad u = g(x) \quad \Rightarrow \quad \frac{dy}{dx} = \frac{dy}{du} \times \frac{du}{dx} \qquad (\text{the chain rule})$$

$$\frac{dy}{dx} = 1 \Big/ \frac{dx}{dy}$$

IMPLICIT DIFFERENTIATION

When y cannot be isolated, each term can be differentiated with respect to x,

e.g. $\quad \dfrac{d}{dx}(y^2) = 2y\dfrac{dy}{dx}, \quad \dfrac{d}{dx}(xy) = y + x\dfrac{dy}{dx} \quad (\text{using the product rule})$

PARAMETRIC DIFFERENTIATION

If $\quad y = f(t) \quad$ and $\quad x = g(t) \quad$ then $\quad \dfrac{dy}{dx} = \dfrac{dy}{dt} \div \dfrac{dx}{dt}$

MULTIPLE CHOICE EXERCISE C

TYPE I

1. $\dfrac{d}{dx}\left(\dfrac{1}{1+x} \right)$ is

    A $\dfrac{-1}{(1+x)^2}$ **B** $\dfrac{1}{1-x}$

    C $\ln(1+x)$ **D** $\dfrac{-1}{1+x^2}$

2. $\dfrac{d}{dx} \ln \left(\dfrac{x+1}{2x} \right)$ is

A $\dfrac{1}{2}$

B $\dfrac{1}{x+1} - \dfrac{1}{2x}$

C $\dfrac{2x}{x+1}$

D $\dfrac{1}{x+1} - \dfrac{1}{x}$

3. $\dfrac{d}{dx} a^x$ is

A xa^{x-1}

B a^x

C $x \ln a$

D $a^x \ln a$

4. If $x = \cos\theta$ and $y = \cos\theta + \sin\theta$, $\dfrac{dy}{dx}$ is

A $1 - \cot\theta$

B $1 - \tan\theta$

C $\cot\theta - 1$

D $\cot\theta + 1$

5. The coefficient of x^3 in the binomial expansion of $(2-x)^8$ is

A 1792

B 56

C -1792

D -2000

E -448

6. If $x^2 + y^2 = 4$ then $\dfrac{dy}{dx}$ is

A $2x + 2y$

B $4 - x^2$

C $-\dfrac{x}{y}$

D $\dfrac{y}{x}$

7. A linear approximation for $(5 - 2x)^3$ when x is small is

A $5 - 2x$

B $15 - 18x$

C $15 - 6x$

D $125 - 150x$

8. $\dfrac{d}{dx} (3 - 5x^2)^3$ is

A $3(3 - 5x^2)^2$

B $-30x(3 - 5x^2)^2$

C $-30x^5$

D $10x(3 - 5x^2)^2$

TYPE II

9. The third term in the binomial expansion of $(2 - 3x)^{10}$ is $(45)(2^7)(3^3)(x^3)$

10. When $y = \cos 2\theta$ and $x = \sin\theta$, $\dfrac{dy}{dx} = -4x\sqrt{(1 - x^2)}$

11. If $y = f(t)$ and $x = g(t)$ then $\dfrac{dy}{dx} = \dfrac{dy}{dt} \div \dfrac{dx}{dt}$

12. $(1+x)^3(1-x)^{20}$ is expanded as a series of ascending powers of x. The first two terms of the series are $1 - 17x$.

13. $\dfrac{d}{dx}(uv) = \dfrac{du}{dx} \times \dfrac{dv}{dx}$

14. $\dfrac{d}{dx}(x^2y^2) = 2xy^2 + 2x^2y$

15. If $y = \ln(\ln x)$ and $x > 1$ then

$$\frac{dy}{dx} = \frac{1}{\ln x}$$

16. The relationship $ay^3 = bx^2$ gives a straight line when $\ln y$ is plotted against $\ln x$, assuming that a and b are constants.

MISCELLANEOUS EXERCISE C

1. Determine the coefficient of x^3 in the binomial expansion of $(1 - 2x)^7$.

(AEB)$_s$

2. The parametric equations of a curve are given by

$$x = 160t - 6t^2, \qquad y = 80t - 8t^2.$$

Find the value of $\dfrac{dy}{dx}$ at each of the points on the curve where $y = 0$.

(UCLES)$_s$

3. The curve with equation $ky = a^x$ passes through the points with coordinates $(7, 12)$ and $(12, 7)$. Find, to 2 significant figures, the values of the constants k and a.

Using your values of k and a, find the value of $\dfrac{dy}{dx}$ at $x = 20$, giving your answer to 1 decimal place.

(ULEAC)$_s$

4. Two variable quantities x and y are related by the equation $y = a(b^x)$, where a and b are constants. When a graph is plotted showing values of $\ln y$ on the vertical axis and values of x on the horizontal axis, the points lie on a straight line having gradient 1.8 and crossing the vertical axis at the point $(0, 4.1)$. Find the values of a and b.

(UCLES)$_s$

5. Expand $(1 + ax)^8$ in ascending powers of x up to and including the term in x^2.

The coefficients of x and x^2 in the expansion of $(1 + bx)(1 + ax)^8$ are 0 and -36 respectively.

Find the values of a and b, given that $a > 0$ and $b > 0$.

(ULEAC)$_s$

6. The curve C has parametric equations

$$x = at, \qquad y = \frac{a}{t}, \qquad t \in \mathbb{R}, \qquad t \neq 0,$$

where t is a parameter and a is a positive constant.

(a) Sketch C.

(b) Find $\dfrac{dy}{dx}$ in terms of t.

The point P on C has parameter $t = 2$.

(c) Show that an equation of the normal to C at P is

$$2y = 8x - 15a.$$

The normal meets C again at the point Q.

(d) Find the value of t at Q.   (ULEAC)$_s$

7. The following measurements of the volume, $V\,\text{cm}^3$, and the pressure, $p\,\text{cm}$ of mercury, of a given mass of gas were taken.

V	10	50	110	170	230
p	1412.5	151.4	50.3	27.4	18.6

By plotting values of $\log_{10} p$ against $\log_{10} V$, verify graphically the relationship $p = kV^n$ where k and n are constants.

Use your graph to find approximate values for k and n, giving your answers to two significant figures.   (AEB)$_p$

8. The Highway Code (1993 edition) gives the braking distances needed (under good conditions) to stop cars travelling at various speeds. For a car travelling at 20 mph the braking distance is stated as 6 metres, and for a car travelling at 40 mph the braking distance is stated as 24 metres. Show that these figures are consistent with the braking distance being proportional to the square of the speed, and find the braking distance that would correspond to a speed of 70 mph.

The total distance required to stop a car in an emergency is greater than the braking distance, because a short time elapses before the driver reacts to the emergency and applies the brakes. The distance that the car travels in this short time before braking starts is the 'thinking distance', and the Highway Code states that, for a car travelling at 40 mph, the thinking distance is 12 metres. Find the total stopping distance for a car travelling at 70 mph.   (UCLES)$_s$

9. Write down the expansion of $(1 + x)^5$.

Hence, by letting $x = z + z^2$, find the coefficient of z^3 in the expansion of $(1 + z + z^2)^5$ in powers of z.   (UCLES)$_s$

10. The parametric equations of a curve are $x = 4t$, $y = \dfrac{4}{t}$, where the parameter t takes all non-zero values. The points A and B on the curve have parameters t_1 and t_2 respectively.

   (i) Write down the coordinates of the mid-point of the chord AB.

   (ii) Given that the gradient of AB is -2, show that $t_1 t_2 = \frac{1}{2}$.

   (iii) Find the coordinates of the points on the curve at which the gradient of the tangent is -2. (UCLES)$_s$

11. The equation of a closed curve is $(x + y)^2 + 2(x - y)^2 = 24$.

   (i) Show, by using differentiation, that the gradient at the point (x, y) on the curve may be expressed in the form $\dfrac{3x - y}{x - 3y}$.

   (ii) Find the coordinates of all the points on the curve at which the tangent is parallel to either the x-axis or the y-axis.

   (iii) Find the exact coordinates of all the points at which the curve crosses the axes, and the gradient of the curve at each of these points. (UCLES)

12. A forest fire spreads so that the number of hectares burnt after t hour is given by
$$h = 30(1.65)^t.$$

   (i) By what constant factor is the burnt area multiplied from time $t = N$ to time $t = N + 1$? Express this as a percentage increase.

   (ii) 1.65 can be written as e^K. Find the value of K.

   (iii) Hence show that $dh/dt = 15 e^{Kt}$.

   (iv) This shows that dh/dt is proportional to h. Find the constant of proportionality. (OCSEB)$_s$

13.

x	2.1	2.8	4.7	6.2	7.3
y	13	32	316	2000	7080

The above table shows corresponding values of variables x and y obtained experimentally. By drawing a suitable graph, show that these values support the hypothesis that x and y are connected by a relationship of the form $y = a^x$, where a is a constant. Use your graph to estimate the value of a to 2 significant figures. (ULEAC)

14. Find the term independent of x in the expansion of
$$\left(x^2 - \frac{2}{x}\right)^6.$$
(UCLES)$_s$

15. (i)   Differentiate with respect to x

       (a) $\ln(x^2)$,   (b) $x^2 \sin 3x$.

   (ii)   Find the gradient of the curve with equation

$$5x^2 + 5y^2 - 6xy = 13$$

      at the point $(1, 2)$.   (ULEAC)

16. Given that

$$(1 + kx)^8 = 1 + 12x + px^2 + qx^3 + \ldots, \quad \text{for all} \ \ x \in \mathbb{R},$$

   (a)   find the value of k, the value of p and the value of q.

   (b)   Using your values of k, p and q find the numerical coefficient of the x^3 term in the expansion of $(1 - x)(1 + kx)^8$ (ULEAC)

17. $f(x) \equiv (1 + x)^4 - (1 - x)^4$.

Expand $f(x)$ as a series in ascending powers of x.   (ULEAC)

CHAPTER 13

INTEGRATION

REVERSING DIFFERENTIATION

The process of finding a function from its derivative is called integration. This process reverses the operation of differentiation.

The result of integrating $f'(x)$, which is called the integral of $f'(x)$, is not unique; any constant can be added to the basic integral because the derivative of any constant is zero. So, for any function $f'(x)$, we have

$$\int f'(x)\,dx = f(x) + K$$

where $\int \ldots dx$ means 'the integral of ... w.r.t. x'

and $\quad K$ is called the constant of integration.

The following results were established in *Module A*.

$$\int x^n\,dx = \frac{1}{(n+1)} x^{n+1} + K, \quad n \neq -1$$

$$\int e^x\,dx = e^x + K$$

$$\int \frac{1}{x}\,dx = \ln x + K$$

A sum or difference of functions can be integrated term by term.

DEFINITE INTEGRATION

Definite integration requires the value of an integral to be calculated between specified values of the variable,

e.g. $\displaystyle\int_a^b x^n \, dx = \left[\frac{1}{(n+1)} x^{n+1} \right]_a^b$

$\displaystyle = \left\{ \frac{1}{(n+1)} x^{n+1} \right\}_{x=b} - \left\{ \frac{1}{(n+1)} x^{n+1} \right\}_{x=a}$

The constant of integration is unnecessary in a definite integral as it disappears upon subtraction.

STANDARD INTEGRALS

Whenever a function $f(x)$ is *recognised* as the derivative of a function $f(x)$ then

$$\frac{d}{dx} f(x) = f'(x) \quad \Rightarrow \quad \int f'(x) \, dx = f(x) + K$$

Thus any function whose derivative is known can be established as a standard integral.

It is already known that $\displaystyle \frac{d}{dx}(ax+b)^n = an(ax+b)^{n-1}$

hence $\displaystyle \int (ax+b)^n \, dx = \frac{1}{a(n+1)}(ax+b)^{n+1}$

e.g. $\displaystyle \int (5x-1)^2 \, dx = \frac{1}{(5)(3)}(5x-1)^3 + K = \frac{1}{15}(5x-1)^3 + K$

and $\displaystyle \int (3-4x)^{-2} \, dx = \frac{1}{(-4)(-1)}(3-4x)^{-1} + K = \frac{1}{4}(3-4x)^{-1} + K$

Further, we have $\dfrac{d}{dx} e^{(ax+b)} = ae^{(ax+b)}$

Hence $\displaystyle\int e^{(ax+b)} = \dfrac{1}{a} e^{(ax+b)} + K$

e.g. $\displaystyle\int 2e^x \, dx = 2e^x + K$ and $\displaystyle\int 4e^{(1-3x)} \, dx = (4)(-\tfrac{1}{3})e^{(1-3x)} + K$

We also know that $\dfrac{d}{dx}(a^x) = (\ln a)a^x$

Therefore $\displaystyle\int a^x \, dx = \dfrac{1}{\ln a} a^x + K$

Example 13a

Write down the integral of e^{3x} w.r.t. x and hence evaluate $\displaystyle\int_0^1 e^{3x} \, dx$

$\displaystyle\int e^{3x} \, dx = \tfrac{1}{3} e^{3x} + K$

The constant of integration disappears when a definite integral is calculated, hence

$\displaystyle\int_0^1 e^{3x} \, dx = \left[\tfrac{1}{3} e^{3x} \right]_0^1 = \tfrac{1}{3} e^3 - \tfrac{1}{3} e^0$

i.e. $\displaystyle\int_0^1 e^{3x} \, dx = \tfrac{1}{3}(e^3 - 1)$

EXERCISE 13a

Integrate each function w.r.t. x.

1. $(4x + 1)^3$
2. $(2 + 3x)^7$
3. $(5x - 4)^5$
4. $(6x - 2)^3$
5. $(4 - x)^2$
6. $(3 - 2x)^3$
7. $(3x + 1)^{-2}$
8. $(2x - 5)^{-4}$
9. $(4 - 3x)^{-2}$
10. $(2x + 3)^{1/2}$
11. $(3x - 1)^{1/3}$
12. $(2x - 1)^{-1/2}$
13. $(8 - 3x)^{-1/2}$
14. $\sqrt{(1 - 4x)}$
15. $(3 - 2x)^{3/2}$

16. $\dfrac{1}{(2-7x)^2}$ 17. $\dfrac{1}{\sqrt{(3-x)}}$ 18. $\dfrac{1}{(1-5x)^{1/3}}$

19. e^{4x} 20. $4e^{-x}$ 21. $e^{(3x-2)}$

22. $2e^{(1-5x)}$ 23. $6e^{-2x}$ 24. $5e^{(x-3)}$

25. $e^{(2+x/2)}$ 26. 2^x 27. $4^{(2+x)}$

28. $e^{2x} + \dfrac{1}{e^{2x}}$ 29. $a^{(1-2x)}$ 30. $2^x + x^2$

Evaluate the following definite integrals.

31. $\displaystyle\int_0^2 e^{2x}\,dx$ 32. $\displaystyle\int_{-1}^1 2e^{(x+1)}\,dx$

33. $\displaystyle\int_2^3 e^{(2-x)}\,dx$ 34. $\displaystyle\int_0^2 -e^x\,dx$

35. $\displaystyle\int_3^8 (1+x)^{1/2}\,dx$ 36. $\displaystyle\int_0^1 (2x-3)^3\,dx$

37. $\displaystyle\int_{-1}^0 \dfrac{1}{(1-x)^2}\,dx$ 38. $\displaystyle\int_{-1}^2 \dfrac{3}{\sqrt{(x+2)}}\,dx$

FUNCTIONS WHOSE INTEGRALS ARE LOGARITHMIC

To Integrate $\dfrac{1}{x}$

At first sight it looks as though we can write $\dfrac{1}{x} = x^{-1}$ and integrate by using

the rule $\displaystyle\int x^n\,dx = \dfrac{1}{n+1}x^{(n+1)} + K$

However, this method fails when $n = -1$ because the resulting integral, $\dfrac{x^0}{0}$, is meaningless.

Taking a second look at $\dfrac{1}{x}$ it can be *recognised* as the derivative of $\ln x$.

It must be remembered, however, that $\ln x$ is defined only when $x > 0$. Hence, provided that $x > 0$ we have

$$\frac{d}{dx}(\ln x) = \frac{1}{x} \quad \Longleftrightarrow \quad \int \frac{1}{x}\,dx = \ln x + K$$

Now if $x < 0$ the statement $\displaystyle\int \frac{1}{x}\,dx = \ln x$ is not valid because the log of a negative number does not exist.

However, $\dfrac{1}{x}$ exists for negative values of x, as the graph of $y = \dfrac{1}{x}$ shows.

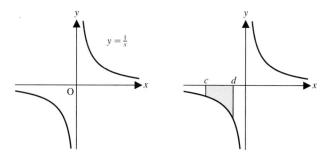

Also, the definite integral $\displaystyle\int_c^d \frac{1}{x}\,dx$, which is represented by the shaded area,

clearly exists. It must, therefore, be possible to integrate $\dfrac{1}{x}$ when x is negative

and we see opposite how to deal with the problem.

If $x < 0$ then $-x > 0$

i.e. $\quad \displaystyle\int \frac{1}{x}\,dx = \int \frac{-1}{(-x)}\,dx = \ln(-x) + K$

Thus, when $x > 0$, $\quad \displaystyle\int \frac{1}{x}\,dx = \ln x + K$

and when $x < 0$, $\quad \displaystyle\int \frac{1}{x}\,dx = \ln(-x) + K$

These two results can be combined so that, for both positive and negative values
of x, we have

$$\int \frac{1}{x} \, dx = \ln |x| + K$$

where $|x|$ denotes the numerical value of x regardless of sign

e.g. $|-1| = 1$ and $|-4| = 4$

The expression $\ln |x| + K$ can be simplified if K is replaced by $\ln A$, where A
is a positive constant, giving

$$\int \frac{1}{x} \, dx = \ln |x| + \ln A = \ln A |x|$$

Further $\dfrac{d}{dx} (\ln x^c) = \dfrac{d}{dx} (c \ln x) = \dfrac{c}{x}$

$\therefore$ $\int \dfrac{c}{x} \, dx = c \ln |x| + K$ or $c \ln A |x|$

e.g. $\int \dfrac{4}{x} \, dx = 4 \ln |x| + K$ or $4 \ln A |x|$

Also $\dfrac{d}{dx} \ln (ax + b) = \dfrac{a}{ax + b}$

$\therefore$ $\int \dfrac{1}{ax + b} \, dx = \dfrac{1}{a} \ln |ax + b| + K = \dfrac{1}{a} \ln A |ax + b|$

e.g. $\int \dfrac{1}{2x + 5} \, dx = \dfrac{1}{2} \ln |2x + 5|$ or $\dfrac{1}{2} \ln A |2x + 5|$

and $\int \dfrac{1}{4 - 3x} \, dx = -\dfrac{1}{3} \ln |4 - 3x| + K$ or $-\dfrac{1}{3} \ln A |4 - 3x|$

$$= \frac{1}{3} \ln \frac{B}{|4 - 3x|}$$

Integrate w.r.t. x giving each answer in a form which

(a) uses K (b) uses $\ln A$ and is simplified.

1. $\dfrac{1}{2x}$ 2. $\dfrac{4}{x}$ 3. $\dfrac{1}{3x+1}$ 4. $\dfrac{3}{1-2x}$

5. $\dfrac{6}{2+3x}$ 6. $\dfrac{3}{4-2x}$ 7. $\dfrac{4}{1-x}$ 8. $\dfrac{5}{6-7x}$

Evaluate

9. $\displaystyle\int_1^2 \dfrac{3}{x}\,dx$ 10. $\displaystyle\int_1^2 \dfrac{1}{2x}\,dx$ 11. $\displaystyle\int_4^5 \dfrac{2}{x-3}\,dx$ 12. $\displaystyle\int_0^1 \dfrac{1}{2-x}\,dx$

INTEGRATING TRIGONOMETRIC FUNCTIONS

Knowing the derivatives of the six trig functions, we can recognise the following integrals.

$$\frac{d}{dx}(\sin x) = \cos x \quad \Longleftrightarrow \quad \int \cos x\,dx = \sin x + K$$

$$\frac{d}{dx}(\cos x) = -\sin x \quad \Longleftrightarrow \quad \int \sin x\,dx = -\cos x + K$$

$$\frac{d}{dx}(\tan x) = \sec^2 x \quad \Longleftrightarrow \quad \int \sec^2 x\,dx = \tan x + K$$

Remembering the derivatives of some variations of the basic trig functions we also have

$$\int c \cos x\,dx = c \sin x + K$$

and

$$\int \cos(ax+b)\,dx = \frac{1}{a}\sin(ax+b) + K$$

There are similar results for the remaining trig integrals,

e.g. $\int 3 \cos x \, dx = 3 \sin x + K$

$\int \sin 4\theta \, d\theta = -\frac{1}{4} \cos 4\theta + K$

$\int \cos\left(3\theta - \frac{1}{2}\pi\right) d\theta = \frac{1}{3} \sin\left(3\theta - \frac{1}{2}\pi\right) + K$

Note that there is no need to *learn* these standard integrals. Knowledge of the standard derivatives is sufficient.

EXERCISE 13c

Integrate each function w.r.t. x

1. $\sin 2x$ 2. $\cos 7x$ 3. $\sec^2 4x$

4. $\sin\left(\frac{1}{4}\pi + x\right)$ 5. $3 \cos\left(4x - \frac{1}{2}\pi\right)$ 6. $\sec^2\left(\frac{1}{3}\pi + 2x\right)$

7. $2 \sin\left(3x - \alpha\right)$ 8. $5 \cos\left(\alpha - \frac{1}{2}x\right)$ 9. $\cos 3x - \cos x$

Evaluate

10. $\int_0^{\pi/6} \sin 3x \, dx$ 11. $\int_{\pi/4}^{\pi/6} \cos\left(2x - \frac{1}{2}\pi\right) dx$

12. $\int_0^{\pi/2} 2 \sin\left(2x - \frac{1}{3}\pi\right) dx$ 13. $\int_0^{\pi/8} \sec^2 2x \, dx$

EXERCISE 13d

This exercise contains a variety of functions. The reader is advised always to check that an integral is correct by differentiating it mentally.

Integrate w.r.t. x

1. $\sin\left(\frac{1}{2}\pi - 2x\right)$ 2. $e^{(4x-1)}$

3. $\sec^2 7x$ 4. $\dfrac{1}{2x - 3}$

5. $\dfrac{1}{\sqrt{(2x - 3)}}$ 6. $\dfrac{1}{(3x - 2)^2}$

7. 5^x

8. $\dfrac{3}{4x-1}$

9. $(3x-5)^2$

10. $e^{(4x-5)}$

11. $\sqrt{(4x-5)}$

12. $\sin\left(4x-\tfrac{1}{3}\pi\right)$

13. $\dfrac{3}{2(1-x)}$

14. $10^{(x+1)}$

15. $\cos\left(3x-\tfrac{1}{3}\pi\right)$

16. $2(3x-4)^5$

Evaluate

17. $\displaystyle\int_{-1/2}^{1/2} \sqrt{(1-2x)}\,dx$

18. $\displaystyle\int_{0}^{2} e^{(x/2+1)}\,dx$

19. $\displaystyle\int_{\pi/4}^{\pi/2} \sin 4x\,dx$

20. $\displaystyle\int_{1}^{3/2} \dfrac{5}{2-x}\,dx$

THE RECOGNITION ASPECT OF INTEGRATION

We have already seen the importance of the recognition aspect of integration in compiling a set of standard integrals.
Recognition is equally important when it is used to avoid serious errors in integration.

Consider, for instance, the derivative of the product $x^2 \sin x$.
Using the product formula gives

$$\frac{d}{dx}(x^2 \sin x) = 2x\sin x + x^2 \cos x$$

Clearly the derivative is the sum of two products therefore the integral of a single product is not itself a simple product

i.e.   **integration is not distributive when applied to a product.**

On the other hand, when we differentiate the function of a function $(1+x^2)^3$ we get

$$\frac{d}{dx}(1+x^2)^3 = 6x(1+x^2)^2$$

This time the derivative *is* a product so clearly the integral of a product *may* be a function of a function.

INTEGRATING PRODUCTS

First consider the function e^u where u is a function of x.

Differentiating as a function of a function gives

$$\frac{d}{dx}(e^u) = \left(\frac{du}{dx}\right)(e^u)$$

Thus any product of the form $\left(\dfrac{du}{dx}\right)e^u$ can be integrated by recognition, since

$$\int \left(\frac{du}{dx}\right) e^u \, dx = e^u + K$$

e.g. $\displaystyle\int 2x\,e^{x^2}\,dx = e^{x^2} + K$ $(u = x^2)$

$\displaystyle\int \cos x\,e^{\sin x}\,dx = e^{\sin x} + K$ $(u = \sin x)$

$\displaystyle\int x^2\,e^{x^3}\,dx = \tfrac{1}{3}\int 3x^2\,e^{x^3}\,dx = \tfrac{1}{3}e^{x^3} + K \quad (u = x^3)$

In these simple cases the substitution of u for $f(x)$ is done mentally. Also all the results can be checked by differentiating them mentally.

Similar, but slightly less simple functions, can also be integrated by changing the variable, but for these the substitution is written down.

Changing the Variable

Consider a general function $g(u)$ where u is a function of x

$$\frac{d}{dx}g(u) = \frac{du}{dx}g'(u) \quad \text{or} \quad g'(u)\frac{du}{dx}$$

Therefore $\displaystyle\int g'(u)\frac{du}{dx}\,dx = g(u) + K$ [1]

We also know that $\displaystyle\int g'(u)\,du = g(u) + K$ [2]

Comparing [1] and [2] gives

$$\int g'(u)\,\frac{du}{dx}\,dx = \int g'(u)\,du$$

i.e. $$\ldots\,\frac{du}{dx}\,dx \equiv \ldots du$$ [3]

i.e.   integrating   (a function of u) $\left(\dfrac{du}{dx}\right)$ w.r.t. x

   is *equivalent to* integrating   (the same function of u)   w.r.t. u

Note that the relationship in [3] is neither an equation nor an identity but is a pair of equivalent operations.

Suppose, for example, that we want to find $\displaystyle\int 2x(x^2+1)^5\,dx$

Writing the integral in the form $\displaystyle\int (x^2+1)^5\,2x\,dx$ and making the substitution $u = x^2+1$ gives

$$\int (x^2+1)^5 2x\,dx = \int u^5(2x)\,dx$$

But $\dfrac{du}{dx} = 2x$ and $\ldots\,\dfrac{du}{dx}\,dx \equiv \ldots du$

Therefore $\ldots 2x\,dx \equiv \ldots du$

i.e. $\displaystyle\int (x^2+1)^5\,2x\,dx = \int u^5\,du$

$$= \tfrac{1}{6}u^6 + K = \tfrac{1}{6}(x^2+1)^6 + K$$

In practice we go direct from $\dfrac{du}{dx} = 2x$ to the equivalent operators $\ldots 2x\,dx \equiv \ldots du$ by 'separating the variables'.

Products which can be integrated by this method are those in which one factor is basically the derivative of the function in the other factor; we choose to substitute u for the function in the other factor.

1. Integrate $x^2 \sqrt{(x^3 + 5)}$ w.r.t. x

In this product x^2 is basically the derivative of $x^3 + 5$ so we choose the substitution $u = x^3 + 5$

If $u = x^3 + 5$ then $\quad \dfrac{du}{dx} = 3x^2$

$\Rightarrow \qquad\qquad \ldots du \equiv \ldots 3x^2 \, dx$

Hence $\displaystyle\int x^2 \sqrt{(x^3 + 5)} \, dx = \tfrac{1}{3} \int (x^3 + 5)^{1/2} (3x^2 \, dx) = \tfrac{1}{3} \int u^{1/2} \, du$

$$= \left(\tfrac{1}{3}\right)\left(\tfrac{2}{3}\right) u^{3/2} + K$$

i.e. $\displaystyle\int x^2 \sqrt{(x^3 + 5)} \, dx = \tfrac{2}{9} (x^3 + 5)^{3/2} + K$

2. Find $\displaystyle\int \cos x \, \sin^3 x \, dx$

Writing the given integral in the form $\cos x (\sin x)^3$ shows that a suitable substitution is $u = \sin x$

If $u = \sin x$ then $\ldots du \equiv \ldots \cos x \, dx$

$\therefore \qquad \displaystyle\int \cos x \, \sin^3 x \, dx = \int (\sin x)^3 \cos x \, dx = \int u^3 \, du$

$$= \tfrac{1}{4} u^4 + K$$

i.e. $\displaystyle\int \cos x \, \sin^3 x \, dx = \tfrac{1}{4} \sin^4 x + K$

Applied generally, the method used above shows that

$$\int \cos x \, \sin^n x \, dx = \frac{1}{n+1} \sin^{(n+1)} x + K$$

and similarly that

$$\int \sin x \, \cos^n x \, dx = \frac{-1}{n+1} \cos^{(n+1)} x + K$$

3. Find $\int \dfrac{\ln x}{x}\,dx$

Initially this looks like a fraction but once it is recognised as the product of $\dfrac{1}{x}$ and $\ln x$, it is clear that

$\dfrac{1}{x} = \dfrac{d}{dx}(\ln x)$ and that we can make the substitution $u = \ln x$

If $u = \ln x$ then $\ldots du \equiv \ldots \dfrac{1}{x}\,dx$

Hence $\int \dfrac{1}{x}\ln x\,dx = \int u\,du = \tfrac{1}{2}u^2 + K$

i.e. $\int \dfrac{\ln x}{x}\,dx = \tfrac{1}{2}(\ln x)^2 + K$

Note that $(\ln x)^2$ is *not* the same as $\ln x^2$

Integrate the following expressions w.r.t. x

1. $4x^3\,e^{x^4}$

2. $\sin x\;e^{\cos x}$

3. $\sec^2 x\;e^{\tan x}$

4. $(2x+1)e^{(x^2+x)}$

5. $\sec^2 x\;e^{(1-\tan x)}$

6. $(1+\cos x)e^{(x+\sin x)}$

7. $2x e^{(1+x^2)}$

8. $(3x^2-2)e^{(x^3-2x)}$

Find the following integrals by making the substitution suggested.

9. $\displaystyle\int x(x^2-3)^4\,dx$ $u = x^2 - 3$

10. $\displaystyle\int x\sqrt{(1-x^2)}\,dx$ $u = 1 - x^2$

11. $\displaystyle\int \cos 2x(\sin 2x + 3)^2\,dx$ $u = \sin 2x + 3$

12. $\displaystyle\int x^2(1-x^3)\,dx$ $u = 1 - x^3$

13. $\displaystyle\int e^x\sqrt{(1+e^x)}\,dx$ $u = 1 + e^x$

14. $\displaystyle\int \cos x \, \sin^4 x \, dx$ $u = \sin x$

15. $\displaystyle\int \sec^2 x \, \tan^3 x \, dx$ $u = \tan x$

16. $\displaystyle\int x^n (1 + x^{n+1})^2 \, dx$ $u = 1 + x^{n+1}$

17. $\displaystyle\int \sin x \, \cos^3 x \, dx$ $u = \cos x$

18. $\displaystyle\int \sqrt{x}\sqrt{(1 + x^{3/2})} \, dx$ $u = 1 + x^{3/2}$

By using a suitable substitution, or by integrating at sight, find

19. $\displaystyle\int x^3 (x^4 + 4)^2 \, dx$

20. $\displaystyle\int e^x (1 - e^x)^3 \, dx$

21. $\displaystyle\int \sin\theta \sqrt{(1 - \cos\theta)} \, d\theta$

22. $\displaystyle\int (x + 1)\sqrt{(x^2 + 2x + 3)} \, dx$

23. $\displaystyle\int x \, e^{(x^2 + 1)} \, dx$

24. $\displaystyle\int \sec^2 x (1 + \tan x) \, dx$

DEFINITE INTEGRATION WITH A CHANGE OF VARIABLE

A definite integral can be evaluated only after the appropriate integration has been performed. Should this require a change of variable, e.g. from x to u, it is usually most convenient also to change the limits of integration from x values to u values.

Example 13f

By using the substitution $u = x^3 + 1$, evaluate $\int_0^1 x^2 \sqrt{(x^3 + 1)}\,dx$

If $u = x^3 + 1$ then $\qquad \ldots du \equiv \ldots 3x^2\,dx$

$$\text{and} \quad \begin{cases} x = 0 & \Rightarrow & u = 1 \\ x = 1 & \Rightarrow & u = 2 \end{cases}$$

Hence $\qquad \displaystyle\int_0^1 x^2\sqrt{(x^3+1)}\,dx = \tfrac{1}{3}\int_1^2 \sqrt{u}\,du$

$$= \tfrac{1}{3}\left[\tfrac{2}{3}u^{3/2}\right]_1^2$$

$$= \tfrac{2}{9}(2\sqrt{2} - 1)$$

EXERCISE 13f

Evaluate

1. $\displaystyle\int_0^1 x\,e^{x^2}\,dx$

2. $\displaystyle\int_0^{\pi/2} \cos x\,\sin^4 x\,dx$

3. $\displaystyle\int_1^2 \frac{1}{x}\ln x\,dx$

4. $\displaystyle\int_1^2 x^2(x^3-1)^4\,dx$

5. $\displaystyle\int_0^{\pi/4} \sec^2 x\,e^{\tan x}\,dx$

6. $\displaystyle\int_1^2 x(1+2x^2)\,dx$

7. $\displaystyle\int_2^3 (x-1)\,e^{(x^2-2x)}\,dx$

8. $\displaystyle\int_0^{\pi/6} \cos x(1+\sin^2 x)\,dx$

9. $\displaystyle\int_1^3 \frac{1}{x}(\ln x)^2\,dx$

10. $\displaystyle\int_0^{\sqrt{3}} x\sqrt{(1+x^2)}\,dx$

11. $\displaystyle\int_0^{\pi/2} \cos x\,e^{\sin x}\,dx$

12. $\displaystyle\int_0^1 x^2(2+x^3)^2\,dx$

INTEGRATION BY PARTS

It is not always possible to express a product in the form $f(u)\dfrac{du}{dx}$ so an alternative approach is needed.

Looking again at the differentiation of a product uv where u and v are both functions of x we have

$$\frac{d}{dx}(uv) = v\frac{du}{dx} + u\frac{dv}{dx} \quad \Rightarrow \quad v\frac{du}{dx} = \frac{d}{dx}(uv) - u\frac{dv}{dx}$$

Now $v\dfrac{du}{dx}$ can be taken to represent a product which is to be integrated w.r.t. x

Thus
$$\int v\frac{du}{dx}\,dx = \int \frac{d}{dx}(uv)\,dx - \int u\frac{dv}{dx}\,dx$$

i.e.
$$\int v\frac{du}{dx}\,dx = uv - \int u\frac{dv}{dx}\,dx$$

At this stage it may appear that the RHS is more complicated than the original product on the LHS.

However, by careful choice of the factor to be replaced by v we can ensure that $u\dfrac{dv}{dx}$ is easier to integrate than $v\dfrac{du}{dx}$

The factor chosen to be replaced by v is usually the one whose derivative is a simpler function. It must also be remembered, however, that the other factor is $\dfrac{du}{dx}$ and therefore it must be possible to integrate it to find u.

This method for integrating a product is called *integrating by parts*.

Examples 13g

1. Integrate xe^x w.r.t. x

Taking $\quad v = x \quad$ and $\quad \dfrac{du}{dx} = e^x$

gives $\quad \dfrac{dv}{dx} = 1 \quad$ and $\quad u = e^x$

Then $\quad \displaystyle\int v \frac{du}{dx} dx = uv - \int u \frac{dv}{dx} dx$

gives $\quad \displaystyle\int x e^x dx = (e^x)(x) - \int (e^x)(1) dx$

$$= x e^x - e^x + K$$

2. Find $\displaystyle\int x^2 \sin x \, dx$

Taking $\quad v = x^2 \quad$ and $\quad \dfrac{du}{dx} = \sin x$

gives $\quad \dfrac{dv}{dx} = 2x \quad$ and $\quad u = -\cos x$

Then $\quad \displaystyle\int v \frac{du}{dx} dx = uv - \int u \frac{dv}{dx} dx$

gives $\quad \displaystyle\int x^2 \sin x \, dx = (-\cos x)(x^2) - \int (-\cos x)(2x) \, dx$

$$= -x^2 \cos x + 2 \int x \cos x \, dx \qquad [1]$$

At this stage the integral on the RHS cannot be found without *repeating* the process of integrating by parts on the term $\displaystyle\int x \cos x \, dx$ as follows.

Taking $\quad v = x \quad$ and $\quad \dfrac{du}{dx} = \cos x$

gives $\quad \dfrac{dv}{dx} = 1 \quad$ and $\quad u = \sin x$

Then $\quad \displaystyle\int x \cos x \, dx = (\sin x)(x) - \int (\sin x)(1) \, dx$

$$= x \sin x + \cos x + K$$

Hence equation [1] becomes

$$\int x^2 \sin x \, dx = -x^2 \cos x + 2x \sin x + 2 \cos x + K$$

3. Find $\displaystyle\int x^4 \ln x \, dx$

Because ln x can be differentiated but *not integrated*, we must use $v = \ln x$

Taking $v = \ln x$ and $\dfrac{du}{dx} = x^4$

gives $\dfrac{dv}{dx} = \dfrac{1}{x}$ and $u = \tfrac{1}{5} x^5$

The formula for integrating by parts then gives

$$\int x^4 \ln x \, dx = \left(\tfrac{1}{5} x^5\right)(\ln x) - \int \left(\tfrac{1}{5} x^5\right)\left(\dfrac{1}{x}\right) dx = \tfrac{1}{5} x^5 \ln x - \tfrac{1}{5}\int x^4 \, dx$$

$\Rightarrow$ $\displaystyle\int x^4 \ln x \, dx = \tfrac{1}{5} x^5 \ln x - \tfrac{1}{25} x^5 + K$

Special Cases of Integration by Parts

An interesting situation arises when an attempt is made to integrate $e^x \cos x$ or $e^x \sin x$

4. Find $\displaystyle\int e^x \cos x \, dx$

Taking $v = e^x$ and $\dfrac{du}{dx} = \cos x$ gives $\dfrac{dv}{dx} = e^x$ and $u = \sin x$

Hence $\displaystyle\int e^x \cos x \, dx = e^x \sin x - \int e^x \sin x \, dx$ [1]

But since $\displaystyle\int e^x \sin x \, dx$ is very similar to $\displaystyle\int e^x \cos x \, dx$ it seems that we have made no progress. However, if we now apply integration by parts to $\displaystyle\int e^x \sin x \, dx$ an interesting situation emerges.

Taking $v = e^x$ and $\dfrac{du}{dx} = \sin x$ gives $\dfrac{dv}{dx} = e^x$ and $u = -\cos x$

so that $\displaystyle\int e^x \sin x \, dx = -e^x \cos x + \int e^x \cos x \, dx$

or $\displaystyle\int e^x \cos x \, dx = e^x \cos x + \int e^x \sin x \, dx$ [2]

Adding [1] and [2] gives $\qquad 2\int e^x \cos x \, dx = e^x(\sin x + \cos x) + K$

i.e. $\qquad \int e^x \cos x \, dx = \tfrac{1}{2} e^x(\sin x + \cos x) + K$

Clearly the same two equations can be used to give

$$2\int e^x \sin x \, dx = e^x(\sin x - \cos x) + K$$

Note that neither of the equations [1] and [2] contains a completed integration process, so the constant of integration is introduced only when these two equations have been combined.

Note also that the same choice of function for v must be made in both applications of integration by parts; in this case we chose $v = e^x$ each time.

Integration of ln x

So far we have found no way to integrate $\ln x$. Now, however, if $\ln x$ is regarded as the product of 1 and $\ln x$ we can apply integration by parts as follows.

Examples 13g (continued)

5. Find $\int \ln x \, dx$

Taking $\qquad v = \ln x \quad$ and $\quad \dfrac{du}{dx} = 1$

gives $\qquad \dfrac{dv}{dx} = \dfrac{1}{x} \quad$ and $\quad u = x$

Then $\qquad \displaystyle\int v\,\frac{du}{dx}\,dx = uv - \int u\,\frac{dv}{dx}\,dx$

becomes $\qquad \displaystyle\int \ln x \, dx = x \ln x - \int x\left(\frac{1}{x}\right) dx$

$$= x \ln x - x + K$$

i.e. $\qquad \displaystyle\int \ln x \, dx = x(\ln x - 1) + K$

Integrate the following functions w.r.t. x

1. $x \cos x$ 2. $x^2 e^x$ 3. $x^3 \ln 3x$

4. $x e^{-x}$ 5. $3x \sin x$ 6. $e^x \sin 2x$

7. $e^{2x} \cos x$ 8. $x^2 e^{4x}$ 9. $e^{-x} \sin x$

10. $\ln 2x$ 11. $e^x (x + 1)$ 12. $x(1 + x)^7$

13. $x \sin (x + \frac{1}{6} \pi)$ 14. $x \cos nx$ 15. $x^n \ln x$

16. $3x \cos 2x$ 17. $2e^x \sin x \cos x$ 18. $x^2 \sin x$

19. $e^{ax} \sin bx$

20. By writing $\cos^3 \theta$ as $(\cos^2 \theta)(\cos \theta)$ use integration by parts to find
    $\int \cos^3 \theta \, d\theta$.

Each of the following products can be integrated either:

(a) by immediate recognition, or

(b) by a suitable change of variable, or

(c) by parts.

Choose the best method in each case and hence integrate each function w.r.t. x.

21. $(x - 1)e^{x^2 - 2x + 4}$ 22. $(x + 1)^2 e^x$ 23. $\sin x (4 + \cos x)^3$

24. $\cos x \, e^{\sin x}$ 25. $x^4 \sqrt{(1 + x^5)}$ 26. $e^x (e^x + 2)^4$

27. $x e^{2x - 1}$ 28. $x(1 - x^2)^9$ 29. $\cos x \sin^5 x$

DEFINITE INTEGRATION BY PARTS

When using the formula $\int v \dfrac{du}{dx} \, dx = uv - \int u \dfrac{dv}{dx} \, dx$

it must be appreciated that the term uv on the RHS is fully integrated. Consequently in a definite integration, uv must be *evaluated between the appropriate boundaries*

i.e. $\displaystyle\int_a^b v \dfrac{du}{dx} \, dx = \left[uv \right]_a^b - \int_a^b u \dfrac{dv}{dx} \, dx$

Example 13h

Evaluate $\displaystyle\int_0^1 x\,e^x\,dx$

$$\int x\,e^x\,dx = \int v\,\frac{du}{dx}\,dx$$

where $\qquad v = x \quad$ and $\quad \dfrac{du}{dx} = e^x$

Hence $\displaystyle\int_0^1 x\,e^x\,dx = \left[\,x\,e^x\,\right]_0^1 - \int_0^1 e^x\,dx$

$$= \left[\,x\,e^x\,\right]_0^1 - \left[\,e^x\,\right]_0^1$$

$$= (e^1 - 0) - (e^1 - e^0)$$

i.e. $\qquad \displaystyle\int_0^1 x\,e^x\,dx = 1$

EXERCISE 13h

Evaluate

1. $\displaystyle\int_0^{\pi/2} x\sin x\,dx$

2. $\displaystyle\int_1^2 x^5\ln x\,dx$

3. $\displaystyle\int_0^1 (x+1)e^x\,dx$

4. $\displaystyle\int_0^\pi e^x\cos x\,dx$

5. $\displaystyle\int_1^2 x\sqrt{(x-1)}\,dx$

6. $\displaystyle\int_0^{\pi/2} x^2\cos x\,dx$

7. $\displaystyle\int_0^1 x\,e^{2x}\,dx$

8. $\displaystyle\int_0^1 x^2\,e^x\,dx$

9. $\displaystyle\int_0^{\pi/2} x\cos 2x\,dx$

MIXED EXERCISE 13

Integrate the following functions, taking care to choose the best method in each case.

1. $x^2\,e^{2x}$

2. $2x\,e^{x^2}$

3. $\sec^2 x\,(3\tan x - 4)$

4. $(x+1)\ln(x+1)$

5. $\sec^2 x\,\tan^3 x$

6. $x^2\cos x$

7. $\sin x\,e^{\cos x}$

8. $x(2x+3)^7$

9. $(1-x)e^{(1-x)^2}$

10. $xe^{(2x-1)}$

11. $\cos x \sin^5 x$

12. $\sin x (4 + \cos x)^3$

13. $(x-1)e^{(x^2-2x+3)}$

14. $x^2(1-x^3)^9$

15. $x \cos 4x$

16. $2x e^{x^2}$

17. $3x(x^2-1)^3$

18. $3x(x-1)^3$

Evaluate each definite integral.

19. $\displaystyle\int_1^3 e^{3x}\,dx$

20. $\displaystyle\int_0^{\pi/8} \cos 4x\,dx$

21. $\displaystyle\int_0^1 \frac{1}{x-2}\,dx$

22. $\displaystyle\int_{\pi/4}^{\pi/2} \operatorname{cosec}^2 x\,dx$

23. $\displaystyle\int_0^1 x^2 e^{3x^3}\,dx$

24. $\displaystyle\int_0^{\pi/4} x \cos 2x\,dx$

25. $\displaystyle\int_0^1 \frac{1}{2-x}\ln(2-x)\,dx$

26. $\displaystyle\int_1^2 x^2 \ln x\,dx$

CHAPTER 14

PARTIAL FRACTIONS

DECOMPOSING A FRACTION

Two separate fractions such as $\dfrac{3}{x+4} + \dfrac{2}{x-5}$ can be expressed as a single fraction with a common denominator,

i.e. $\dfrac{3}{x+4} + \dfrac{2}{x-5} = \dfrac{3(x-5)+2(x+4)}{(x+4)(x-5)} = \dfrac{5x-7}{(x+4)(x-5)}$

Later on in the course it is necessary to reverse this operation, i.e. to take an expression such as $\dfrac{x-2}{(x+3)(x-4)}$ and express it as the sum of two separate fractions.

This process is called splitting up, or decomposing, into *partial fractions*.

Consider again $\qquad \dfrac{x-2}{(x+3)(x-4)}$

This fraction is a *proper fraction* because the highest power of x in the numerator (1 in this case) is less than the highest power of x in the denominator (2 in this case when the brackets are expanded).

Therefore its separate (or partial) fractions also will be proper,

i.e. $\dfrac{x-2}{(x+3)(x-4)}$ can be expressed as $\dfrac{A}{x+3} + \dfrac{B}{x-4}$

where A and B are numbers. The worked example which follows shows how the values of A and B can be found.

Express $\dfrac{x-2}{(x+3)(x-4)}$ in partial fractions.

$$\frac{x-2}{(x+3)(x-4)} = \frac{A}{x+3} + \frac{B}{x-4}$$

Expressing the separate fractions on the RHS as a single fraction over a common denominator gives

$$\frac{x-2}{(x+3)(x-4)} = \frac{A(x-4)+B(x+3)}{(x+3)(x-4)}$$

This is not an equation because the RHS is just another way of expressing the LHS. It follows that, as the denominators are identical the numerators also are identical.

i.e. $\qquad x-2 = A(x-4)+B(x+3)$

Remembering that this is *not* an equation but two ways of writing the same expression, it follows that LHS = RHS for any value that we choose to give to x.

Choosing to substitute 4 for x (to eliminate A) gives

$$2 = A(0)+B(7)$$

$\Rightarrow \qquad B = \frac{2}{7}$

Choosing to substitute -3 for x (to eliminate B) gives

$$-5 = A(-7)+B(0)$$

$\Rightarrow \qquad A = \frac{5}{7}$

Therefore $\qquad \dfrac{x-2}{(x+3)(x-4)} = \dfrac{\frac{5}{7}}{x+3} + \dfrac{\frac{2}{7}}{x-4}$

$$= \frac{5}{7(x+3)} + \frac{2}{7(x-4)}$$

Express the following fractions in partial fractions.

1. $\dfrac{x-2}{(x+1)(x-1)}$

2. $\dfrac{2x-1}{(x-1)(x-7)}$

3. $\dfrac{4}{(x+3)(x-2)}$

4. $\dfrac{7x}{(2x-1)(x+4)}$

5. $\dfrac{2}{x(x-2)}$

6. $\dfrac{2x-1}{x^2-3x+2}$

7. $\dfrac{3}{x^2 - 9}$ 8. $\dfrac{6x + 7}{3x(x + 1)}$

9. $\dfrac{9}{2x^2 + x}$ 10. $\dfrac{x + 1}{3x^2 - x - 2}$

The Cover-up Method

There is a quicker method for finding the numerators of the separate fractions.

Consider $f(x) = \dfrac{x}{(x - 2)(x - 3)} = \dfrac{A}{x - 2} + \dfrac{B}{x - 3}$

$$= \dfrac{A(x - 3) + B(x - 2)}{(x - 2)(x - 3)}$$

$\Rightarrow$ $x \equiv A(x - 3) + B(x - 2)$

When $x = 2$, $A = \dfrac{2}{2 - 3} = -2$,

which is the value of $\dfrac{x}{\rule{1cm}{0.4pt}(x - 3)}$ when $x = 2$

i.e. $A = f(2)$ with the factor $(x - 2)$ 'covered up'.

Using this fact, A can be written down at sight.

Similarly when $x = 3$, $B = \dfrac{3}{3 - 2} = 3$

which is the value of $\dfrac{x}{(x - 2)\rule{1cm}{0.4pt}}$ when $x = 3$

i.e. $B = f(3)$ with the factor $(x - 3)$ covered up.

Hence $\dfrac{x}{(x - 2)(x - 3)} \equiv \dfrac{-2}{x - 2} + \dfrac{3}{x - 3}$

Note that this method can be used only for linear factors.

Example 14b

Express $\dfrac{1}{(2x-1)(x+3)}$ in partial fractions.

$$\frac{1}{(2x-1)(x+3)} = \frac{f(\frac{1}{2}) \text{ with } (2x-1) \text{ covered up}}{(2x-1)} + \frac{f(-3) \text{ with } (x+3) \text{ covered up}}{(x+3)}$$

$$\frac{1}{(2x-1)(x+3)} = \frac{2}{7(2x-1)} - \frac{1}{7(x+3)}$$

EXERCISE 14b

Express these fractions in partial fractions.

1. $\dfrac{x+4}{(x+3)(x-5)}$

2. $\dfrac{3x+1}{(2x-1)(x-1)}$

3. $\dfrac{2x-3}{(x-2)(4x-3)}$

4. $\dfrac{3-x}{(x+1)(2x-1)}$

5. $\dfrac{3}{x(2x+1)}$

6. $\dfrac{4}{x^2-7x-8}$

7. $\dfrac{4x}{4x^2-9}$

8. $\dfrac{4x-2}{x^2+2x}$

9. $\dfrac{3x}{2x^2-2x-4}$

10. $\dfrac{3x+2}{2x^2-4x}$

Quadratic Factors in the Denominator

It is also possible to decompose fractions with quadratic or higher degree factors.

Consider $\dfrac{x^2+1}{(x^2+2)(x-1)}$

This is a proper fraction, so its partial fractions are also proper,

i.e. $\dfrac{x^2+1}{(x^2+2)(x-1)}$ can be expressed in the form $\dfrac{Ax+B}{x^2+2} + \dfrac{C}{x-1}$

Using the cover-up method gives $C = \frac{2}{3}$, but to find A and B the earlier method of expressing the partial fraction form as a single fraction must be used giving

$$x^2 + 1 \equiv (Ax + B)(x - 1) + \tfrac{2}{3}(x^2 + 2)$$

The values of A and B can then be found by substituting any suitable values for x.

We will choose $x = 0$ and $x = -1$ as these are simple values to handle. Choosing $x = 1$ is not suitable as it eliminates A and B and it was used to find C.

$$x = 0 \quad \text{gives} \quad 1 = B(-1) + \tfrac{2}{3}(2) \qquad \Rightarrow \quad B = \tfrac{1}{3}$$

$$x = -1 \quad \text{gives} \quad 2 = (-A + \tfrac{1}{3})(-2) + \tfrac{2}{3}(3) \quad \Rightarrow \quad A = \tfrac{1}{3}$$

$$\therefore \qquad \frac{x^2 + 1}{(x^2 + 2)(x - 1)} \equiv \frac{x + 1}{3(x^2 + 2)} + \frac{2}{3(x - 1)}$$

A Repeated Factor in the Denominator

Consider the fraction $\dfrac{2x - 1}{(x - 2)^2}$

This is a proper fraction, and it is possible to express this as two fractions with numerical numerators as we can see if we adjust the numerator,

i.e. $\qquad \dfrac{2x - 1}{(x - 2)^2} \equiv \dfrac{2(x - 2) + 4 - 1}{(x - 2)^2} \equiv \dfrac{2}{x - 2} + \dfrac{3}{(x - 2)^2}$

Any fraction whose denominator is a repeated linear factor can be expressed as separate fractions with numerical numerators, for example,

$$\frac{2x^2 - 3x + 4}{(x - 1)^3} \quad \text{can be expressed as} \quad \frac{A}{x - 1} + \frac{B}{(x - 1)^2} + \frac{C}{(x - 1)^3}$$

In the general case the values of the numerators can be found using the method in the next worked example.

To summarise, a proper fraction can be decomposed into partial fractions and the form of the partial fractions depends on the form of the factors in the denominator where

a linear factor gives a partial fraction of the form $\dfrac{A}{ax+b}$

a quadratic factor gives a partial fraction of the form $\dfrac{Ax+B}{ax^2+bx+c}$

a repeated factor gives two partial fractions of the form

$$\frac{A}{ax+b}+\frac{B}{(ax+b)^2}$$

Examples 14c

1. Express $\dfrac{x-1}{(x+1)(x-2)^2}$ in partial fractions.

$$\underbrace{\frac{x-1}{(x+1)(x-2)^2}}_{x+1+\frac{B}{(x-2)}+\frac{C}{(x-2)^2}} \equiv -\frac{2}{9}$$

$$\Rightarrow \qquad x-1 \equiv \left(-\tfrac{2}{9}\right)(x-2)^2 + B(x+1)(x-2) + C(x+1)$$

$$x=2 \quad \text{gives} \quad C = \tfrac{1}{3}$$

Comparing coefficients of x^2 gives $0 = -\tfrac{2}{9} + B \quad \Rightarrow \quad B = \tfrac{2}{9}$

$$\therefore \qquad \frac{x-1}{(x+1)(x-2)^2} \equiv -\frac{2}{9(x+1)} + \frac{2}{9(x-2)} + \frac{1}{3(x-2)^2}$$

Note that C can be found by the cover-up method, but B cannot.

2. Express $\dfrac{x^3}{(x+1)(x-3)}$ in partial fractions.

This fraction is improper and it must be divided out to obtain a mixed fraction before it can be expressed in partial fractions.

$$
\begin{array}{r}
x+2 \\
x^2-2x-3\overline{)x^3} \\
\underline{x^3-2x^2-3x} \\
2x^2+3x \\
\underline{2x^2-4x-6} \\
7x+6
\end{array}
$$

$$\therefore \quad \frac{x^3}{(x+1)(x-3)} \equiv x+2+\frac{7x+6}{(x+1)(x-3)}$$

$$\equiv x+2+\frac{1}{4(x+1)}+\frac{27}{4(x-3)}$$

EXERCISE 14c

Express in partial fraction,

1. $\dfrac{2}{(x-1)(x^2+1)}$

2. $\dfrac{x^2+1}{x(2x^2+1)}$

3. $\dfrac{x^2+3}{x(x^2+2)}$

4. $\dfrac{2x^2+x+1}{(x-3)(2x^2+1)}$

5. $\dfrac{x^3-1}{(x+2)(2x+1)(x^2+1)}$

6. $\dfrac{x^2+1}{x(2x^2-1)(x-1)}$

7. $\dfrac{x}{(x-1)(x-2)^2}$

8. $\dfrac{x^2-1}{x^2(2x+1)}$

9. $\dfrac{3}{x(3x-1)^2}$

10. $\dfrac{x^2}{(x+1)(x-1)}$

11. $\dfrac{x^2-2}{(x+3)(x-1)}$

12. $\dfrac{x^3+3}{(x-1)(x+1)}$

13. $\dfrac{2}{(x+1)(x-1)}$

14. $\dfrac{3}{(x-2)(x+1)}$

15. $\dfrac{1}{x(x-3)}$

16. $\dfrac{1}{x^2-1}$

17. Express in partial fractions

   (a) $\dfrac{2}{x^2(x-1)}$ (b) $\dfrac{1}{x(x^2+1)}$ (c) $\dfrac{x^2}{x-1}$ (d) $\dfrac{x}{(x-1)^2(2x+1)}$

18. Given that $y = \dfrac{1}{x(x-1)}$, express y in partial fractions and hence find $\dfrac{dy}{dx}$.

19. Express $f(x) = \dfrac{2x}{(x-1)(x+1)}$ in partial fractions and hence find $f'(x)$. Show that $f(x)$ has no stationary values.

CHAPTER 15

FURTHER INTEGRATION

INTEGRATING FRACTIONS

Some expressions have an integral that is a function but there are many others for which an exact integral cannot be found. In this book we are concerned mainly with expressions which *can* be integrated but even with this proviso the reader should be aware that, while the methods suggested usually work, they are not infallible.

There are several different methods for integrating fractions, the appropriate method in a particular case depending upon the form of the fraction. Consequently it is very important that each fraction be categorised carefully to avoid embarking on unnecessary and lengthy working.

Method 1   Using Recognition

Consider the function $\ln u$ where $u = f(x)$

Differentiating with respect to x gives

$$\frac{d}{dx} \ln u = \left(\frac{1}{u}\right)\left(\frac{du}{dx}\right) \quad \text{i.e.} \quad \frac{du/dx}{u}$$

i.e. $\quad \dfrac{d}{dx} \ln f(x) = \dfrac{f'(x)}{f(x)}$

Hence $\qquad \displaystyle\int \frac{f'(x)}{f(x)}\, dx = \ln|f(x)| + K$

Thus all fractions of the form $f'(x)/f(x)$ can be integrated *immediately* by recognition, e.g.

$$\int \frac{\cos x}{1 + \sin x}\,dx = \ln|1 + \sin x| + K \quad \text{as} \quad \frac{d}{dx}(1 + \sin x) = \cos x$$

$$\int \frac{e^x}{e^x + 4}\,dx = \ln|e^x + 4| + K \quad \text{as} \quad \frac{d}{dx}(e^x + 4) = e^x$$

Note, however, that $\displaystyle\int \frac{x}{\sqrt{(1 + x)}}\,dx$ is *not* $\ln|\sqrt{(1 + x)}| + K$

because $\qquad\qquad \dfrac{d}{dx}\sqrt{(1 + x)}$ is not x

Method 1 applies only to an integral whose numerator is basically the derivative of *the complete denominator*.

An integral whose numerator is the derivative, not of the complete denominator but of a function *within* the denominator, belongs to the next type.

Method 2   Using Substitution

Consider the integral $\displaystyle\int \frac{2x}{\sqrt{(x^2 + 1)}}\,dx$

Noting that $2x$ is the derivative of $x^2 + 1$ we make the substitution $u = x^2 + 1$, i.e.

if $\qquad u = x^2 + 1 \quad$ then $\quad \ldots du \equiv \ldots 2x\,dx$

By this change of variable the given integral is converted into the simple form

$$\int \frac{1}{\sqrt{u}}\,du$$

Examples 15a

1. Find $\displaystyle\int \frac{x^2}{1+x^3}\,dx$

$$\int \frac{x^2}{1+x^3}\,dx = \frac{1}{3}\int \frac{3x^2}{1+x^3}\,dx$$

This integral is of the form $\displaystyle\int \frac{f'(x)}{f(x)}\,dx$ so we use recognition

$$= \tfrac{1}{3}\ln|1+x^3| + K$$

2. By writing $\tan x$ as $\dfrac{\sin x}{\cos x}$, find $\displaystyle\int \tan x\,dx$

$$\int \tan x\,dx = \int \frac{\sin x}{\cos x}\,dx = -\int \frac{f'(x)}{f(x)}\,dx \quad \text{where} \quad f(x) = \cos x$$

so

$$\int \frac{\sin x}{\cos x}\,dx = -\ln|\cos x| + K$$

$$\therefore \qquad \int \tan x\,dx = K - \ln|\cos x|$$

Note that, similarly, $\displaystyle\int \cot x\,dx = \ln|\sin x| + K$

3. Find $\displaystyle\int \frac{e^x}{(1-e^x)^2}\,dx$

e^x is basically the derivative of $1-e^x$ but not of $(1-e^x)^2$ so we make the substitution $u = 1 - e^x$

If $\qquad u = 1 - e^x \quad$ then $\quad \ldots du \equiv \ldots -e^x\,dx$

So $\qquad \displaystyle\int \frac{e^x}{(1-e^x)^2}\,dx = \int \frac{-1}{u^2}\,du = \frac{1}{u} + K$

$$\therefore \qquad \int \frac{e^x}{(1-e^x)^2}\,dx = \frac{1}{1-e^x} + K$$

In Questions 1 to 18 integrate each function w.r.t. x

1. $\dfrac{\cos x}{4 + \sin x}$

2. $\dfrac{e^x}{3e^x - 1}$

3. $\dfrac{x}{(1 - x^2)^3}$

4. $\dfrac{\sin x}{\cos^3 x}$

5. $\dfrac{x^3}{1 + x^4}$

6. $\dfrac{2x + 3}{x^2 + 3x - 4}$

7. $\dfrac{x^2}{\sqrt{(2 + x^3)}}$

8. $\dfrac{\cos x}{(\sin x - 2)^2}$

9. $\dfrac{1}{x \ln x}$ i.e. $\dfrac{1/x}{\ln x}$

10. $\dfrac{\cos x}{\sin^6 x}$

11. $\dfrac{e^x}{\sqrt{(1 - e^x)}}$

12. $\dfrac{x - 1}{3x^2 - 6x + 1}$

13. $\dfrac{\cos x}{\sin^n x}$

14. $\dfrac{\sin x}{\cos^n x}$

15. $\dfrac{\sin x}{(3 + \cos x)^2}$

Evaluate

16. $\displaystyle\int_1^2 \dfrac{2x + 1}{x^2 + x}\, dx$

17. $\displaystyle\int_0^1 \dfrac{x}{x^2 + 1}\, dx$

18. $\displaystyle\int_2^3 \dfrac{2x}{(x^2 - 1)^3}\, dx$

19. $\displaystyle\int_0^1 \dfrac{e^x}{(1 + e^x)^2}\, dx$

20. $\displaystyle\int_{\pi/6}^{\pi/3} \dfrac{\sin 2x}{\cos(2x - \pi)}\, dx$

21. $\displaystyle\int_2^4 \dfrac{1}{x(\ln x)^2}\, dx$

USING PARTIAL FRACTIONS

If a fraction has not fallen into any of the previous categories, it may be that it is easy to integrate when expressed in partial fractions.
Remember however that only proper fractions can be converted directly into partial fractions; an improper fraction must first be divided out until it comprises non-fractional terms and a proper fraction.

It is not very often that actual long division is needed. Usually a simple adjustment in the numerator is all that is required, as the following examples show. When such an adjustment is not obvious, however, long division can always be used.

Examples 15b

1. Integrate $\dfrac{2x-3}{(x-1)(x-2)}$ w.r.t. x

Using the cover-up method gives

$$\frac{2x-3}{(x-1)(x-2)} = \frac{1}{x-1} + \frac{1}{x-2}$$

$$\therefore \quad \int \frac{2x-3}{(x-1)(x-2)}\, dx = \int \frac{1}{x-1}\, dx + \int \frac{1}{x-2}\, dx$$

$$= \ln|x-1| + \ln|x-2| + \ln A$$

$$= \ln A|(x-1)(x-2)|$$

2. Find $\displaystyle\int \frac{x^2+1}{x^2-1}\, dx$

This fraction is improper so, before we can factorise the denominator and use partial fractions, we must adjust the given fraction as follows.

$$\frac{x^2+1}{x^2-1} = \frac{(x^2-1)+2}{x^2-1} = 1 + \frac{2}{x^2-1} = 1 + \frac{2}{(x-1)(x+1)}$$

Then

$$\int \frac{x^2+1}{x^2-1}\, dx = \int 1\, dx + \int \frac{1}{x-1}\, dx - \int \frac{1}{x+1}\, dx$$

$$= x + \ln|x-1| - \ln|x+1| + \ln A$$

$$= x + \ln \frac{A|x-1|}{|x+1|}$$

Even when improper fractions do not need conversion into partial fractions, it is still essential to reduce to proper form before attempting to integrate, i.e.

$$\int \frac{2x+4}{x+1}\, dx = \int \frac{2(x+1)+2}{x+1}\, dx = \int 2\, dx + \int \frac{2}{x+1}\, dx$$

$$= 2x + 2\ln A|x+1|$$

The reader should not fall into the trap of thinking that, whenever the denominator of a fraction factorises, integration will involve partial fractions. Careful scrutiny is vital, as fractions requiring quite different integration techniques often *look* very similar. The following example shows this clearly.

3. Integrate w.r.t. x,

(a) $\dfrac{x+1}{x^2+2x-8}$ (b) $\dfrac{x+1}{(x^2+2x-8)^2}$ (c) $\dfrac{x+2}{x^2+2x-8}$

(a) This fraction is basically of the form $f'(x)/f(x)$

$$\int \frac{x+1}{x^2+2x-8}\,dx = \frac{1}{2}\int \frac{2x+2}{x^2+2x-8}\,dx$$

$$= \tfrac{1}{2}\ln A\,|x^2+2x-8|$$

(b) This time the numerator is basically the derivative of the function *within* the denominator so we use

$$u = x^2+2x-8 \quad\Rightarrow\quad \ldots du \equiv \ldots (2x+2)\,dx \equiv \ldots 2(x+1)\,dx$$

$$\therefore \quad \int \frac{x+1}{(x^2+2x-8)^2}\,dx = \frac{1}{2}\int \frac{1}{u^2}\,du = -\frac{1}{2u}+K$$

$$= K - \frac{1}{2(x^2+2x-8)}$$

(c) In this fraction the numerator is not related to the derivative of the denominator so, as the denominator factorises, we use partial fractions.

$$\int \frac{x+2}{x^2+2x-8}\,dx = \int \frac{\frac{1}{3}}{x+4}\,dx + \int \frac{\frac{2}{3}}{x-2}\,dx$$

$$= \tfrac{1}{3}\ln|x+4| + \tfrac{2}{3}\ln|x-2| + \ln A$$

$$= \tfrac{1}{3}\ln A\,|(x+4)(x-2)^2|$$

EXERCISE 15b

Integrate each of the following functions w.r.t. x

1. $\dfrac{2}{x(x+1)}$ 2. $\dfrac{4}{(x-2)(x+2)}$ 3. $\dfrac{x}{(x-1)(x+1)}$

4. $\dfrac{x-1}{x(x+2)}$ 5. $\dfrac{x-1}{(x-2)(x-3)}$ 6. $\dfrac{1}{(x-1)(x+1)}$

7. $\dfrac{x}{x+1}$ 8. $\dfrac{x+4}{x}$ 9. $\dfrac{x}{x+4}$

10. $\dfrac{3x-4}{x(1-x)}$ 11. $\dfrac{x^2-2}{x^2-1}$ 12. $\dfrac{x^2}{(x+1)(x+2)}$

Choose the best method to integrate each function.

13. $\dfrac{x}{x^2 - 1}$ 14. $\dfrac{2x}{(x^2 - 1)^2}$ 15. $\dfrac{2}{x^2 - 1}$

16. $\dfrac{2x - 5}{x^2 - 5x + 6}$ 17. $\dfrac{2x}{x^2 - 5x + 6}$ 18. $\dfrac{2x - 3}{x^2 - 5x + 6}$

Evaluate

19. $\displaystyle\int_0^4 \dfrac{x+2}{x+1}\, dx$ 20. $\displaystyle\int_{-1}^1 \dfrac{5}{x^2 + x - 6}\, dx$ 21. $\displaystyle\int_1^2 \dfrac{x+2}{x(x+4)}\, dx$

22. $\displaystyle\int_0^1 \dfrac{2}{3 + 2x}\, dx$ 23. $\displaystyle\int_{1/2}^3 \dfrac{2}{(3 + 2x)^2}\, dx$ 24. $\displaystyle\int_1^2 \dfrac{2x}{3 + 2x}\, dx$

The Use of Partial Fractions in Differentiation

Rational functions with two or more factors in the denominator are sometimes easier to differentiate if expressed in partial fractions.

The use of partial fractions is of particular benefit when a second derivative is required.

Example 15c

Find the first and second derivatives of $\dfrac{x}{(x-1)(x+1)}$

Taking $y = \dfrac{x}{(x-1)(x+1)} = \dfrac{\frac{1}{2}}{(x-1)} + \dfrac{\frac{1}{2}}{(x+1)}$

$$= \tfrac{1}{2}(x-1)^{-1} + \tfrac{1}{2}(x+1)^{-1}$$

gives $\dfrac{dy}{dx} = -\tfrac{1}{2}(x-1)^{-2} - \tfrac{1}{2}(x+1)^{-2}$

$$= \dfrac{-1}{2(x-1)^2} - \dfrac{1}{2(x+1)^2}$$

and $\dfrac{d^2y}{dx^2} = (-2)\left(-\tfrac{1}{2}\right)(x-1)^{-3} - (-2)\left(\tfrac{1}{2}\right)(x+1)^{-3}$

$$= \dfrac{1}{(x-1)^3} + \dfrac{1}{(x+1)^3}$$

In each question express the given function in partial fractions and hence find its first and second derivatives.

1. $\dfrac{2}{(x-2)(x-1)}$

2. $\dfrac{3x}{(2x-1)(x-3)}$

3. $\dfrac{x}{(x+2)(x-4)}$

4. $\dfrac{5}{(x+2)(x-3)}$

5. $\dfrac{x}{(2x+3)(x+1)}$

6. $\dfrac{3}{(3x-1)(x-1)}$

SPECIAL TECHNIQUES FOR INTEGRATING SOME TRIGOMETRIC FUNCTIONS

To Integrate a Function Containing an Odd Power of sin x or cos x

When $\sin x$ or $\cos x$ appear to an odd power other than 1, the identity $\cos^2 x + \sin^2 x \equiv 1$ may convert the given function to an integrable form,

e.g. $\sin^3 x$ is converted to $(\sin^2 x)(\sin x)$ $\Rightarrow$ $(1 - \cos^2 x)(\sin x)$

$\Rightarrow$ $\sin x - \cos^2 x \sin x$

Examples 15d

1. Integrate w.r.t. x,   (a) $\cos^5 x$ (b) $\sin^3 x \cos^2 x$

(a) $\cos^5 x = (\cos^2 x)^2 \cos x$

$= (1 - \sin^2 x)^2 \cos x$

$= (1 - 2\sin^2 x + \sin^4 x) \cos x$

$\therefore$ $\displaystyle\int \cos^5 x \, dx = \int \cos x \, dx - 2 \int \sin^2 x \cos x \, dx + \int \sin^4 x \cos x \, dx$

For any value of n we know that $\displaystyle\int \sin^n x \cos x \, dx = \dfrac{1}{n+1} \sin^{n+1} x + K$

$\therefore$ $\displaystyle\int \cos^5 x \, dx = \sin x - 2\left(\tfrac{1}{3}\right)\sin^3 x + \left(\tfrac{1}{5}\right)\sin^5 x + K$

$= \sin x - \tfrac{2}{3}\sin^3 x + \tfrac{1}{5}\sin^5 x + K$

(b) $\qquad \sin^3 x \cos^2 x = \sin x (1 - \cos^2 x) \cos^2 x$

$$= \cos^2 x \sin x - \cos^4 x \sin x$$

$\therefore \qquad \int \sin^3 x \cos^2 x \, dx = \int \cos^2 x \sin x \, dx - \int \cos^4 x \sin x \, dx$

$$= -\tfrac{1}{3} \cos^3 x + \tfrac{1}{5} \cos^5 x + K$$

To Integrate a Function Containing only Even Powers of $\sin x$ or $\cos x$

This time the double angle identities are useful,

e.g. $\qquad \cos^4 x$ becomes $(\cos^2 x)^2 = \{\tfrac{1}{2}(1 + \cos 2x)\}^2$

$$= \tfrac{1}{4}\{1 + 2\cos 2x + \cos^2 2x\}$$

then we can use a double angle identity again

$$= \tfrac{1}{4}(1 + 2\cos 2x) + \tfrac{1}{4}\{\tfrac{1}{2}(1 + \cos 4x)\}$$

$$= \tfrac{3}{8} + \tfrac{1}{2}\cos 2x + \tfrac{1}{8}\cos 4x$$

Now each of these terms can be integrated.

Examples 15d (continued)

2. Integrate w.r.t. x, (a) $\sin^2 x$ (b) $16\sin^4 x \cos^2 x$

(a) $\qquad \int \sin^2 dx = \int \tfrac{1}{2}(1 - \cos 2x) \, dx$

$$= \int \tfrac{1}{2} \, dx - \tfrac{1}{2} \int \cos 2x \, dx$$

$$= \tfrac{1}{2} x - \tfrac{1}{4} \sin 2x + K$$

(b) $\qquad 16 \int \sin^4 x \cos^2 x \, dx = 16 \int \{\tfrac{1}{2}(1 - \cos 2x)\}^2 \{\tfrac{1}{2}(1 + \cos 2x)\} \, dx$

$$= 2 \int (1 - \cos 2x - \cos^2 2x + \cos^3 2x) \, dx$$

$$= 2x - \sin 2x - 2 \int \cos^2 2x \, dx + 2 \int \cos^3 2x \, dx$$

Now $\quad 2 \int \cos^2 2x \, dx = \int (1 + \cos 4x) \, dx = x + \frac{1}{4} \sin 4x$

and $\quad 2 \int \cos^3 2x \, dx = 2 \int \cos 2x \, (1 - \sin^2 2x) \, dx = \sin 2x - \frac{1}{3} \sin^3 2x$

$\therefore \quad 16 \int \sin^4 x \cos^2 x \, dx = x - \frac{1}{4} \sin 4x - \frac{1}{3} \sin^3 2x + K$

Note that for a product with an odd power in one term and an even power in the other, the method for an odd power is usually best (see Example 1b on page 218).

It is important to appreciate that the techniques used in these examples, although they are of most general use, are by no means exhaustive. Because there are so many trig identities there is always the possibility that a particular integral can be dealt with in several different ways,

e.g. to integrate $\sin^2 x \cos^2 x$ w.r.t. x the best conversion would use

$2 \sin x \cos x \equiv \sin 2x$, so that

$$\int \sin^2 x \cos^2 x \, dx = \int \left(\tfrac{1}{2} \sin 2x \right)^2 dx = \tfrac{1}{4} \int \tfrac{1}{2} (1 - \cos 4x) \, dx$$

It is always advisable to look for the identity which will make the given function integrable as quickly and simply as possible.

Further, as mentioned earlier, it must be remembered that there are many expressions whose integrals cannot be found as a function at all.
(In examination papers, of course, any integral asked for *can* be found!)

EXERCISE 15d

Integrate each function w.r.t. x

1. $\cos^2 x$
2. $\cos^3 x$
3. $\sin^5 x$
4. $\tan^2 x$

5. $\sin^4 x$
6. $\tan^3 x$
7. $\cos x$
8. $\sin^3 x$

Find

9. $\displaystyle\int \sin^2 \theta \cos^3 \theta \, d\theta$

10. $\displaystyle\int \sin^{10} \theta \cos^3 \theta \, d\theta$

11. $\displaystyle\int \sin^n\theta \, \cos^3\theta \, d\theta$ 12. $\displaystyle\int \sin^2\theta \, \cos^2\theta \, d\theta$

13. $\displaystyle\int \tan^2\theta \, d\theta$ (Hint: Change $\tan^2\theta$ into $\sec^2\theta - 1$)

14. $\displaystyle\int \tan^3\theta \, d\theta$ (Hint: Change $\tan^2\theta$ into $\sec^2\theta - 1$)

SYSTEMATIC INTEGRATION

At this stage it is possible to classify most of the integrals which the reader is likely to meet.

Once correctly classified, a given expression can be integrated using the method best suited to its category.

The simplest category comprises the quotable results listed below.

Standard Integrals

Function	Integral
x^n	$\dfrac{1}{n+1} x^{n+1}$ $(n \neq -1)$
e^x	e^x
$\dfrac{1}{x}$	$\ln|x|$
$\cos x$	$\sin x$
$\sin x$	$-\cos x$
$\sec^2 x$	$\tan x$

Each of these should be recognised equally readily when x is replaced by ax or $(ax+b)$, e.g. for e^{ax+b} the standard integral is $\dfrac{1}{a} e^{ax+b}$ and for $\cos ax$ the standard integral is $\dfrac{1}{a} \sin ax$

Classification

When attempting to classify a particular function the following questions should
be asked, *in order*, about the form of the integral.

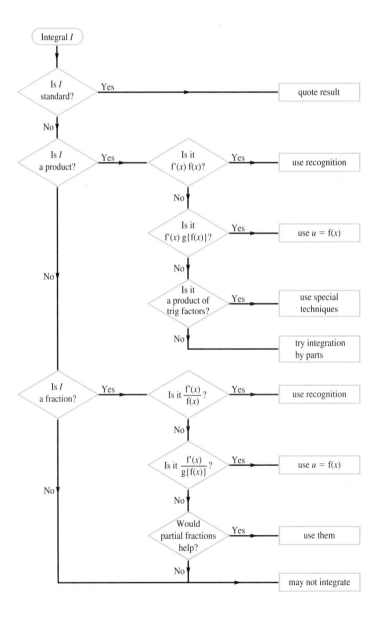

Other Techniques

Although this systematic approach deals successfully with most integrals at this level, inevitably the reader will encounter some integrals for which no method is obvious. A fraction, for instance, may be such that its *numerator* can be separated, thus producing *two* (or more) fractions of different types, e.g.

$$\int \frac{2x+1}{1-x^2}\,dx = \int \frac{2x}{1-x^2}\,dx + \int \frac{1}{1-x^2}\,dx$$

Further, many expressions other than products and fractions can be integrated by making a suitable substitution. Because at this stage the reader cannot always be expected to 'spot' an appropriate change of variable, a substitution is suggested in all but the simplest of cases. The resulting integral must be converted so that it is expressed in terms of the original variable *except in the case of a definite integral* when it is usually much easier to change the limits.

The following example illustrates an integral which responds to a change of variable.

Example 15e

Use the substitution $u = 1 + 2x$ to find $\displaystyle\int x(1+2x)^{11}\,dx$

$$u = 1 + 2x \quad \Rightarrow \quad \ldots\, du \,\ldots \equiv \,\ldots\, 2dx \quad \Rightarrow \quad \ldots \tfrac{1}{2}\, du \equiv \,\ldots\, dx$$

Hence $\displaystyle\int x(1+2x)^{11}\,dx = \int \tfrac{1}{2}(u-1)(u^{11})(\tfrac{1}{2}\,du)$

$$= \tfrac{1}{4}\int (u^{12} - u^{11})\,du$$

$$= \tfrac{1}{4}\left(\tfrac{1}{13}\,u^{13} - \tfrac{1}{12}\,u^{12}\right) + K$$

$$= \tfrac{1}{624}\,u^{12}(12u - 13) + K$$

i.e. $\displaystyle\int x(1+2x)^{11}\,dx = \tfrac{1}{624}(1+2x)^{12}(24x - 1) + K$

Find the following integrals using the suggested substitution.

1. $\displaystyle\int (x+1)(x+3)^5\,dx$; $\qquad x+3 = u$

2. $\displaystyle\int \frac{x}{\sqrt{(3-x)}}\,dx$; $\qquad 3-x = u^2$

3. $\displaystyle\int x\sqrt{(x+1)}\,dx$; $\qquad x+1 = u^2$

4. $\displaystyle\int \frac{2x+1}{(x-3)^6}\,dx$; $\qquad x-3 = u$

5. $\displaystyle\int 2x\sqrt{(3x-4)}\,dx$; $\qquad 3x-4 = u^2$

Devise a suitable substitution and hence find:

6. $\displaystyle\int 2x(1-x)^7\,dx$ 　　　　　　7. $\displaystyle\int \frac{x+3}{(4-x)^5}\,dx$

Use the flow chart to classify each of the following integrals. Hence perform each integration using an appropriate method.

1. $\displaystyle\int x(x-1)\,dx$ 　　　　　　　2. $\displaystyle\int (x-2)(x^2)\,dx$

3. $\displaystyle\int e^{2x+3}\,dx$ 　　　　　　　4. $\displaystyle\int x\sqrt{(2x^2-5)}\,dx$

5. $\displaystyle\int x\,e^x\,dx$ 　　　　　　　　6. $\displaystyle\int \ln x\,dx$

7. $\displaystyle\int \sin^2 3x\,dx$ 　　　　　　8. $\displaystyle\int x\,e^{-x^2}\,dx$

9. $\displaystyle\int \cos x\sin^2 x\,dx$ 　　　　10. $\displaystyle\int u(u+7)^9\,du$

11. $\displaystyle\int \frac{x^2}{(x^3+9)^5}\,dx$ 　　　　12. $\displaystyle\int \frac{\sin 2y}{1-\cos 2y}\,dy$

13. $\displaystyle\int \frac{1}{2x+7}\,dx$

14. $\displaystyle\int \sin 3x\sqrt{(1+\cos 3x)}\,dx$

15. $\displaystyle\int x\sin 4x\,dx$

16. $\displaystyle\int \frac{x+2}{x^2+4x-5}\,dx$

17. $\displaystyle\int \frac{x+1}{x^2+4x-5}\,dx$

18. $\displaystyle\int \frac{x+2}{(x^2+4x-5)^3}\,dx$

19. $\displaystyle\int 3y\sqrt{(9-y^2)}\,dy$

20. $\displaystyle\int e^{2x}\cos 3x\,dx$

21. $\displaystyle\int \ln 5x\,dx$

22. $\displaystyle\int \cos^3 2x\,dx$

23. $\displaystyle\int \cos x\,e^{\sin x}\,dx$

24. $\displaystyle\int \frac{\sin y}{\sqrt{(7+\cos y)}}\,dy$

25. $\displaystyle\int x^2 e^x\,dx$

26. $\displaystyle\int \frac{x}{x^2-4}\,dx$

27. $\displaystyle\int \frac{x^2}{x^2-4}\,dx$

28. $\displaystyle\int \frac{1}{x^2-4}\,dx$

29. $\displaystyle\int x\ln x\,dx$

30. $\displaystyle\int \cos^2 u\sin^3 u\,du$

31. $\displaystyle\int \tan^2\theta\,d\theta$

32. $\displaystyle\int \frac{2-x}{1-x}\,dx$

33. $\displaystyle\int \frac{\sec^2 x}{1-\tan x}\,dx$

34. $\displaystyle\int x\sqrt{(7+x^2)}\,dx$

35. $\displaystyle\int x(1+x^2)^4\,dx$

36. $\displaystyle\int x\,e^{-3x}\,dx$

37. $\displaystyle\int \frac{x+3}{x+2}\,dx$

38. $\displaystyle\int \frac{3}{(x-4)(x-1)}\,dx$

39. $\displaystyle\int \frac{x+1}{x(2x+1)}\,dx$

40. $\displaystyle\int \frac{\sin x}{\sqrt{(\cos x)}}\,dx$

41. $\displaystyle\int \sin(5\theta-\pi/4)\,d\theta$

42. $\displaystyle\int \sec^2 u\,e^{\tan u}\,dx$

CHAPTER 16

DIFFERENTIAL EQUATIONS

An equation in which at least one term contains $\dfrac{dy}{dx}$, $\dfrac{d^2y}{dx^2}$ etc., is called a *differential equation*. If it contains only $\dfrac{dy}{dx}$ it is of the first order whereas if it contains $\dfrac{d^2y}{dx^2}$ it is of the second order, and so on.

For example, $x + 2\dfrac{dy}{dx} = 3y$ is a first order differential equation and $\dfrac{d^2y}{dx^2} - 5\dfrac{dy}{dx} + 4y = 0$ is a second order differential equation.

Each of these examples is a *linear* differential equation because none of the differential coefficients $\left(\text{i.e. } \dfrac{dy}{dx}, \dfrac{d^2y}{dx^2}\right)$ is raised to a power higher than 1.

A differential equation represents a relationship between two variables. The same relationship can often be expressed in a form which does not contain a differential coefficient,

e.g. $\dfrac{dy}{dx} = 2x$ and $y = x^2 + K$ express the same relationship between x and y, but $\dfrac{dy}{dx} = 2x$ is a differential equation whereas $y = x^2 + K$ is not.

Converting a differential equation into a direct one is called *solving the differential equation*. This clearly involves some form of integration.

There are many different types of differential equation, each requiring a specific technique for its solution.

At this stage however we are going to deal with only one simple type, i.e. linear differential equations of the first order.

FIRST ORDER DIFFERENTIAL EQUATIONS WITH SEPARABLE VARIABLES

Consider the differential equation $\quad 3y \dfrac{dy}{dx} = 5x^2$ [1]

Integrating both sides of the equation gives

$$\int 3y \frac{dy}{dx}\, dx = \int 5x^2\, dx$$

We know that $\quad \ldots \dfrac{dy}{dx}\, dx \equiv \ldots dy$

so $$\int 3y\, dy = \int 5x^2\, dx$$ [2]

Temporarily removing the integral signs from this equation gives

$$3y\, dy = 5x^2\, dx$$ [3]

This can be obtained direct from equation [1] by *separating the variables*, i.e. by *separating* dy *from* dx *and collecting on one side all the terms involving* y *together with* dy, *while all the* x *terms are collected, along with* dx, *on the other side.*

It is vital to appreciate that what is given in [3] above does not, in itself, have any meaning and it *should not be written down as a step in the solution.* It simply provides a way of making a quick mental conversion from the differential equation [1] to the form [2] which is ready for two separate integrations.

Now returning to equation [2] and integrating each side we have

$$\tfrac{3}{2} y^2 = \tfrac{5}{3} x^3 + A$$

Note that it is unnecessary to introduce a constant of integration on both sides. It is sufficient to have a constant on one side only.

When solving differential equations, the constant of integration is usually denoted by A, B, etc. and is called the *arbitrary constant.*

The solution of a differential equation that includes the arbitrary constant is called *the general solution,* or, very occasionally, *the complete primitive.* It represents a family of straight lines or curves, each member of the family corresponding to one value of A.

$$\frac{1}{x}\frac{dy}{dx} = \frac{2y}{x^2+1}$$

$$\frac{1}{x}\frac{dy}{dx} = \frac{2y}{x^2+1} \quad \Rightarrow \quad \frac{1}{y}\frac{dy}{dx} = \frac{2x}{x^2+1}$$

So, after separating the variables we have,

$$\int \frac{1}{y}\,dy = \int \frac{2x}{x^2+1}\,dx$$

$$\Rightarrow \quad \ln|y| = \ln A\,|x^2+1|$$

(Using $\ln A$ as the arbitrary constant.)

i.e. $\quad y = A(x^2+1)$

Note that, whenever we solve a differential equation some integration has to be done so the systematic classification of each integral involved is an essential part of solving differential equations.

EXERCISE 16a

Find the general solution of each differential equation

1. $y\dfrac{dy}{dx} = \sin x$

2. $x^2\dfrac{dy}{dx} = y^2$

3. $\dfrac{1}{x}\dfrac{dy}{dx} = \dfrac{1}{y^2-2}$

4. $\sin y\,\dfrac{dy}{dx} = \dfrac{1}{x}$

5. $\dfrac{dy}{dx} = y^2$

6. $\dfrac{1}{x}\dfrac{dy}{dx} = \dfrac{1}{1-x^2}$

7. $(x-3)\dfrac{dy}{dx} = y$

8. $\cos y\,\dfrac{dy}{dx} = 4$

9. $u\dfrac{du}{dv} = v+2$

10. $\dfrac{y^2}{x^3}\dfrac{dy}{dx} = \ln x$

11. $e^x\dfrac{dy}{dx} = \dfrac{x}{y}$

12. $\sec x\,\dfrac{dy}{dx} = e^y$

13. $r\dfrac{dr}{d\theta} = \sin^2\theta$

14. $\dfrac{dv}{du} = \dfrac{v+1}{u+2}$

15. $xy \dfrac{dy}{dx} = \ln x$ 16. $y(x+1) = (x^2 + 2x) \dfrac{dy}{dx}$

17. $v^2 \dfrac{dv}{dt} = (2+t)^3$ 18. $x \dfrac{dy}{dx} = \dfrac{1}{y} + y$

19. $r \dfrac{d\theta}{dr} = \cos^2 \theta$ 20. $y \sin^3 x \dfrac{dy}{dx} = \cos x$

21. $\dfrac{uv}{u-1} = \dfrac{du}{dv}$ 22. $e^x \dfrac{dy}{dx} = e^{y-1}$

23. $\tan x \dfrac{dy}{dx} = 2y^2 \sec^2 x$ 24. $\dfrac{dy}{dx} = \dfrac{x(y^2 - 1)}{(x^2 + 1)}$

CALCULATION OF THE ARBITRARY CONSTANT

We saw on p. 226 that

$$3y \frac{dy}{dx} = 5x^2 \iff \tfrac{3}{2} y^2 = \tfrac{5}{3} x^3 + A$$

The equation $\tfrac{3}{2} y^2 = \tfrac{5}{3} x^3 + A$ represents a family of curves with similar characteristics. Each value of A gives one particular member of the family, i.e. a *particular solution*.

The value of A cannot be found from the differential equation alone; further information is needed.

Suppose that we require the equation of a curve that satisfies the differential equation $2 \dfrac{dy}{dx} = \dfrac{\cos x}{y}$ and which passes through the point $(0, 2)$.

We want one member of the family of curves represented by the differential equation, i.e. the particular value of the arbitrary constant must be found.

The general solution has to be found first so, separating the variables, we have

$$\int 2y \, dy = \int \cos x \, dx \quad \Rightarrow \quad y^2 = A + \sin x$$

In order to find the required curve we need the value of A such that the general solution is satisfied by $x = 0$ and $y = 2$

i.e. $4 = A + 0 \quad \Rightarrow \quad A = 4$

Hence the equation of the specified curve is $y^2 = 4 + \sin x$

1. Describe the family of curves represented by the differential equation $y = x \dfrac{dy}{dx}$

and sketch any three members of this family.

Find the particular solution for which $y = 2$ when $x = 1$ and sketch this member of the family on the same axes as before.

By separating the variables,

$$y = x \frac{dy}{dx} \quad \Rightarrow \quad \int \frac{1}{y}\,dy = \int \frac{1}{x}\,dx$$

$$\Rightarrow \quad \ln|y| = \ln|x| + \ln A$$

i.e. the general solution is $y = Ax$

This equation represents a family of straight lines through the origin, each line having a gradient A, as shown in the following diagram.

If $y = 2$ when $x = 1$ then $A = 2$ and the corresponding member of the family is the line $y = 2x$.

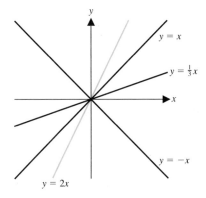

2. A curve is such that the gradient is proportional to the product of the x and y coordinates. If the curve passes through the points $(2, 1)$ and $(4, e^2)$, find its equation.

First find the general solution

$$\frac{dy}{dx} = kxy \quad \text{where } k \text{ is a constant of proportion.}$$

$$\therefore \quad \int \frac{1}{y}\,dy = \int kx\,dx \quad \Rightarrow \quad \ln y = \tfrac{1}{2}kx^2 + A$$

There are *two* unknown constants this time so we need two extra pieces of information; these are

(i) $y = 1$ when $x = 2$ $\Rightarrow$ $\ln 1 = 2k + 4$

$\ln 1 = 0$ so $A + 2k = 0$

(ii) $y = e^2$ when $x = 4$ $\Rightarrow$ $\ln e^2 = 8k + A$

$\ln e^2 = 2$ so $A + 8k = 2$

Solving these equations for A and k we get $k = \frac{1}{3}$ and $A = -\frac{2}{3}$

$\therefore$ the equation of the specified curve is $\ln y = \frac{1}{6}x^2 - \frac{2}{3}$

$$= \tfrac{1}{6}(x^2 - 4)$$

or $y^6 = e^{(x^2 - 4)}$

EXERCISE 16b

Find the particular solution of each of the following differential equations.

1. $y^2 \dfrac{dy}{dx} = x^2 + 1$ and $y = 1$ when $x = 2$

2. $e^t \dfrac{ds}{dt} = \sqrt{s}$ and $s = 4$ when $t = 0$

3. $\dfrac{y}{x}\dfrac{dy}{dx} = \dfrac{y^2 - 1}{x^2 - 1}$ and $y = 3$ when $x = 2$

4. A curve passes through the origin and its gradient function is $2x - 1$. Find the equation of the curve and sketch it.

5. A curve for which $e^{-x}\dfrac{dy}{dx} = 1$, passes through the point $(0, -1)$. Find the equation of the curve.

6. A curve passes through the points $(1, 2)$ and $(\frac{1}{5}, -10)$ and its gradient is inversely proportional to x^2. Find the equation of the curve.

7. If $y = 2$ when $x = 1$, find the coordinates of the point where the curve represented by $\dfrac{2y}{3}\dfrac{dy}{dx} = e^{-3x}$ crosses the y-axis.

8. Find the equation of the curve whose gradient function is $\dfrac{y + 1}{x^2 - 1}$ and which passes through the point $(-3, 1)$.

9. The gradient function of a curve is proportional to $x + 3$. If the curve passes through the origin and the point $(2, 8)$, find its equation.

10. Solve the differential equation $(1 + x^2)\dfrac{dy}{dx} - y(y + 1)x = 0$, given that $y = 1$ when $x = 0$

11. Solve the differential equation $\dfrac{dy}{dx} = 3x^2 y^2$ given that $y = 1$ when $x = 0$.

12. If $\dfrac{dy}{dx} = x(y + 1)$ and $y = 1$ when $x = 2$ find the particular solution of the differential equation.

*13. Find the equation of the curve which passes through the point $\left(\frac{1}{2}, 1\right)$ and is defined by the differential equation $y e^{y^2}\dfrac{dy}{dx} = e^{2x}$.

Show that the curve also passes through the point $(2, 2)$ and sketch the curve.

GENERAL RATES OF INCREASE

We have already seen that

$$\dfrac{dy}{dx} \text{ represents the rate at which } y \text{ increases compared with } x$$

Whenever the variation in one quantity, p say, depends upon the changing value of another quantity, q, then the rate of increase of p compared with q can be expressed as $\dfrac{dp}{dq}$.

There are many every-day situations where such relationships exist, e.g.

- liquid expands when it is heated so, if V is the volume of a quantity of liquid and T is the temperature, then the rate at which the volume increases with temperature can be written $\dfrac{dV}{dT}$

- if the profit, P, made by a company selling radios depends upon the number, n, of radios sold, then $\dfrac{dP}{dn}$ represents the rate of increase of profit compared with the increase in sales.

NATURAL OCCURRENCE OF DIFFERENTIAL EQUATIONS

Differential equations often arise when a physical situation is interpreted mathematically (i.e. when a mathematical model is made of the physical situation).

Consider the following examples.

1) Suppose that a body falls from rest in a medium which causes the velocity to decrease at a rate proportional to the velocity.

   As the velocity is *decreasing* with time, its rate of increase is *negative*.

   Using v for velocity and t for time, the rate of change of velocity can be written as $-\dfrac{dv}{dt}$.

   Thus the motion of the body satisfies the differential equation

   $$-\frac{dv}{dt} = kv$$

2) During the initial stages of the growth of yeast cells in a culture, the number of cells present increases in proportion to the number already formed.

   Thus n, the number of cells at a particular time t, can be found from the differential equation

   $$\frac{dn}{dt} = kn$$

3) Suppose that a chemical mixture contains two substances A and B whose weights are W_A and W_B and whose combined weight remains constant. B is converted into A at a rate which is inversely proportional to the weight of B and proportional to the square of the weight of A in the mixture at any time t. The weight of B present at time t can be found using

   $$\frac{d}{dt}(W_B) = \frac{k}{W_B} \times (W_A)^2$$

   But $W_A + W_B$ is constant, W say

   Hence $\quad \dfrac{d}{dt}(W_B) = \dfrac{k(W - W_B)^2}{W_B}$

   This differential equation now relates W_B and t.

   Note. In forming (and subsequently solving) differential equations from naturally occuring data, it is not actually necessary to understand the background of the situation or experiment.

In Questions 1 to 4 form, but *do not solve,* the differential equation representing the given data.

1. A body moves with a velocity v which is inversely proportional to its displacement s from a fixed point.

2. The rate at which the height h of a certain plant increases is proportional to the natural logarithm of the difference between its present height and its final height H.

3. The manufacturers of a certain brand of soap powder are concerned that the number, n, of people buying their product at any time t has remained constant for some months. They launch a major advertising programme which results in the number of customers increasing at a rate proportional to the square root of n. Express as differential equations the progress of sales

   (a) before advertising

   (b) after advertising.

4. In an isolated community, the number, n, of people suffering from an infectious disease is N_1 at a particular time. The disease then becomes epidemic and spreads so that the number of sick people increases at a rate proportional to n, until the total number of sufferers is N_2. The rate of increase then becomes inversely proportional to n until N_3 people have the disease. After this, the total number of sick people decreases at a constant rate. Write down the differential equation governing the incidence of the disease

   (a) for $N_1 \leqslant N_2$

   (b) for $N_2 \leqslant N_3$

   (c) for $n \geqslant N_3$.

5. Two chemicals, P and Q, are involved in a reaction. The masses of P and Q present at any time t, are p and q respectively. The rate at which p is increasing at time t is k times the product of the two masses. If the masses of P and Q have a constant sum s, find the differential equation expressing $\dfrac{dp}{dt}$ in terms of p, s and k.

SOLVING NATURALLY OCCURRING DIFFERENTIAL EQUATIONS

We have seen that when one naturally occurring quantity varies with another, the relationship between them often involves a constant of proportion. Consequently, a differential equation that represents the relationship contains a constant of proportion whose value is not necessarily known. So the initial solution of the differential equation contains both this constant and the arbitrary constant. Extra given information may allow either or both constants to be evaluated.

1. A particle moves in a straight line with an acceleration that is inversely proportional to its velocity. (Acceleration is the rate of increase of velocity.)

   (a) Form a differential equation to represent this data.

   (b) Given that the acceleration is $2\,\text{m/s}^2$ when the velocity is $5\,\text{m/s}$, solve the differential equation.

(a) Using $\dfrac{dv}{dt}$ for acceleration we have

$$\frac{dv}{dt} \propto \frac{1}{v} \quad \Rightarrow \quad \frac{dv}{dt} = \frac{k}{v}$$

(b) If $v = 5$ when $\dfrac{dv}{dt} = 2,$

then $2 = \dfrac{k}{5} \quad \Rightarrow \quad k = 10$

$\therefore \qquad \dfrac{dv}{dt} = \dfrac{10}{v}$

Separating the variables gives $\displaystyle\int v\,dv = \int 10\,dt$

$\Rightarrow \qquad \tfrac{1}{2}v^2 = 10t + A$

NATURAL GROWTH AND DECAY

There is a particularly important type of naturally occurring relationship of which there are many examples in real life. These arise in situations where the rate of change of a quantity Q is proportional to the value of Q. Very often (though not invariably) the rate of change is taken with respect to time and then the relationship can be expressed as

$\dfrac{dQ}{dt}$ varies with Q

This relationship can be expressed as the differential equation

$\dfrac{dQ}{dt} = kQ$ where k is a constant of proportion

Solving this differential equation by separating the variables gives

$$\int \frac{1}{Q} \, dQ = \int k \, dt \quad \Rightarrow \quad \ln AQ = kt$$

i.e. $AQ = e^{kt}$ or $Q = Be^{kt}$ (where B replaces $1/A$)

This equation shows that Q varies exponentially with time and a sketch of the corresponding graph is

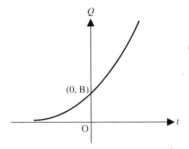

Quantities that behave in this way are said to undergo exponential or *natural growth*. For example, a yeast undergoes natural growth when the rate of increase of the number of cells of yeast is proportional to the number of cells present.

Now if it is the rate of *decrease* of Q that is proportional to Q, then we have

$$\frac{-dQ}{dt} = kQ \quad \Rightarrow \quad Q = Be^{-kt}$$

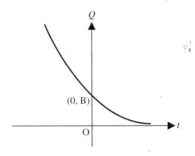

This graph is typical of a quantity undergoing exponential or *natural decay*.
If, when $t = 0$, the value of Q is Q_0, the equations representing natural growth and decay becomes

$$Q = Q_0 e^{kt} \quad \text{and} \quad Q = Q_0 e^{-kt}$$

HALF-LIFE

When a substance is decaying naturally, the time taken for one-half of the original quantity to decay is called the *half-life* of the substance. So if the original amount is Q_0, the half-life is given by

$$\tfrac{1}{2}Q_0 = Q_0 e^{-kt} \quad \Rightarrow \quad e^{kt} = 2$$

i.e.   the value of the half-life is $\dfrac{1}{k}\ln 2$

Examples 16d (continued)

2. When a uniform rod is heated it expands so that the rate of increase of its length, l, with respect to the temperature, $\theta\,^\circ$C, is proportional to the length. When the temperature is $0\,^\circ$C the length of the rod is L.

(a) Form and solve the differential equation that models this data; express l as a function of θ and illustrate this with a sketch.

(b) Given that the length of the rod has increased by 1% when the temperature is 20 $^\circ$C, find the value of θ at which the length of the rod has increased by 5%.

(c) Give a possible reason why the model may not be appropriate for very high temperatures.

(a) From the given data $\dfrac{dl}{d\theta} = kl \quad \Rightarrow \quad \displaystyle\int \frac{1}{l}\,dl = \int k\,d\theta$

Hence $\ln Al = k\theta \quad \Rightarrow \quad l = B\,e^{k\theta}$

When $\theta = 0,\; l = L,$ so $L = B$

$\therefore \quad\quad l = L\,e^{k\theta}$

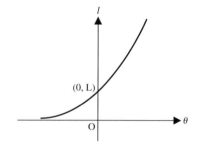

(b) When the length has increased by 1%, $l = L + 0.01L$

Then $\qquad 1.01L = Le^{20k}$

$\therefore \qquad\qquad e^{20k} = 1.01 \quad \Rightarrow \quad 20k = \ln 1.01$

i.e. $\qquad\qquad k = 0.000498 \quad (3\,\text{sf})$

$\therefore \qquad\qquad l = Le^{0.000498\,\theta} \quad \Rightarrow \quad 0.000498\,\theta = \ln(l/L)$

When $l = L + 0.05L = 1.05L$,

$\qquad\qquad 0.000498\,\theta = \ln 1.05 \quad \Rightarrow \quad \theta = 98 \text{ (nearest degree)}$

(c) The rod might distort or melt at high temperatures.

3. The rate at which the atoms in a mass of radioactive material are disintegrating is proportional to N, the number of atoms present at any time. Initially the number of atoms is M.

(a) Form and solve the differential equation that represents this data.

(b) Given that half of the original mass disintegrates in 152 days, evaluate the constant of proportion in the differential equation.

(c) Sketch the graph of N against time.

(a) The rate at which the atoms are disintegrating is $-\mathrm{d}N/\mathrm{d}t$

$\therefore \qquad -\mathrm{d}N/\mathrm{d}t = kN$

Separating the variables gives $\displaystyle \int \frac{1}{N}\,\mathrm{d}N = -\int k\,\mathrm{d}t$

Hence $\quad \ln AN = -kt \quad \Rightarrow \quad N = Be^{kt}$

When $t = 0$, $N = M \quad \Rightarrow \quad M = B$

$\therefore \qquad N = Me^{-kt}$

(b) When $N = \frac{1}{2}M$, $t = 152$

$\therefore \qquad \frac{1}{2}M = Me^{-152k} \quad \Rightarrow \quad \ln\left(\frac{1}{2}\right) = -152k$

$\therefore \qquad 152k = \ln 2 \quad \Rightarrow \quad k = 0.00456 \quad (3\,\text{sf})$

(c)

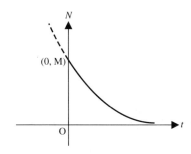

1. Grain is pouring from a hopper on to a barn floor where it forms a conical pile whose height h is increasing at a rate that is inversely proportional to h^3. The initial height of the pile is h_0 and the height doubles after a time T. Find, in terms of T, the time after which its height has grown to $3h_0$.

2. The gradient at any point of a curve is proportional to the square root of the x–coordinate. Given that the curve passes through the point $(1, 2)$ and at that point the gradient is 0.6, form and solve the differential equation representing the given relationship.
   Show that the curve passes through the point $(4, 4.8)$ and find the gradient at this point.

3. A colony of micro-organisms in a liquid is growing at a rate proportional to the number of organisms present at any time. Initially there are N organisms.

   (a) Form a differential equation that models the growth in the size of the colony.

   (b) Given that the colony increases by 50% in T hours, find the time that elapses from the start of the reaction before the size of the colony doubles.

   (c) Under what conditions might the model be inappropriate?

4. If the half-life of a radioactive element that is decaying naturally is 500 years, find how many years it will be before the original mass of the element is reduced by 75%.

5. In a certain chemical reaction, a substance is transformed into a compound. The mass of the substance after any time t is m and the substance is being transformed at a rate that is proportional to the mass of the substance at that time. Given that the original mass is 50 g and that 20 g is transformed after 200 seconds,

   (a) form and solve the differential equation relating x and t

   (b) find the mass of the substance transformed in 300 seconds.

6. The rate of decrease of the temperature of a liquid is proportional to the amount by which this temperature exceeds the temperature of its surroundings. This is known as Newton's law of Cooling. Taking θ as the excess temperature at any time t, and θ_0 as the initial excess,

   (a) show that $\theta = \theta_0 e^{-kt}$

A pan of water at $65°$ is standing in a kitchen whose temperature is a steady $15°$.

(b) Show that, after cooling for t minutes, the water temperature, φ, can be modelled by the equation

$$\varphi = 15 + 50\,e^{-kt} \quad \text{where } k \text{ is a constant.}$$

(c) Given that after 10 minutes the temperature of the water has fallen to $50°$, find the value of k.

(d) Find the temperature after 15 minutes and sketch the graph relating φ and t.

(e) Do you think that this model would be appropriate for a cooling time of 24 hours?

*7. Use the knowledge gained from the last question to undertake the following piece of detective work.
You are a forensic doctor called to a murder scene. When the victim was discovered, the body temperature was measured and found to be $20\,°C$. You arrive one hour later and find the body temperature at that time to be $18\,°C$. Assuming that the ambient temperature remained constant in that intervening hour, give the police an estimate of the time of death. (Take $37\,°C$ as normal body temperature.)

CHAPTER 17

APPLICATIONS OF CALCULUS

COMPARATIVE RATES OF CHANGE

Some problems involving the rate of change of one variable compared with another do not provide a direct relationship between these two variables. Instead, each of them is related to a third variable.

The identity $\dfrac{dy}{dx} = \dfrac{dy}{dt} \times \dfrac{dt}{dx}$ is useful in solving problems of this type.

Suppose, for instance, that the radius, r, of a circle is increasing at a rate of 1 mm per second. This means that $\dfrac{dr}{dt} = 1$.

The rate at which the area, A, of the circle is increasing is $\dfrac{dA}{dt}$.

We do not know A as a function of t but we do know that $A = \pi r^2$

and that $\dfrac{dA}{dt} = \dfrac{dA}{dr} \times \dfrac{dr}{dt}$

Then $\dfrac{dA}{dt}$ can be calculated, since $\dfrac{dr}{dt}$ is given and $\dfrac{dA}{dr}$ can be found from $A = \pi r^2$.

In some cases, more than three variables may be involved but the same approach is used with a relationship of the form

$$\frac{dy}{dx} = \frac{dy}{dp} \times \frac{dp}{dq} \times \frac{dq}{dx}$$

Examples 17a

1. A spherical balloon is being blown up so that its volume increases at a constant rate of $1.5\,\text{cm}^3/\text{s}$. Find the rate of increase of the radius when the volume of the balloon is $56\,\text{cm}^3$.

If, at time t, the radius of the balloon is r and the volume is V then

$$V = \tfrac{4}{3}\pi r^3 \quad \Rightarrow \quad \frac{dV}{dr} = 4\pi r^2$$

We are looking for $\dfrac{dr}{dt}$ and we are given $\dfrac{dV}{dt} = 1.5$ so we use

$$\frac{dr}{dt} = \frac{dr}{dV} \times \frac{dV}{dt} = \frac{dV}{dt} \div \frac{dV}{dr} \quad \Rightarrow \quad \frac{dr}{dt} = \frac{1.5}{4\pi r^2} = \frac{3}{8\pi r^2}$$

Now substituting $V = 56$ in $V = \tfrac{4}{3}\pi r^3$ gives $r = 2.373$ to 4 sf.

Therefore, when $V = 56$, $\quad \dfrac{dr}{dt} = \dfrac{3}{8\pi\,(2.373)^2} = 0.021\,20$

i.e. the radius is increasing at a rate of $0.0212\,\text{cm/s}$ (correct to 3 sf).

2. A funnel holding liquid has the shape of an inverted cone with a semi-vertical angle of $30°$. The liquid is running out of a small hole at the vertex.

(a) If it is assumed that the hole is small enough to be ignored when finding the volume, show that the volume, $V\ \text{cm}^3$, of liquid left in the funnel when the depth of the liquid is h cm, is given by

$$V = \tfrac{1}{9}\pi h^3$$

(b) If it is further assumed that the liquid is running out at a constant rate of $3\ \text{cm}^3/\text{s}$, find the rate of change of h when $V = 81\pi\ \text{cm}^3$.

(c) If the rate of flow of the liquid is not constant, but starts off at $3\ \text{cm}^3/\text{s}$ when the funnel is full and then decreases as h decreases, state with a reason the nature of the error in the answer to part (b).

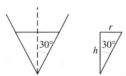

(a)   At any time t, the volume V of liquid is given by $V = \frac{1}{3}\pi r^2 h$.

From the diagram, $r = h\tan 30° = \frac{1}{\sqrt{3}}h \quad \Rightarrow \quad r^2 = \frac{1}{3}h^2$

$\therefore \qquad V = \frac{1}{3}\pi\left(\frac{1}{3}h^2\right)h = \frac{1}{9}\pi h^3$ [1]

(b)   As liquid is running out at a constant rate of $3\,\text{cm}^3/\text{s}$, V is decreasing by $3\,\text{cm}^3/\text{s}$,

i.e. $\dfrac{\mathrm{d}V}{\mathrm{d}t} = -3$

We want $\dfrac{\mathrm{d}h}{\mathrm{d}t}$,   so we use $\dfrac{\mathrm{d}h}{\mathrm{d}t} = \dfrac{\mathrm{d}V}{\mathrm{d}t} \times \dfrac{\mathrm{d}h}{\mathrm{d}V}$

$\qquad\qquad = \dfrac{\mathrm{d}V}{\mathrm{d}t} \div \dfrac{\mathrm{d}V}{\mathrm{d}h}$,

and from [1], $\dfrac{\mathrm{d}V}{\mathrm{d}h} = \dfrac{1}{3}\pi h^2$

$\therefore \qquad \dfrac{\mathrm{d}h}{\mathrm{d}t} = (-3)\div\left(\dfrac{1}{3}\pi h^2\right) = -\dfrac{9}{\pi h^2}$

When $V = 81\pi$, using [1] gives $81\pi = \frac{1}{9}\pi h^3 \quad \Rightarrow \quad h = 9$

$\therefore$ at this instant, $\dfrac{\mathrm{d}h}{\mathrm{d}t} = -\dfrac{1}{9\pi}$

i.e. the depth is decreasing at the rate of $\dfrac{1}{9\pi}\,\text{cm/s}$

(c)   When $V = 81\pi$,   the rate of decrease in volume is less than $3\,\text{cm}^3/\text{s}$,

i.e. $\left|\dfrac{\mathrm{d}V}{\mathrm{d}t}\right| < 3$,

hence $\left|\dfrac{\mathrm{d}h}{\mathrm{d}t}\right|\left(= \left|\dfrac{\mathrm{d}V}{\mathrm{d}t} \times \dfrac{\mathrm{d}h}{\mathrm{d}V}\right|\right)$ is less than $\dfrac{1}{9\pi}$

$\therefore$ the rate of decrease of h is less than $\dfrac{1}{9\pi}$.

So the answer for (b) contains a positive error, i.e. it is larger than the true value.

If we also consider the effect of neglecting the hole, i.e. the missing part of the funnel at the vertex, when calculating V, then we are using values of V, and hence h, that are larger than their true values. Since $\mathrm{d}V/\mathrm{d}h = (1/3)\pi h^2$, this increases the value of $\mathrm{d}V/\mathrm{d}h$, but as we *divide* by it to find $\mathrm{d}h/\mathrm{d}t$, the error in the numerical value of $\mathrm{d}h/\mathrm{d}t$ is reduced. Depending on the relative sizes of the errors introduced by the assumptions made, to some extent they cancel each other out.

1. Ink is dropped on to blotting paper forming a circular stain which increases in area at a rate of $2.5\,\text{cm}^2/\text{s}$. Find the rate at which the radius is changing when the area of the stain is 16π cm^2.

2. The surface area of a cube is increasing at a rate of 10 cm^2/s.
Show that, when the edge is r cm, the surface area, A cm^2, can be modelled by the equation

$$A = 6r^2$$

Find the rate of increase of the volume of the cube when the edge is of length 12 cm.

3. The circumference of a circular patch of oil on the surface of a pond is assumed to be increasing at the constant rate of 2 m/s.

   (a)   When the radius is 4 m, at what rate is the area of the oil changing?

   (b)   If the circumference is actually increasing more quickly than 2 m/s when the radius is 4 m, is your answer for (a) too large or too small?

4. A container in the form of a right circular cone of height 16 cm and base radius 4 cm is held vertex downward and filled with liquid. If the liquid leaks out from the vertex at a rate of 4 cm^3/s, find the rate of change of the depth of the liquid in the cone when half of the liquid has leaked out.

5. A right circular cone has a constant volume. The height h and the base radius r can both vary. Find the rate at which h is changing with respect to r at the instant when r and h are equal.

6. The radius of a hemispherical bowl is a cm. The bowl is being filled with water at a steady rate of $3\,\pi a^3$ cm^3 per minute. Find, in terms of a, the rate at which the water is rising when the depth of water in the bowl is $\frac{1}{2}a$ cm.

The volume of the shaded part of this hemisphere is $\frac{1}{3}\pi h^2(3a - h)$

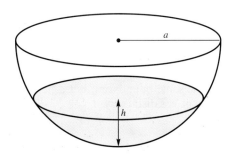

AREA FOUND BY INTEGRATION

The area bounded by part of a curve $y = f(x)$, the x-axis and the lines $x = a$ and $x = b$ is found by summing the areas of vertical strips and taking the limiting value of that sum.

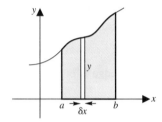

$$\text{Shaded area} = \lim_{\delta x \to 0} \sum_a^b y \, \delta x = \int_a^b y \, dx$$

Similarly the area bounded by a curve $x = f(y)$, the y-axis and the lines $y = a$ and $y = b$ is found by summing the areas of horizontal strips.

The limit of this sum is $\displaystyle\int_a^b x \, dy$.

In *Module A*, curves with algebraic equations only were used as boundaries of required areas. The methods used there apply equally well to the graphs of other functions and to areas where *two* of the boundaries are curves, provided that an element can be found,

- which has the same format throughout, i.e. the ends of all the elements are on the same boundaries;

- whose length and width are measured parallel to the x- and y-axes.

Examples 17b

1. A plane region is defined by the line $y = 4$, the x and y axes and part of the curve $y = \ln x$. Find the area of the region.

A vertical element is unsuitable in this case as the top and bottom are not always on the same boundaries, but a horizontal element is satisfactory.

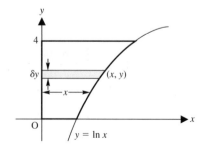

The area, δA, of a typical horizontal element is given by $\delta A \approx x\,\delta y$.
Because the width of our element is δy we will have to integrate w.r.t. y, so we
need the equation of the curve in the form $x = f(y)$

$$y = \ln x \quad \Rightarrow \quad x = e^y$$

$\therefore \quad \delta A \approx e^y\,\delta y$

$$\Rightarrow \qquad A = \lim_{\delta y \to 0} \sum_{y=0}^{y=4} e^y\,\delta y = \int_0^4 e^y\,dy$$

$$= \left[e^y \right]_0^4 = e^4 - e^0$$

The defined area is $(e^4 - 1)$ square units.

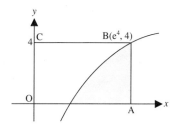

Note that this area can also be found
by subtracting the shaded region
from the area of the rectangle OABC
but this alternative is not always suitable.

2. Find the area between the curve $y = x^2$ and the line $y = 3x$

The line and curve meet where $x^2 = 3x$,   i.e. where $x = 0$ and $x = 3$

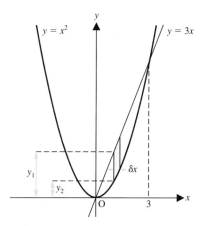

A vertical strip always has its top on the line and its foot on the curve so it is a
suitable element. It is approximately a rectangle whose width is δx and whose
height is the vertical distance between the line and the curve.

The area of the element, δA, is given by

$$\delta A \approx (y_1 - y_2)\,\delta x = (3x - x^2)\,\delta x$$

$$\therefore \qquad A = \lim_{\delta x \to 0} \sum_{x=0}^{x=3} (3x - x^2)\,\delta x = \int_0^3 (3x - x^2)\,dx$$

$$= \left[\tfrac{3}{2}x^2 - \tfrac{1}{3}x^3 \right]_0^3$$

$$= 4\tfrac{1}{2}$$

The required area is $4\tfrac{1}{2}$ square units.

The area bounded partly by a curve whose equation is given parametrically can also be found by summing the areas of suitable elements.

3. Find the area bounded in the first quadrant by the x-axis, the line $x = 2$ and part of the curve with parametric equations $x = 2t^2$, $y = 2t$

The sketch of this curve need not be an accurate shape but it is important to realise that it goes through the origin because when $t = 0$, both x and y are zero.

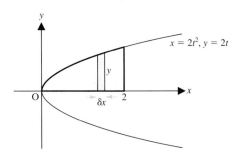

A suitable element is a vertical strip which is approximately a rectangle of height y, width δx and area δA where $\delta A \approx y\,\delta x$

Considering $y = 2t$ gives

$$\delta A \approx 2t\,\delta x$$

Then $\qquad A = \lim_{\delta x \to 0} \sum_{x=0}^{x=2} 2t\,\delta x = \int_{x=0}^{x=2} 2t\,dx$

But $\qquad x = 2t^2$

$$\therefore \qquad \frac{dx}{dt} = 4t \quad \Rightarrow \quad \dots dx = \dots 4t\,dt$$

Also, when $x = 0$, $t = 0$, and when $x = 2$, $t = 1$

$$\therefore \qquad A = \int_0^1 2t(4t)\,dt = \int_0^1 8t^2\,dt = \left[\tfrac{8}{3}t^3 \right]_0^1 = \tfrac{8}{3}$$

The required area is $\tfrac{8}{3}$ square units.

1. Calculate the area bounded by the curve $y = \sqrt{x}$, the y-axis and the line $y = 3$.

2. Find, by integration, the area bounded by

   (a) the x-axis, the line $x = 2$ and the curve $y = x^2$
   (b) the y-axis, the line $y = 4$ and the curve $y = x^2$

   Sketch these two areas on the same diagram and hence check the sum of the answers to (a) and (b).

3. A region in the xy plane is bounded by the lines $y = 1$ and $x = 1$, and the curve $y = e^x$. Find its area.

4. Find the area bounded by the inequalities $y \leqslant 1 - x^2$ and $y \geqslant 1 - x$.

5. Calculate the area bounded by the curve $y = \sin x$ and the lines $y = \frac{1}{2}$ and $x = \frac{1}{2}\pi$.

6. Find the area of the region of the xy plane defined by

   (a) $y \geqslant e^x$, $x \geqslant 0$, $y \leqslant e$

   (b) $1 \geqslant y \geqslant \dfrac{1}{x+1}$, $x \leqslant 2$.

7. The equations of a curve are $x = 2t$, $y = 2/t$. Find the area bounded by this curve, the x-axis and the ordinates at $x = 1$, $x = 4$.

8. Calculate the area in the first quadrant between the curve $y^2 = x$ and the line $x = 9$.

9. Find the area between the y-axis and the curve $y^2 = 1 - x$.

*10. Evaluate the area between the line $y = x - 1$ and the curve

   (a) $y = x(1 - x)$
   (b) $y = (2x + 1)(x - 1)$.

*11. Calculate the area of the region of the xy plane defined by the inequalities $y \geqslant (x+1)(x-2)$ and $y \leqslant x$.

VOLUME OF REVOLUTION

If an area is rotated about a straight line, the three-dimensional object so formed is called a *solid of revolution,* and its volume is a *volume of revolution.*

The line about which rotation takes place is always an axis of symmetry for the solid of revolution. Also, any cross-section of the solid which is perpendicular to the axis of rotation, is circular.

Consider the solid of revolution formed when the area shown in the diagram is rotated about the *x*-axis.

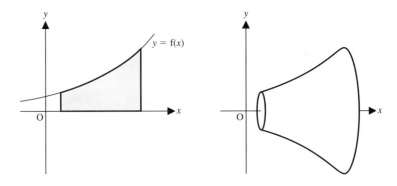

To calculate the volume of this solid we can divide it into 'slices' by making cuts perpendicular to the axis of rotation.

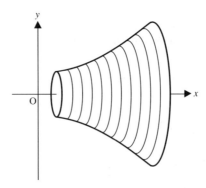

If the cuts are reasonably close together, each slice is approximately cylindrical and the approximate volume of the solid can be found by summing the volumes of these cylinders.

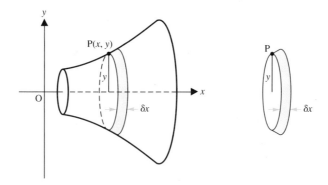

Consider an element formed by one cut through the point $P(x, y)$ and the other cut distant δx from the first.

The volume, δV, of this element is approximately that of a cylinder of radius y and 'height' δx

i.e. $$\delta V \approx \pi y^2 \delta x$$

Then the total volume of the solid is V, where

$$V \approx \sum \pi y^2 \delta x$$

The smaller δx is, the closer is this approximation to V, i.e.

$$V = \lim_{\delta x \to 0} \sum \pi y^2 \delta x = \int \pi y^2 \, dx$$

Now if the equation of the rotated curve is given, this integral can be evaluated and the volume of the solid of revolution found,

e.g. to find the volume generated when the area between part of the curve $y = e^x$ and the x-axis is rotated about the x-axis, we use

$$\int \pi (e^x)^2 \, dx = \pi \int e^{2x} \, dx$$

When an area rotates about the y-axis we can use a similar method based on slices perpendicular to the y-axis, giving

$$V = \int \pi x^2 \, dy$$

1. Find the volume generated when the area bounded by the x- and y-axes, the line $x = 1$ and the curve $y = e^x$ is rotated through one revolution about the x-axis.

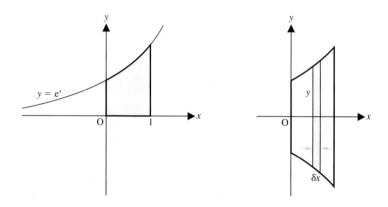

The volume, δV, of the element shown is approximately that of a cylinder of radius y and thickness δx, therefore

$$\delta V \approx \pi y^2 \, \delta x$$

∴ the total volume is V, where $\displaystyle V \approx \sum_{x=0}^{x=1} \pi y^2 \, \mathrm{d}x$

$\Rightarrow \qquad \displaystyle V = \lim_{\delta x \to 0} \sum_{x=0}^{x=1} \pi y^2 \, \delta x = \int_0^1 \pi y^2 \, \mathrm{d}x$

$$= \pi \int_0^1 (e^x)^2 \, \mathrm{d}x$$

$$= \pi \int_0^1 e^{2x} \, \mathrm{d}x$$

$$= \pi \left[\tfrac{1}{2} e^{2x} \right]_0^1$$

$$= \tfrac{1}{2} \pi (e^2 - e^0)$$

i.e. the specified volume of revolution is $\tfrac{1}{2} \pi (e^2 - 1)$ cubic units.

2. The area defined by the inequalities $y \geqslant x^2 + 1$, $x \geqslant 0$, $y \leqslant 2$, is rotated completely about the y-axis. Find the volume of the solid generated.

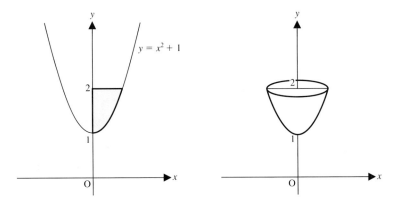

Rotating the shaded area about the y-axis gives the solid shown.

This time we use horizontal cuts to form elements which are approximately cylinders with radius x and thickness δy

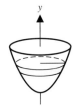

$$\delta V \approx \pi x^2 \, \delta y \quad \Rightarrow \quad V \approx \sum_{y=1}^{y=2} \pi x^2 \, \delta y$$

$$\therefore \quad V = \lim_{\delta y \to 0} \sum_{y=1}^{y=2} \pi x^2 \, \delta y = \int_1^2 \pi x^2 \, dy$$

Using the equation $y = x^2 + 1$ gives $x^2 = y - 1$

$$\therefore \quad V = \pi \int_1^2 (y-1) \, dy = \pi \left[\tfrac{1}{2} y^2 - y \right]_1^2$$

$$= \pi \left\{ (2-2) - (\tfrac{1}{2} - 1) \right\}$$

i.e. the volume of the specified solid is $\tfrac{1}{2}\pi$ cubic units.

3. The area enclosed by the curve $y = 4x - x^2$ and the line $y = 3$ is rotated about the line $y = 3$. Find the volume of the solid generated.

The line $y = 3$ meets the curve $y = 4x - x^2$ at the points $(1, 3)$ and $(3, 3)$, therefore the volume generated is as shown in the diagram.

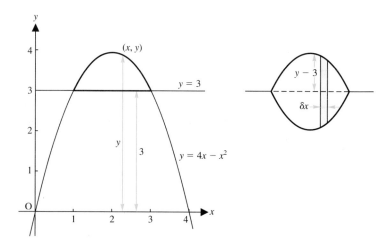

The element shown is approximately a cylinder with radius $(y - 3)$ and thickness δx, so its volume, δV, is given by $\delta V \approx \pi (y - 3)^2 \delta x$

i.e.
$$V = \lim_{\delta x \to 0} \sum_{x=1}^{x=3} \pi (y - 3)^2 \delta x = \pi \int_1^3 (y - 3)^2 \, dx$$

$$\Rightarrow \qquad V = \pi \int_1^3 (4x - x^2 - 3)^2 \, dx$$

$$= \pi \int_1^3 (9 - 24x + 22x^2 - 8x^3 + x^4) \, dx$$

$$= \pi \left[9x - 12x^2 + \tfrac{22}{3}x^3 - 2x^4 + \tfrac{1}{5}x^5 \right]_1^3$$

$$= \tfrac{16}{15} \pi$$

$\therefore$ the required volume is $\tfrac{16}{15} \pi$ cubic units.

4. Find the volume generated when the area between the curve $y^2 = x$ and the line $y = x$ is rotated completely about the x-axis.

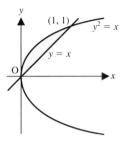

The defined area is shown in the diagram.

When this area rotates about Ox, the solid generated is bowl-shaped on the outside, with a conical hole inside.
The cross-section this time is not a simple circle but is an annulus, i.e. the area between two concentric circles.

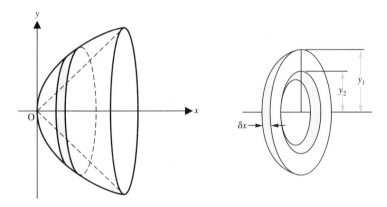

For a typical element the area of cross-section is $\pi y_1^2 - \pi y_2^2$

Therefore the volume of an element is given by $\delta V \approx \pi \{ y_1^2 - y_2^2 \} \delta x$

$$\therefore \quad V \approx \sum_{x=0}^{x=1} \pi \{ y_1^2 - y_2^2 \} \delta x = \pi \int_0^1 (y_1^2 - y_2^2) \, dx$$

Now $y_1 = \sqrt{x}$ and $y_2 = x$

$$\therefore \quad V = \pi \int_0^1 (x - x^2) \, dx = \pi \left[\tfrac{1}{2} x^2 - \tfrac{1}{3} x^3 \right]_0^1 = \tfrac{1}{6} \pi$$

The volume generated is $\tfrac{1}{6} \pi$ cubic units.

Note that the volume specified in Example 4 could be found by calculating separately

1) the volume given when the curve $y^2 = x$ rotates about the x-axis;

2) the volume, by formula, of a cone with base radius 1 and height 1 and subtracting from (1).

The method in which an annulus element is used, however, applies whatever the shape of the hollow interior.

EXERCISE 17c

In each of the following questions, find the volume generated when the area defined is rotated completely about the x-axis.

1. The area between the curve $y = x(4 - x)$ and the x-axis.

2. The area bounded by the x- and y-axes, the curve $y = e^x$ and the line $x = 3$.

3. The area bounded by the x-axis, the curve $y = 1/x$ and the lines $x = 1$ and $x = 2$.

4. The area defined by the inequalities $0 \leqslant y \leqslant x^2$ and $-2 \leqslant x \leqslant 2$.

5. The area between the curve $y^2 = x$ and the line $x = 2$.

In each of the following questions, the area bounded by the curve and line(s) given is rotated about the y-axis to form a solid. Find the volume generated.

6. $y = x^2$, $y = 4$.

7. $y = 4 - x^2$, $y = 0$.

8. $y = x^3$, $y = 1$, $y = 2$, for $x \geqslant 0$.

9. $y = \ln x$, $x = 0$, $y = 0$, $y = 1$.

10. Find the volume generated when the area enclosed between $y^2 = x$ and $x = 1$ is rotated about the line $x = 1$.

11. The area defined by the inequalities

$$y \geqslant x^2 - 2x + 4, \ y \leqslant 4$$

is rotated about the line $y = 4$. Find the volume generated.

12. The area enclosed by $y = \sin x$ and the x-axis for $0 \leqslant x \leqslant \pi$ is rotated about the x-axis. Find the volume generated.

13. An area is bounded by the line $y = 1$, the x-axis and parts of the curve $y = 3 - x^2$. Find the volume generated when this area rotates completely about the y-axis.

14. The area enclosed between the curves $y = x^2$ and $y^2 = x$ is rotated about the x-axis. Find the volume generated.

MIXED EXERCISE 17

1. The surface area of an expanding sphere is increasing at a constant rate of 0.02 cm^2 per second. Find the rate of increase of the volume of the sphere when the radius is 15 cm.

2. A region of the xy plane is defined by the inequalities $0 \leqslant x \leqslant 4$ and $0 \leqslant y \leqslant e^x$. Find the area of the region.

3. (a) Find the area of the region in the first quadrant bounded by the y-axis, the line $y = 6$ and the curve $y = x^2 + 2$

   (b) If this area is rotated completely about Oy to form a solid, find the volume of the solid.

4. (a) Find the area enclosed between the x-axis and the curve $y = x^2 - 3$

   (b) Find the volume generated when this area is rotated completely about
   (i) the x-axis   (ii) the y-axis.

5. A region of the xy plane is bounded by the curve $y = e^x$, the x-axis and the lines $x = -1$ and $x = 1$

   (a) Find the area of the region.

   (b) Find the volume generated when this area is rotated completely about the x-axis.

6. A plane region is bounded by the curve $y = 6 - x^2$ and the line $y = 2$. Find

   (a) the area of the region

   (b) the volume generated when this area rotates through $360°$ about the line $y = 2$.

7. The equation of a curve is $y = \sin x$. Find the area bounded by this curve and the x-axis between $x = 0$ and $x = \pi$.
   Find also the volume generated when this area rotates through one revolution about the x-axis.

8. Find the area enclosed between the curves $y = x^2$ and $y^2 = x$.

*9. The diagram represents the printout from a tachograph fitted inside an experimental car. It shows the speed of the car over a 60-second time interval.

Speed is the rate of change of distance, i.e. $\text{speed} = \dfrac{\mathrm{d}}{\mathrm{d}t}(\text{distance})$

Hence $\text{distance} = \displaystyle\int (\text{speed})\,\mathrm{d}t$, so the distance covered is represented by the area under this graph.

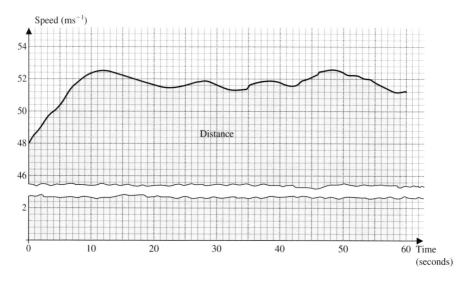

(a) The area under this curve can be found by counting squares. This is very time-consuming when using small squares. Another method for finding an approximate value for the area is to divide the area into vertical strips and to join the tops of the vertical edges with straight lines to form a set of trapeziums, and then to find the areas of these trapeziums. Use this method with six strips to find an approximate value for the distance covered by the car.

(b) Is the value found an over- or under-estimate?

(c) Find an better estimate and explain how you obtained it.

Partial Fractions

A proper fraction with a denominator that factorises can be expressed in partial fractions as follows.
The numerators which can be found by the cover-up method are green.

$$\frac{f(x)}{(x-a)(x-b)} = \frac{A}{(x-a)} + \frac{B}{(x-b)}$$

$$\frac{f(x)}{(x-a)(x-b)^2} = \frac{A}{(x-a)} + \frac{B}{(x-b)} + \frac{C}{(x-b)^2}$$

$$\frac{f(x)}{(x-a)(x^2+b)} = \frac{A}{(x-a)} + \frac{Bx+C}{(x^2+b)}$$

Integration

STANDARD RESULTS

$f(x)$	$\int f(x)\,dx$
x^n	$\dfrac{1}{n+1}x^{n+1}$
e^x	e^x
$1/x$	$\ln x$
$\cos x$	$\sin x$
$\sin x$	$-\cos x$
$\sec^2 x$	$\tan x$
$(ax+b)^n$	$\dfrac{1}{a(n+1)}(ax+b)^{n+1}$
$\cos(ax+b)$	$\dfrac{1}{a}\sin(ax+b)$

INTEGRATING PRODUCTS CAN BE DONE BY

- **recognition**, in particular
$$\int \sin nx \cos x \, dx = \frac{1}{n+1} \sin^{n+1} x + K$$

$$\int f'(x) e^{f(x)} \, dx = e^{f(x)} + K$$

- **change of variable**, suitable for
$$\int f'(x) g \{f(x)\} \, dx$$

- **by parts**
$$\int v \frac{du}{dx} \, dx = uv - \int u \frac{dv}{dx} \, dx$$

INTEGRATING FRACTIONS CAN BE DONE BY

- **recognition**, in particular
$$\int \frac{f'(x)}{f(x)} \, dx = \ln |f(x)| + K$$

- **change of variable**, suitable for
$$\int \frac{f'(x)}{g\{f(x)\}} \, dx$$

- **using partial fractions**

DIFFERENTIAL EQUATIONS

A first order linear differential equation is a relationship between x, y and dy/dx. It can be solved, when the variables can be separated, by collecting all the x terms, along with dx on one side and all the y terms along with dy on the other side. Then each side is integrated with respect to its own variable.

A constant of integration, called an arbitrary constant, is introduced on one side only to give a general solution which is a family of curves or lines. If extra information is available from which the value of the constant can be found, we have a particular solution, i.e. one member of the family.

Applications of Calculus

COMPARATIVE RATES OF CHANGE

If a quantity p depends on a quantity q and the rate at which q increases with time is known, then

$$\frac{dp}{dt} = \frac{dp}{dq} \times \frac{dq}{dt}$$

AREA

The area bounded by the x-axis, the two lines $x = a$ and $x = b$, and the curve $y = f(x)$ can be found by summing the areas of vertical strips of width δx and using

$$\text{area} = \lim_{\delta x \to 0} \sum_{x=a}^{x=b} y\, \delta x = \int_a^b y\, dx$$

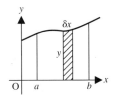

Similarly for horizontal strips,

$$\text{area} = \lim_{\delta y \to 0} \sum_{y=a}^{y=b} x\, \delta y = \int_a^b x\, dy$$

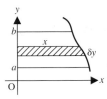

VOLUME

When the area bounded by the x-axis, the ordinates at $x = a$ and $x = b$ and the curve $y = f(x)$ is rotated about the x-axis, the volume formed is given by

$$\text{volume} = \lim_{\delta x \to 0} \sum_{x=a}^{x=b} \pi y^2\, \delta x = \int_a^b \pi y^2\, dx$$

For rotation about the y-axis, $\quad \text{volume} = \lim_{\delta y \to 0} \sum_{y=a}^{y=b} \pi x^2\, \delta y = \int_a^b \pi x^2\, dy$

MULTIPLE CHOICE EXERCISE D

Type I

1. The fraction $\dfrac{1}{(x+1)(x-1)}$ can be expressed as

A $\dfrac{1}{2(x-1)} - \dfrac{1}{2(x+1)}$

B $\dfrac{1}{x+1} + \dfrac{1}{x-1}$

C $\dfrac{2}{x+1} - \dfrac{2}{x-1}$

D $\dfrac{1}{x^2} - \dfrac{1}{1}$

E $\dfrac{1}{x+1} - \dfrac{1}{x-1}$

2.

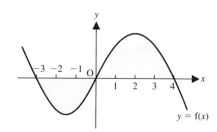

$y = f(x)$

The shaded area in the diagram is rotated about the x-axis. The volume formed is given by

A $\quad \pi \displaystyle\int_{-3}^{4} [f(x)]^2 \, dx$

B $\quad \pi \displaystyle\int_{-3}^{4} f(x) \, dx$

C $\quad \pi \displaystyle\int_{-3}^{0} f(x) \, dx + \pi \displaystyle\int_{0}^{4} f(x) \, dx$

D $\quad \pi^2 \displaystyle\int_{-3}^{4} f(x) \, dx$

3. e^{x^2} could be the integral w.r.t. x of

A $\quad e^{2x}$

B $\quad 2x \, e^{x^2}$

C $\quad \dfrac{e^{x^2}}{2x}$

D $\quad x^2 e^{x^2} - 1$

E $\quad$ none of these

4. If $\displaystyle\int_{1}^{5} \dfrac{dx}{2x - 1} = \ln K$, the value of K is

A 9 $\qquad$ **B** 3 $\qquad$ **C** undefined $\qquad$ **D** 81 $\qquad$ **E** 8

5. $\dfrac{2}{(x+1)(x-1)} \equiv \dfrac{A}{x+1} + \dfrac{B}{x-1}$ corresponds to

A $\quad A = 1, \ B = 1$

B $\quad A = -1, \ B = 1$

C $\quad A = x, \ B = 1$

D $\quad A = 0, \ B = 2$

E $\quad A = x - 1, \ B = x + 1$

6. $I = \displaystyle\int_{1}^{2} x\sqrt{(x^2 - 1)} \, dx$ is found as follows. Where does an error first occur?

A $\quad$ Let $u \equiv x^2 - 1$

B $\quad \ldots du \equiv \ldots 2x \, dx$

C $\quad I = \frac{1}{2} \displaystyle\int_{1}^{2} u^{1/2} \, du$

D $\quad I = \frac{3}{4} \left[u^{3/2} \right]_{1}^{2}$

7. The value of $\int_0^2 2e^{2x}\,dx$ is

   A e^4 **B** $e^4 - 1$ **C** ∞ **D** $4e^4$ **E** $\frac{1}{2}e^4$

8. If $\dfrac{x+p}{(x-1)(x-3)} \equiv \dfrac{q}{x-1} + \dfrac{2}{x-3}$, the values of p and q are

   A $p = -2,\ q = 1$ **B** $p = 2,\ q = 1$

   C $p = 1,\ q = -2$ **D** $p = 1,\ q = 1$

   E $p = 1,\ q = -1$

Type II

9. Values of A and B can be found such that $\dfrac{x}{(x-2)(x+1)} \equiv \dfrac{A}{x-2} + \dfrac{B}{x+1}$

10. The area between the curve $y = 1 - x^2$ and the x-axis is given by $\int_{-1}^{1} y\,dx$

11. $\int \tan x\,dx = \sec^2 x + K$

12. $\int_0^a f(y)\,dy = \lim\limits_{\delta y \to 0} \sum\limits_{y=0}^{y=a} f(y)\,\delta y$

13. $\left[f(x) \right]_0^a = f(a) - 0$

14. A differential equation must contain $\dfrac{dy}{dx}$.

15. $\sum\limits_{x=a}^{x=b} y\,\delta x = \int_a^b y\,dx$

16. $f(x) = 2x^2 + 3x - 2$ can be expressed as the sum of two partial fractions.

17. Integration by parts can be used to find $\int \ln x\,dx$

18. $\int_1^2 x e^x\,dx = x e^x - e^x$

19. The differential equation $e^{x+y} = y\dfrac{dy}{dx}$ can be solved by separating the variables.

1. (a) Differentiate $(1 + x^3)^{1/2}$ with respect to x.

(b) Use the result from (a), or an appropriate substitution, to find the value of

$$\int_0^2 \frac{x^2}{\sqrt{(1 + x^3)}}\, dx.$$ (AEB)$_s$

2. (i) Use the chain rule to differentiate $(3 + 2x)^4$ with respect to x.

(ii) Hence evaluate $\displaystyle\int_0^1 (3 + 2x)^3\, dx.$

(iii) Expand $(3 + 2x)^3$ in powers of x, and use your answer to check the value of the integral in (ii). (OCSEB)$_s$

3. (i) Write $1/\{y(3 - y)\}$ in partial fractions.

(ii) Find $\displaystyle\int \frac{1}{y(3 - y)}\, dy.$

(iii) Solve the differential equation $x\dfrac{dy}{dx} = y(3 - y)$

where $x = 2$ when $y = 2$, giving y as a function of x. (MEI)$_s$

4. The region R in the first quadrant is bounded by the curve $y = e^{-x}$, the x-axis, the y-axis, and the line $x = 2$. Show that the volume of the solid formed when R is completely rotated about the x-axis is

$$\pi \int_0^2 e^{-2x}\, dx,$$

and evaluate this volume, giving your answer in terms of e and π. (UCLES)$_s$

5. Find

(a) $\displaystyle\int x \cos x\, dx,$

(b) $\displaystyle\int \cos^2 y\, dy.$

Hence find the general solution of the differential equation

$$\frac{dy}{dx} = x \cos x \sec^2 y, \qquad 0 < y < \frac{x}{2}.$$ (ULEAC)$_s$

6. $f(x) \equiv \dfrac{x^2 + 6x + 7}{(x+2)(x+3)}, \quad x \in \mathbb{R}.$

Given that $f(x) \equiv A + \dfrac{B}{x+2} + \dfrac{C}{x+3}$,

(a) find the values of the constants A, B and C.

(b) show that $\displaystyle\int_0^2 f(x)\,dx = 2 + \ln\left(\tfrac{25}{18}\right).$ (ULEAC)$_s$

7. Find $\displaystyle\int x \ln x \, dx.$ (UCLES)$_s$

8. Express $\dfrac{1}{x(2x-1)^2}$ in the form $\dfrac{A}{x} + \dfrac{B}{(2x-1)} + \dfrac{C}{(2x-1)^2}.$ (AEB)$_p$

9. The region R is bounded by the x-axis and the part of the curve $y = \sin 2x$ between $x = 0$ and $x = \tfrac{1}{2}\pi$. Use integration to find the exact values of

(i) the area of R,

(ii) the volume of the solid formed when R is rotated completely about the x-axis.   (UCLES)

10. A circular patch of oil on the surface of water has radius r metres at time t minutes. When $t = 0$, $r = 1$ and when $t = 10$, $r = 2$. It is desired to predict the value T of t when $r = 4$.

(i) In a simple model the rate of increase of r is taken to be constant. Find T for this model.

(ii) In a more refined model, the rate of increase of r is taken to be proportional to $\dfrac{1}{r}$. Express this statement as a differential equation, and find the general solution.
Find T for this second model.   (UCLES)$_s$

11. The diagram shows a sketch of the curve defined for $x > 0$ by the equation $y = x^2 \ln x$.

The curve crosses the x-axis at A and has a local minimum at B.

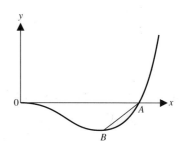

(a) State the coordinates of A and calculate the gradient of the curve at A.

(b) Calculate the coordinates of B in terms of e and determine the value of $\dfrac{d^2y}{dx^2}$ at B.

(c) The region bounded by the line segment AB and an arc of the curve is R, as shaded in the diagram. Show that the area of R is

$$\left(\frac{4 - e^{-3/2} - 9e^{-1}}{36}\right).$$

(AEB)ₛ

12. Express as the sum of partial fractions

$$\frac{2}{x(x+1)(x+2)}.$$

Hence show that

$$\int_2^4 \frac{2}{x(x+1)(x+2)} \, dx = 3 \ln 3 - 2 \ln 5.$$

(ULEAC)

13.

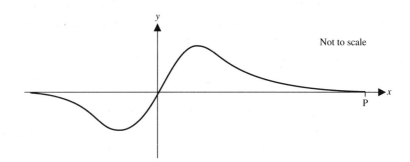

Not to scale

The diagram shows a sketch of the curve $y = \dfrac{x}{x^2+1}$.

(i) Differentiate $y = \dfrac{x}{x^2+1}$.

(ii) Hence find the coordinates of the two turning points of the curve.

(iii) By substituting $t = x^2 + 1$, or otherwise, find $\displaystyle\int \frac{x}{x^2+1} \, dx$.

(iv) Hence find the x-coordinate of the point P, given that the area between the curve and the x-axis, from the origin to P is 10 units². (MEI)ₛ

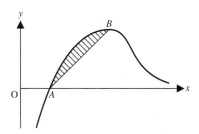

14. The diagram shows a sketch of the curve with equation

$$y = \frac{\ln x}{x} \qquad (x > 0)$$

(a) State the coordinates of A, where the curve crosses the x-axis.

(b) Calculate, in terms of e, the coordinates of the maximum point B.

(c) Calculate, in terms of e, the value of $\dfrac{d^2 y}{dx^2}$ at B.

(d) Using the substitution $u = \ln x$, or otherwise, find

$$\int \frac{\ln x}{x} \, dx.$$

Hence determine, in terms of e, the area of the shaded region bounded by the curve and the chord AB. (AEB)

15. (a) Sketch the graph of $y = e^x$ and $y = 1 + 2e^{-x}$ on the same diagram.

Give the coordinates of any points of intersection with the axes and state the equation of the asymptote for each graph.

(b) By solving an appropriate equation, show that the only point where the two graphs meet is $(\ln 2, 2)$.

(c) Show that the area of the finite region defined by

$$y \leqslant e^x, \ y \leqslant 1 + 2e^{-x}, \ y \geqslant 0, \ x \geqslant 0, \ x \leqslant 2 \ln 2,$$

is $\ln 2 + \frac{3}{2}$ (ULEAC)

CHAPTER 18

COLLECTING AND REPRESENTING INFORMATION

STATISTICS

The word statistics can mean simply collections of numerical data, such as unemployment figures or sales figures. It can also mean the processes used to collect, analyse and interpret the data.

Populations and Samples

Most statistical investigations use data from just a small part (a *sample*) of all the relevant data (the *population*). There are several reasons why this is necessary, some of which are listed below.

- It is often impossible to collect all the relevant data. This is the case when the population is infinite; e.g. all possible throws of a dice. A population can be finite and yet be impossible to list; e.g. the number of red squirrels in the U.K.

- It is too time-consuming. It can be that the time needed to collect and analyse all the data is such that any information gained is out of date or too late for its purpose.

- It is expensive. The cost of collecting all the information can outweigh its benefit.

- It is self-defeating. Finding how long each light bulb coming off the production line lasts before it burns out would not leave any to sell.

The word population has a precise definition in relation to statistical investigations.

The population is the complete set of data being investigated.

CENSUS AND SURVEY

An investigation that uses the whole population is called a *census*.

An investigation based on a sample is often called a *survey*.

For example, a cattle census in Surrey requires information about all cattle in the county whereas a cattle survey requires information about some of the cattle.

SAMPLING FRAME

When a sample is needed, individual items are chosen from the whole population. In a few specialised cases, the sample chooses itself; if you are investigating the proportion of sixes that turn up when an ordinary die is rolled, the sample is the number of rolls of the die actually made, out of the infinite number of possible rolls. In most cases a positive choice of individual items has to be made but this can only be done if we have a list of the population from which they will be chosen. This is called the *sampling frame*.

For example, a university may want to survey the attitudes of its students to a suggestion that some lectures take place in the evenings. The sampling frame would be the list of all students enrolled at the university.

There is usually a list that can act as the sampling frame for any well defined group of people, e.g. employees of a company, practising solicitors, and so on.

Ideally this list should be complete, up to date and contain no duplication, but such a list does not always exist. If the population is 'all adults living in the UK', the Electoral Register is usually used as the sampling frame, but it does not satisfy any of the criteria above. It is not complete (some people are omitted for a variety of reasons), it is out of date (it is published annually four months after the forms have to be returned and some people will have moved into and out of the area) and it contains duplications (some people are on more than one register). However it is the best list available.

Choosing a Sample

Once a population has been identified and the sampling frame determined, both the size of the sample and the items forming the sample have to be chosen. As the error in the results from a sample of a given size can be quantified, it is possible to choose the size of the sample so that this error is within acceptable limits. However, the reasons for deciding in the first place that a sample is necessary will largely dictate its size: the reasons are often the availability of time and money.

The aim is to select a sample with the following two characteristics.

**The sample must be representative of the population
and the selection procedure must avoid bias.**

A sample can be unrepresentative of its population by chance. For example, if a fair coin is tossed five times, it could, by chance show a head on all five tosses. On the other hand, five heads could show for a reason other than pure chance (for example, unreliable tossing) and, in this case, the sample is *biased*.

Samples can be biased for several reasons. One common source of bias is the use of data from responses to questions. For example, if information about the weights of a group of people is gathered by asking them to fill in a questionnaire, overweight people may enter less than their true weight (probably from embarrassment). In fact people can and do lie, for many reasons, in response to questionnaires.

VARIABLES

Any quantity that can vary is called a *variable*. Statistics tries to make sense of variables such as voting intentions, attitudes towards police, numbers of people waiting for a bus, numbers of heads from three tosses of a coin, heights, weights, and so on.

Variables such as attitudes, opinions, gender, and so on, are attributes which are described using words. These are *qualitative* variables.

Variables like shoe size, height, number of broken eggs, and so on, have numerical values. These are *quantitative* variables.

There are two categories of quantitative data.

The number of broken eggs can be 0, 1, 2, 3, ... and a shoe size can be ... 4, $4\frac{1}{2}$, 5, $5\frac{1}{2}$, 6, ... It is not possible to have 3.75 broken eggs and there is no shoe size of 3.75. Variables that can only have distinct values, like shoe size and the number of broken eggs, are called *discrete* variables.

The weight of an apple may be ... 40 g, 55 g, 62 g, 120 g, ... to the nearest gram, but could be anywhere on a continuous scale:

A variable that can have a value anywhere on a continuous scale is called a *continuous* variable.

COLLECTING DATA

Once a sample has been selected, the information required from each member of the sample needs to be collected.

If the quantity of information required is large and related to human activity, it may be possible to use outside sources, such as the vast quantity of statistics produced by Government Departments. If figures gathered by an outside agency are used, the source must be stated.

When information is collected directly, the following points need to be considered.

- How is the information to be recorded? This is largely determined by the nature of the data. If it involves a variable that can be measured or counted, or a clearly defined attribute such as gender, the information can be recorded directly on an observation sheet. If it involves opinion, a questionnaire will be needed.

- If the variable is continuous, it cannot be measured exactly. In this case a decision about the accuracy required. For example, when recording times with an ordinary stop watch, it should be possible to do so correct to the nearest second. On the other hand, when measuring the heights of people, it may be decided to do so correct to the nearest centimetre, even if greater accuracy than this were possible. Whatever the decision, it is essential that the accuracy of continuous data is stated.

Observation Sheets

If one category of continuous variables, e.g. the lengths of rods, is to be recorded,, a simple list is enough. If more than one category of information is involved, e.g. the quantity of each type of sweet sold, it is sensible to draw up a table with a column for each category.

Questionnaires

When information is required from people, and that information cannot be directly observed or measured, questions have to be asked. The answers can be obtained by asking the questions personally (i.e. by interview) or by asking people to fill in answers to questions on a form. In either case, each person must obviously answer the same questions. This means that the questions need to be prepared, i.e. a questionnaire is needed.

When preparing a questionnaire, questions should be worded so that they

- are short and unambiguous,
- can be answered yes or no, or by choosing one of a set of predetermined answers, e.g. agree, no opinion, disagree,
- are not leading or provocative (i.e. do not indicate a 'correct answer'),
- are likely to be answered truthfully.

It is not easy to predict what other people will find ambiguous or embarrassing, so before finalising the questions, try them out on a few people, i.e. do a *pilot survey.*

The questionnaire can be completed by several methods of which the following are the most common.

- Direct interview.

This involves a trained interviewer asking each individual the questions directly. This method produces the best results because any ambiguities or misunderstandings can be dealt with on the spot. It is, however, expensive and time consuming.

- Telephone interview.

This has the advantages of direct interview and it is less expensive and usually quicker. The disadvantages are that it is more difficult to get a representative sample because not everyone has a telephone and of those that do, some are ex-directory. Also a telephone number is usually listed under only one name in a household so that other members of that household cannot be included in the sample.

- Post.

Questionnaires are distributed by post and the recipients are asked to complete and return them. This method is the cheapest and easiest to administer but has several disadvantages. The response rate is usually very poor and those that do return completed questionnaires often do so because they are particularly interested in the subject matter and are therefore not representative of the general population.

EXERCISE 18a

1. For each of the following variables, state whether it is qualitative or quantitative and in the latter case state also whether it is discrete or continuous.

   (a) Voting intentions in the Christchurch constituency at the next general election.

   (b) The number of people passing through the ticket barrier in one minute at Euston Underground Station.

(c) The weight of an apple from a particular tree.

(d) The age of each child in year 9 at Mill School.

(e) The language spoken at home by each child in year 9 at Mill School.

(f) The number of heads showing when three coins are tossed.

(g) The suit of a card drawn from an ordinary pack of 52 playing cards.

(h) The mass of nitrate in a one litre sample of river water.

(i) The speed of a car passing a check-point on a motorway.

2. A canvasser for a political party intends to ask people which candidate they will vote for at the forthcoming council by-election in Heath ward.

(a) Is the number of people that he could question finite or infinite?

(b) What kind of variable is he dealing with?

(c) If he intends to use a sample, what would be a suitable sampling frame?

3. The number of heads that appear when three one-pound coins are tossed is being investigated.

(a) What values can the variable have?

(b) What is the population that is being investigated?

(c) Is this population finite or infinite?

(d) Which of these observation sheets is more efficient for recording the information?

(i)

Number of heads	0	1	2	3
Tally				
Totals				

(ii)

Number of heads	Tally	Totals
0		
1		
2		
3		

4. The governors of a secondary school want to change the hours of the school day to start at 8 a.m. and end at 1 p.m. Before deciding to implement this change, they want information on attitudes to the proposed hours.

(a) Is it possible to ask everyone affected by the change? Give reasons for your answer.

(b) It is decided to use a sample. Suggest a possible sampling frame.

5. Suggest a suitable sampling frame for gathering information on the following variables.

   (a) The heights now of children born on or between 1.9.86 and 31.8.87.

   (b) The salaries of practising solicitors.

   (c) The weights of adult men.

   (d) The number of pupils in school for whom English is not the language spoken at home.

   (e) The weights of adult mice reared in a laboratory.

   For questions 6 to 9, list at least one reason why the conclusion may be suspect.

6. This year, the published figures for new registrations for all road vehicles showed an increase of 20% over the previous year. These figures were used to conclude that there are 20% more cars on the road now than there were last year.

7. This table shows the number of accident claims made by different age groups on car insurance policies issued by one insurance company.

Age of driver	17–20	21–60	61–
Total number of claims	325	4788	1076

   These figures show that people under 21 years old are safer drivers than older people.

8. An opinion poll was conducted on behalf of a motoring organisation by asking people whether or not they were in favour of charging motorists for use of roads in the town centre. Those questioned were people leaving a city centre car park after parking their cars. The results were published in the form 'Only 1 in 50 people agreed that motorists should be charged for using roads in the town centre.'

9. An opinion poll on voting intentions was conducted by telephone. The results showed that party A would receive 55% of the vote.

10. A questionnaire was used to gather information from some year 11 school children. The first draft contained these two questions.

    Do you own a games machine   Yes ... No

    How much money do you get each week? ......

    A pilot study revealed problems with both questions.

    (a) Give these questions to a few year 7 pupils and identify any problems.

    (b) Reword the questions so that these problems are avoided. Try your version again with a few more year 7 pupils.

11. (a) Design a questionnaire to be used to collect information on the following items from school children aged about 11 to 15.

   Age. Sex. Name. Opinion on having to wear uniform.

   (b) Try it out on a few children (about five) and make notes about any difficulties.

   (c) Use the results of your pilot survey to amend your questionnaire and try it out again on about five (different) children. Write a short report on any problems that you encountered.

12. Give an example of the use of (a) a census (b) a survey.

FREQUENCY TABLES

Figures that have been collected, but not organised in any way are called *raw data*. Raw data needs sorting before any sense can be made of it. The first step is usually to list it in tabular form showing the different items and the number of times they occur. A table giving this information is called a *frequency table*.

Frequency Tables for Discrete Data

Frequency means the number of times that an event occurs.

If four coins are tossed, and the variable is the number of heads that show, the event is the value the variable can have; in this case it is 0, 1, 2, 3 or 4. If the four coins are tossed several times, and the results listed on an observation sheet, the tally marks can be added in each column to give the number of times each event occurred. The result can be displayed in a table, e.g.

Number of heads, x	0	1	2	3	4	
Frequency, f	6	18	38	23	5	Total: 90

(Adding the frequencies gives the total number of times the four coins were tossed.)

This table gives a 'picture' of the distribution of the number of occurrences for each value of the variable; for example, we can see that the number of times that 2 heads showed was roughly twice the number of times that 1 or 3 heads showed.

Column Graphs

The table on the previous page can be illustrated by a column graph in which the frequency of each item is represented by a line whose length is proportional to the frequency. This gives a visual representation of the frequencies which makes it easier to appreciate their relative sizes.

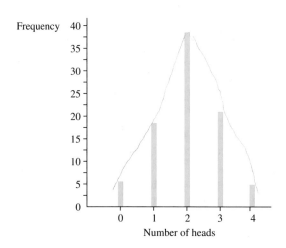

Column graphs can only be used to illustrate frequencies of distinct items, i.e. the distribution of a discrete variable.

It is usual to place the possible values of the variable along the horizontal axis and the frequency along the vertical axis, but column graphs are sometimes drawn with the 'lines' horizontal.

(This type of diagram is also called a bar-line graph.)

1. A 50 metre length of cloth was inspected for flaws. This list gives the number of flaws in each length of 1 metre.

   2, 1, 3, 0, 0, 2, 0, 1, 3, 1, 0, 1, 1, 1, 3, 2, 0, 1, 1, 2, 1, 1, 2, 0, 0,
   2, 1, 0, 0, 2, 4, 1, 0, 0, 2, 1, 1, 1, 5, 2, 1, 1, 0, 1, 0, 0, 1, 2, 3, 0.

   (a) List the information in a frequency table and illustrate it with a column graph.

   (b) What does the graph show to be the most likely number of flaws to occur in a length of cloth?

2. The graph shows the number of immunisation injections given by a local general practice in one year.

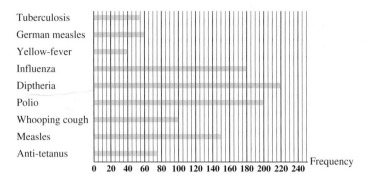

(a) Make a frequency table from this graph.

(b) What is the total number of immunisations given in the year?

(c) Is there any query that you would raise concerning these figures?

GROUPED FREQUENCY TABLES

Discrete Variables

When, as is the case above, a variable has just a few discrete values, tabulating each event together with its frequency shows the information clearly and concisely without losing any of it. When a variable can have many discrete values, making a frequency table for each of these values may not add much clarity to the raw data.

Consider, for example, this frequency table, where x is the number of vehicles per minute passing a road survey point.

x	0	1	2	3	4	5	6	7	8	9	10	11	12	13	14	15	16	17	18	19	20	21	22	23	24
f	4	3	2	2	0	4	2	5	2	1	4	0	2	3	3	3	1	2	1	0	2	0	1	0	1

If this distribution has a 'shape', it is not very obvious from the table. This is partly because there is too much information available for much sense to be made of it. A clearer 'picture' can be seen if values of the variable are grouped together and the frequency for each group is considered.

Taking 0 to 4 vehicles/minute as the first group, 5 to 9 vehicles/minute as the second group, and so on, gives this frequency table for the grouped data.

x	0–4	5–9	10–14	15–19	20–24
f	11	14	12	7	4

The set of data is clearer in this grouped form and from it we can get a feel for the traffic flow past the survey point.

The disadvantage of grouping data is that information is lost; from the grouped table, for instance, it is not possible to tell how many times 6 vehicles passed the survey point in a minute.

Continuous Variables

Consider this frequency table, which shows the distribution of the masses of some apples.

Mass in grams (to nearest gram)	86	87	88	89	90	91	92	93	94	95	96	97
Frequency	8	5	7	9	10	4	12	15	9	5	8	3

Continuous variables, such as the mass of an apple, cannot be measured exactly. If an apple is recorded as having a mass of 95 grams correct to the nearest gram, then its mass, m grams, is in the range $94.5 \leqslant m < 95.5$. Another apple with a recorded mass of 95 g, also has its mass in the same range, but the two apples are unlikely to have the same masses.

Therefore continuous data is already in groups, or *classes* as they are called in this context.

The values at the extreme ends of the classes are called the *class boundaries*.

The table above can now be presented in this form, which shows the true nature of the information.

Mass in grams, m	Frequency
$85.5 \leqslant m < 86.5$	8
$86.5 \leqslant m < 87.5$	5
$87.5 \leqslant m < 88.5$	7
$88.5 \leqslant m < 89.5$	9
$89.5 \leqslant m < 90.5$	10
$90.5 \leqslant m < 91.5$	4
$91.5 \leqslant m < 92.5$	12
$92.5 \leqslant m < 93.5$	15
$93.5 \leqslant m < 94.5$	9
$94.5 \leqslant m < 95.5$	5
$95.5 \leqslant m < 96.5$	8
$96.5 \leqslant m < 97.5$	3

More sense can be made of the figures if we group them into fewer but bigger classes. The following tables comes from doubling the width of the classes.

class width = ucb - lcb

Mass in grams, m	Frequency
$85.5 \leqslant m < 87.5$	13
$87.5 \leqslant m < 89.5$	16
$89.5 \leqslant m < 91.5$	14
$91.5 \leqslant m < 93.5$	27
$93.5 \leqslant m < 95.5$	14
$95.5 \leqslant m < 97.5$	11

Notice that the upper class boundary of one group is the same as the lower class boundary of the next group. Notice also that the widths of all classes are the same.

When deciding how to group data:

- Decide how many groups to use; the number chosen depends on the nature of the data and the accuracy of the information that is required (remember that grouping loses some of the information), but about five to ten groups is sensible for most purposes. Ideally the groups should be the same width, but there are situations where this is either not possible or not desirable.

- Make sure that the groups do not overlap, i.e. that there is no uncertainty about the group into which an item should be placed. (Notice the use of inequalities in defining the groups above.)

- Make sure that the groups span all the values; there must not be any gaps. If there are no items in a group it can be left as a group with zero frequency or it can be combined with an adjacent group.

- If there are a few items that straggle well outside the main groupings, they must not be ignored but can be included by making the first and/or the last group wider, or even open ended. In the example above, if there were a couple of apples whose masses were 100 g and 120 g, we could add another group, $97.5 \leqslant m < 120.5$. This is considerably wider than the other groups, but has the advantage of keeping a reasonable amount of detail for the bulk of the distribution without increasing the number of groups unreasonably.

Arrange each of the following sets of data in a grouped frequency table, using five to ten groups.

1. Heights, measured correct to the nearest tenth of a millimetre.

    5.5, 5.5, 5.6, 5.6, 5.6, 5.7, 5.7, 5.7, 5.8, 5.8, 5.8, 5.8, 5.9, 5.9,
    5.9, 6.0, 6.0, 6.0, 6.0, 6.0, 6.1, 6.1, 6.1, 6.1, 6.2, 6.2, 6.2, 6.3,
    6.3, 6.3, 6.3, 6.4, 6.4, 6.4, 6.5, 6.5, 6.6, 6.7, 6.7, 6.8, 6.8, 6.9

2. Lengths, measured correct to the nearest five centimetres.

    100, 100, 105, 105, 110, 110, 110, 115, 115, 115, 120, 125, 125, 125, 130,
    130, 135, 135, 140, 140, 140, 140, 145, 145, 145, 145, 150, 150, 150, 150,
    150, 155, 160, 160, 160, 165, 165, 165, 165, 170, 170, 170, 170, 175, 175,
    180, 180, 180, 185, 185, 185, 185, 190, 190, 190, 200

3. Times, measured to the nearest second.

    1, 3, 3, 4, 4, 5, 5, 5, 5, 6, 6, 6, 6, 7, 7, 7, 7, 7, 8, 8, 8, 8,
    9, 9, 9, 9, 9, 9, 10, 10, 10, 10, 10, 10, 11, 11, 11, 11, 11,
    11, 12, 12, 12, 13, 13, 14, 15, 20, 25, 35

4. Number of goals scored in each match by a hockey team.

    0, 0, 0, 0, 0, 0, 1, 1, 1, 1, 1, 1, 1, 1, 2, 2, 2, 2, 2, 2, 3, 3, 3,
    3, 3, 3, 4, 4, 4, 4, 5, 5, 5, 6, 6, 7, 7, 8, 9, 10, 12, 15, 18

5. Suggest two further groupings for the data in question 4 other than the one you have already used. Give any advantages and disadvantages for each of the three possible groupings chosen.

HISTOGRAMS

Grouped data (and this includes all continuous data) can be illustrated with a histogram. A histogram is drawn so that

- a bar represents each group where the *area* of the bar represents the frequency of the items in the group,

- the width of a bar is the width of the class,

- the height of the bar is given by $\dfrac{\text{frequency}}{\text{class width}}$. This is called the *frequency density*.

This histogram illustrates the grouped frequency distribution of the masses of apples given on p. 277.

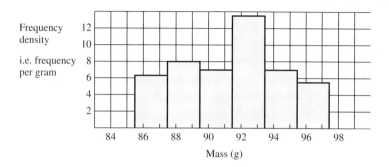

Notice that the class width is 2 units for each bar, so the heights of the bars are half their respective frequencies.

Notice how the vertical axis is labelled. This is because the unit chosen for the class width is a matter of choice. (In the example above we could use 2 grams as the unit class width, in which case the bars would be twice as high.)

For a grouped frequency distribution of car prices, the width of the classes is likely to be in thousands of pounds; in this case it is sensible to use £1000 as the class width; the vertical axis would then be labelled 'frequency per £1000 of price'.
If 'frequency density' only is used as a label, a key describing the unit of density is essential.

Histograms are also used to illustrate grouped frequency distributions of discrete variables. Consider again the grouped frequency distribution given on p. 276.

x	0–4	5–9	10–14	15–19	20–24
f	11	14	9	7	4

Each group contains five integers, so the bar representing each group must be five units wide. If we drew the vertical edges of the bar for the first group at 0 and 4 on the horizontal axis, the bar would be only four units wide and there would be a gap between this bar and the next one which starts at 5. To overcome this problem we use class boundaries that are midway between the end values of adjacent groups. For example, the lower class boundary of the third group is midway between 9 and 10, i.e. 9.5, and the upper class boundary is 14.5. (To get the lower class boundary of the first group and the upper class boundary of the last group, imagine groups beyond these.)

We can now add two rows to the table, one for the class boundaries and one for the frequency density.

x	0–4	5–9	10–14	15–19	20–24
f	11	14	9	7	4
Class	−0.5–4.5	4.5–9.5	9.5–14.5	14.5–19.5	19.5–24.5
Frequency density	2.2	2.8	1.8	1.4	0.8

This histogram illustrates the distribution.

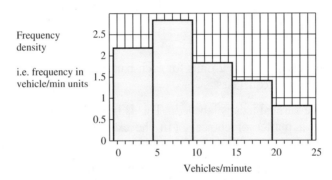

As the area of each bar represents the number of items in that group,

> **the total area of a histogram represents
> the total number of items in the distribution.** or frequency

It also follows that the sum of the areas of the last two bars on the right represents the number of times 15 or more vehicles/minute passed the survey point. Hence

> **the area of a histogram between two values on the horizontal axis
> represents the number of items whose values are in that range.**

Note that if these values do not coincide with the ends of bars, the area between the values can only give an approximation for the number of items. This is illustrated in the examples below.

Examples 18d

1. Eggs are graded by size. This frequency table shows the distribution of the masses of some Size 2 eggs. Draw a histogram to illustrate this distribution.

Mass, w grams	$60 \leqslant w < 63$	$63 \leqslant w < 64$	$64 \leqslant w < 65$	$65 \leqslant w < 66$	$66 \leqslant w < 68$	$68 \leqslant w < 72$
Frequency	9	12	15	17	10	8

The class widths are not all the same. We will add another two lines to the table to show the class widths and frequency densities.

Mass, w grams	$60 \leqslant w < 63$	$63 \leqslant w < 64$	$64 \leqslant w < 65$	$65 \leqslant w < 66$	$66 \leqslant w < 68$	$68 \leqslant w < 72$
Frequency	9	12	15	17	10	8
Class width	3	1	1	1	2	4
Frequency density	3	12	15	17	5	2

We can now draw the histogram

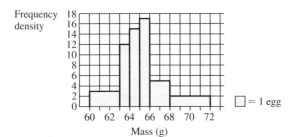

2. This histogram illustrates the distribution of the masses of another batch of Size 2 eggs.

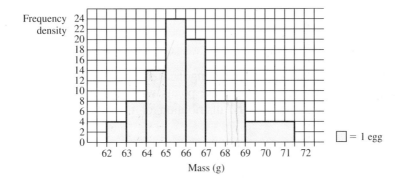

Find the number of eggs with a mass that is

(a) less than 66 grams

(b) more than 68.5 grams.

(a) As the area of each bar represents the frequency, we can get a value for the number of eggs with mass less than 66 g by finding the area of the bars to the left of 66 on the horizontal axis.

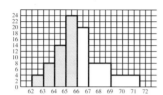

The area required is
$1 \times 4 + 1 \times 8 + 1 \times 14 + 1 \times 24 = 50$
Therefore there are 50 eggs with mass less than 66 g.

(b) 68.5 g is in the middle of a bar. By drawing a vertical line through this value, we can find the area of the histogram to the right of this line to give an approximate value for the number of eggs with a mass more than 68.5 g.

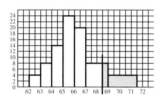

The area required is
$0.5 \times 8 + 2.5 \times 4 = 14$
Therefore there are approximately 14 eggs whose mass is more than 68.5 g.

Notice that the answer to part (b) above is only approximate because, although we know there are 16 eggs with masses in the range $67 \leqslant m < 69$, we do not know how many of those eggs weigh more than 68.5 g; by taking the line three-quarters of the way through the bar representing the class, we have assumed that a quarter of them do.
Looking at the shape of the histogram in the example above, it would be more reasonable to assume that less than a quarter of the eggs in the class have a mass greater than 68.5 g. Therefore the answer given is probably an overestimate.

Frequency Polygons

A frequency polygon is drawn by plotting the frequency density at the midpoint of each class interval and joining these points with straight lines. This is the frequency polygon drawn for the distribution of masses in the first worked example above.

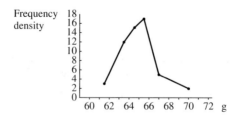

When two distributions are being compared, frequency polygons give a clearer picture of the situation as one polygon can be superimposed on another without obstructing either polygon.

This shows the distribution of the masses of a second batch of eggs superimposed on the first polygon.

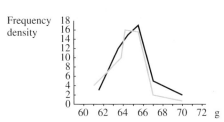

EXERCISE 18d

1. Draw a histogram to illustrate the frequency table drawn up for question 1 in Exercise 18c.

2. Illustrate the frequency table made for question 4 in Exercise 18c with a histogram.

3. This histogram illustrates the distribution of goals scored by another hockey team.

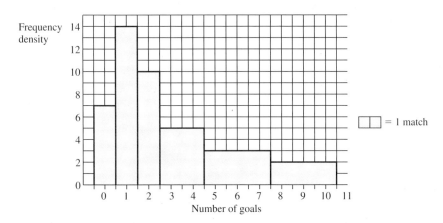

(a) Estimate the number of matches in which the team scored more than 5 goals.

(b) In how many matches did the team score three goals or less?

(c) Draw up a frequency table from this histogram.

4. (a) On the same set of axes, draw frequency polygons for questions 2 and 3. Make sure that you use the same frequency density for each graph.

(b) From the polygons, can you deduce which team had the higher goal average? Give reasons for your answer.

5. The table shows the distribution of share ownership in Forbus Ltd.

Value of shares held (£)	Number of individuals holding shares
1–50	200
51–150	450
151–500	564
501–1000	45
1001–10 000	20
10 001–50 000	8
50 001–100 000	4
100 001–	2

Explain why it would be difficult to illustrate this distribution with a histogram.

CHAPTER 19

SUMMARISING DATA

The Need to Summarise Data

If you are considering various ways of earning a living, one factor of interest is almost certainly the amount of money that you are likely to earn. At the initial stage of enquiry, you would probably not want to see a full, or even a grouped, distribution of earnings. It would be enough to know a single figure giving some central value together with an idea of the spread of earnings about this central figure.

There are several situations where data needs to be summarised into a single representative value together with a measure of its spread.

In this chapter we are going to look at the main ways of locating a representative value and most of them involve a great deal of number crunching. You will need a calculator with statistical functions *and the manual* that goes with it.

MEASURES OF CENTRAL TENDENCY

There are various methods for finding a central value to represent a set of values. These are collectively known as measures of *central tendency.*

The most common measure of central tendency is the *arithmetic mean,* usually called just the mean, and it is defined as follows.

The mean of a set of values is equal to $\dfrac{\textbf{the sum of all the values}}{\textbf{the number of values}}$

If we have a set of n values, $x_1, \ x_2, \ x_3, \ \dots, \ x_n$ then the mean, $\bar{x}$, of these values is given by $\quad \bar{x} = \dfrac{x_1 + x_2 + x_3 + \dots + x_n}{n}$

Using the symbol Σ to mean 'the sum of terms such as' we can define the mean more concisely as $\dfrac{\Sigma x_i}{n}$ where i takes all integer values from 1 to n,

or even more briefly, we can write

$$\bar{x} = \frac{\Sigma x}{n}$$

When a set of values has been arranged in order of size, the value of the middle item is sometimes used to represent the set. This is called the *median* and is defined as follows.

When n values have been arranged in order of size,

the median is the value of the $\dfrac{n+1}{2}$ th item.

If n is an even number, e.g. 24, then $\frac{n+1}{2}$ is $12\frac{1}{2}$. As there is no such item, the median in this case is the mean of the 12th and 13th items.

One other measure of central tendency that is used is the *mode,* where:

The mode of a set of values is the value with the greatest frequency.

Every distribution has a mean and a median, but not necessarily a single mode; a distribution may have more than one mode or even no mode.

Example 19a

This table shows the quarterly sales figures (to the nearest £1000) of a shop for a three-year period.
Find (a) the mean quarterly sales figure
(b) the median   (c) the mode.

Quarter	Sales (£)
1990–1	12 000
2	15 000
3	18 000
4	13 000
1991–1	13 000
2	16 000
3	18 000
4	14 000
1992–1	15 000
2	17 000
3	20 000
4	17 000

(a)   There are 12 quarters so the mean quarterly sales figure is the sum of the sales figures divided by 12.

Mean quarterly sales $\approx$ £188 000 ÷ 12 = £15 666.6...

$\qquad\qquad\qquad\qquad\quad$ = £16 000   (nearest £1000)

The figures in the table are not exact so the calculated mean can only be an estimate.

(b)   We now need the sales figures listed in order of size.

£12 000, 13 000, 13 000, 14 000, 15 000, 15 000,
16 000, 17 000, 17 000, 18 000, 18 000, 20 000

There are 12 figures here so the median is the $(12 + 1)/2$ th, i.e. the $6\frac{1}{2}$ th figure in the list.

Therefore the median is the mean of the 6th and 7th sales figures,

i.e.   £$\frac{1}{2}$(15 000 + 16 000) = £15 500

(c)   There is no single mode; there are four figures that each occur twice, i.e. £13 000, £15 000, £17 000, £18 000

EXERCISE 19a

Find the mean, median and mode of the sets of data in questions 1 to 3.

1. This is the set of (unordered) class marks in an end-of-session maths test.

47, 54, 43, 50, 61, 36, 65, 54, 50, 43, 62, 59,
36, 45, 45, 33, 53, 67, 21, 45, 50, 36, 58

2. The number of packets of fancy napkins made each month by a company is shown in the table.

Jan	Feb	Mar	Apr	May	Jun	Jly	Aug	Sep	Oct	Nov	Dec
430	560	450	550	760	430	525	110	635	450	800	950

3. This list is the sorted values of the lengths, in centimetres correct to 1 dp, of some cucumbers.

32.5, 40.1, 40.3, 44.5, 49.8, 50.6, 51.2, 53.4, 55.5, 56.1, 56.4, 57.2,
57.4, 58.0, 58.7, 58.8, 58.9, 59.1, 59.3, 59.4, 60.1, 60.3, 60.5, 62.8

4. In the six months following those listed in question 2, the mean number of fancy napkins made each month was 520.
Find the mean number of packets made each month for the whole 18-month period.

5. This table is taken from the *Statistical Abstract* for the United Kingdom (which included Ireland) published in 1886.

Population in the United Kingdom	
Year	Estimate
1871	31 556 000
1872	31 874 000
1873	33 178 000
1874	33 502 000
1875	33 839 000
1876	33 200 000
1877	33 576 000
1878	33 949 000
1879	34 303 000
1880	34 623 000

(a) Find the mean population from the data given in this table.

(b) Plot this information on a graph and join the points with straight lines. (Use values to the nearest 100 000 for the population)

Now calculate the average population for 1871 to 1874, 1872 to 1875, 1873 to 1876, and so on. (These are called moving averages.) Plot this information on your graph and join the points with straight lines. Compare the second graph with the first. Give a reason why a plot showing moving averages for figures that vary over time may have advantages over using the raw figures.

(c) The population in the UK is now about 55 million. Is it reasonable to use these figures to compare the populations then and now?

6. A girl is given a short test each week in mathematics and after each one, she updates the values of the mean, median and mode for all the tests taken so far in the year. Each test is marked out of 10 and after 10 tests the mean is 6.8, the median is 6 and the mode is 5. Her score in the next test is 9 which is her highest mark yet. What effect does this mark have on the mean, median and mode?

Finding the Mean from a Frequency Table

This frequency table, which appears in Chapter 18, gives the frequency of various numbers of heads that showed when four coins were tossed.

Number of heads, x	0	1	2	3	4
Frequency, f	6	18	38	23	5

The information in this table is not grouped so we can find the exact number of heads that showed for any value of x, e.g. 2 heads appeared 38 times, giving 2×38 heads.

Thus we can find the total number of heads by multiplying each value of x by its frequency and summing the results, i.e. Σfx. The number of tosses is the sum of the frequencies, i.e. Σf.

Then $\bar{x}$, the mean number of heads per toss, is given by $\dfrac{\Sigma fx}{\Sigma f}$

i.e. $\quad \bar{x} = \dfrac{\Sigma fx}{\Sigma f} = \dfrac{0 \times 6 + 1 \times 18 + 2 \times 38 + 3 \times 23 + 4 \times 5}{6 + 18 + 38 + 23 + 5} = 2.03\ldots$

The mean value is accurate in this case because all the information is available from the table.

The mean value should normally be calculated by using the statistical functions in a calculator. Enter the data directly from the table (consult your manual to find out how this is done); the $\bar{x}$ button gives the mean.

Now consider this grouped frequency table, also from Chapter 18, showing the distribution of masses of some apples.

Mass in grams, m	Frequency
$85.5 \leqslant m < 87.5$	13
$87.5 \leqslant m < 89.5$	16
$89.5 \leqslant m < 91.5$	14
$91.5 \leqslant m < 93.5$	27
$93.5 \leqslant m < 95.5$	14
$95.5 \leqslant m < 97.5$	11

We know that there are 13 apples whose masses are between 85.5 g and 87.5 g but we do not know the mass of any one of these apples. If we take the average mass of these 13 apples as being halfway between the class boundaries 85.5 g and 87.5 g, which is 86.5 g, then we can *estimate* the total mass of these 13 apples as

$$13 \times 86.5 \text{ g} = 1124.5 \text{ g}$$

Repeating this for the other classes and summing the results gives an estimate for the total mass of all the apples. Dividing by the sum of the frequencies then gives an estimate of the mean mass.

Hence for a grouped frequency distribution, $\bar{x} \approx \dfrac{\Sigma fx}{\Sigma f}$ where x is the mid-class value.

To sum up,

the mean value of a frequency distribution is given by

$$\bar{x} = \frac{\Sigma fx}{\Sigma f}$$

When the frequency distribution is grouped, x is the mid-class value, and the result is approximate.

Example 19b

Find the mean mass of the eggs in this distribution.

Mass, w grams	$60 \leqslant w < 63$	$63 \leqslant w < 64$	$64 \leqslant w < 65$	$65 \leqslant w < 66$	$66 \leqslant w < 68$	$68 \leqslant w < 72$
Frequency	9	12	15	17	10	8

This is a grouped distribution so we need the mid-class values, and we add another row to the table to display these.

Mass, w grams	$60 \leqslant w < 63$	$63 \leqslant w < 64$	$64 \leqslant w < 65$	$65 \leqslant w < 66$	$66 \leqslant w < 68$	$68 \leqslant w < 72$
Frequency	9	12	15	17	10	8
Mid-class value	61.5	63.5	64.5	65.5	67	70

The frequencies and mid-class values can be entered into a calculator in statistics mode and the mean found directly. Alternatively they can be calculated as follows, again using a calculator but in ordinary computation mode.

$$\bar{x} \approx \frac{\Sigma fx}{\Sigma f}$$

$$= \frac{9 \times 61.5 + 12 \times 63.5 + 15 \times 64.5 + 17 \times 65.5 + 10 \times 67 + 8 \times 70}{9 + 12 + 15 + 17 + 10 + 8} = 65.16\ldots$$

The mean mass is approximately 65.2 g

1. The lengths of some leaves are given in the table.

Length, l mm	Frequency
$5.45 \leqslant l < 5.85$	5
$5.85 \leqslant l < 6.25$	9
$6.25 \leqslant l < 6.65$	15
$6.65 \leqslant l < 7.05$	19
$7.05 \leqslant l < 7.45$	16
$7.45 \leqslant l < 7.85$	8
$7.85 \leqslant l < 8.25$	2

Find the mean length.

2. Find the mean number of pupils on a school roll from the figures given in the table.

Number on school roll	Number of schools
0–199	20
200–399	45
400–599	34
600–799	53
800–999	21
1000–1999	10
2000–3999	4

3. This histogram, from Example 18d, shows the distribution of the masses of some Size 2 eggs.

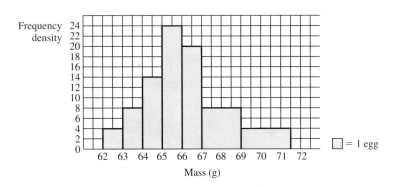

(a) How many eggs are represented in this distribution?

(b) Find the mean mass of an egg in this group of eggs.

4. An analysis of the ages of people taking the fixed price menu on offer at the Cartorelli restaurant one weekday evening is shown in the table.

Age profile of customers	
Age group (years)	Number of people
1–20	10
21–40	48
41–60	53
61 and over	25

(a) Explain why it is difficult to find the mean age from these figures.

(b) Estimate the mean age, stating clearly any assumptions that you make.

The Mode of a Grouped Frequency Distribution

Finding the mode of an ungrouped frequency distribution is easy: the item with the highest frequency is the mode.

It is not possible to identify the mode from grouped data; all we can do is to give the modal class, or classes. However, which group forms the modal class will depend on how the groups are chosen, especially if groups of uneven widths are used, so modal class is useless as a measure of central tendency for grouped data. However the words 'modal', 'bimodal', etc., are useful when describing the shape of a distribution.

Finding the Median from a Grouped Frequency Table

Finding the median of ungrouped data from a frequency table is straightforward. Here is the table from Chapter 18 again.

Number of heads showing when 4 coins were tossed, x	0	1	2	3	4	
Frequency, f	6	18	38	23	5	Total: 90

Since there are 90 tosses, the median is the 45.5th number of heads; counting the frequencies shows this is in the list of 2 heads, so the median number of heads is 2.

Now consider this grouped distribution of the masses of apples (from Chapter 18, p. 277)

Mass in grams, m	Frequency	
$85.5 \leqslant m < 87.5$	13	
$87.5 \leqslant m < 89.5$	16	
$89.5 \leqslant m < 91.5$	14	
$91.5 \leqslant m < 93.5$	27	
$93.5 \leqslant m < 95.5$	14	
$95.5 \leqslant m < 97.5$	11	Total $= 95$

The median mass is the 48th mass, and from the table, it lies in the group $91.5 \leqslant m < 93.5$. If we *assume* that the masses are evenly distributed throughout the class, we can find an estimate for the median by calculating how far through the class the 48th mass lies.

Now there are 43 masses below the lower class boundary (l.c.b.), so the 48th mass is the 5th mass through the 27 masses in the class,

i.e. it is $\frac{5}{27}$ th of the way through the 2 gram interval $91.5 \leqslant m < 93.5$

so its mass is $(91.5 + \frac{5}{27} \times 2)$ grams, i.e. $91.8 \ldots$ g $= 92$ g (nearest g)

This method of calculating an estimate for a median is called *linear interpolation*. It can be summarised as follows.

**If the median is the k th item of m items in a class,
the value of the median is approximately**

$$(\text{l.c.b.}) + \frac{k}{m} \times (\text{class width})$$

CUMULATIVE FREQUENCY

An easier way of estimating the median is to read its value from a cumulative frequency graph, but first we have to define what we mean by cumulative frequencies.

**A cumulative frequency gives the total number of items
from the lowest in the distribution
up to the upper class boundary of each group.**

Here is the grouped frequency table for the masses of apples again. Next to it is the cumulative frequency table for the same distribution.

Mass, m grams	Frequency
$85.5 \leqslant m < 87.5$	13
$87.5 \leqslant m < 89.5$	16
$89.5 \leqslant m < 91.5$	14
$91.5 \leqslant m < 93.5$	27
$93.5 \leqslant m < 95.5$	14
$95.5 \leqslant m < 97.5$	11

Mass (grams)	Cumulative frequency
< 85.5	0
< 87.5	13
< 89.5	29
< 91.5	43
< 93.5	70
< 95.5	84
< 97.5	95

A cumulative frequency table can be illustrated by plotting the cumulative frequency against the upper class boundary and joining the points with straight lines. It is called a *cumulative frequency polygon*.

This is the cumulative frequency polygon for the distribution above.

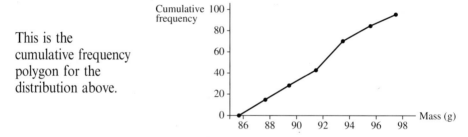

Note that the construction of the cumulative frequency polygon assumes that the items are evenly distributed within each group.

Finding the Median from a Cumulative Frequency Polygon

Now we can find an estimate for the value of the median from the graph, by finding 48 (i.e. half way between 1 and 95) on the vertical scale and reading its corresponding value off the horizontal scale.

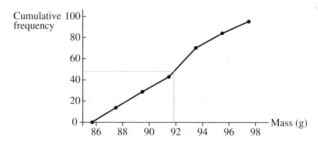

This gives the median as slightly less than 92 g, i.e. 92 g to 2 sf.

Note that this method is equivalent to linear interpolation, which we can see from the relevant part of the graph shown in the diagram below.

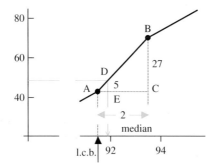

Using similar triangles,

$$\frac{AE}{AC} = \frac{DE}{BC} \quad \Rightarrow \quad AE = \frac{5}{27} \times (\text{class width})$$

$$\therefore \quad \text{median} \approx \text{l.c.b.} + \frac{5}{27} \times (\text{class width})$$

Cumulative Frequency Curves

If instead of using straight lines, we draw a smooth curve through the points on the cumulative frequency diagram, we produce a cumulative frequency curve.

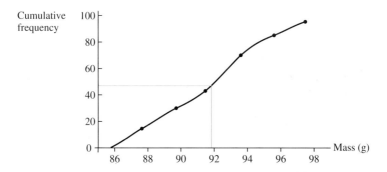

From the curve, the median is seen to be about 91.9 g, which agrees well with the value given by linear interpolation. In the context of this distribution, however, a measurement to the nearest gram is as accurate as is reasonable because the original data was given only to this degree of accuracy.

The advantage of a curve is that it *probably* gives a more realistic representation of the distribution than the polygon, as it does not assume that the items are evenly spread throughout a class and so avoids the jerkiness.

Note also that, for a grouped frequency distribution, the median is usually found by using $\dfrac{n}{2}$ $\left(\text{rather than } \dfrac{n+1}{2}\right)$

EXERCISE 19c

Questions 1 to 4 in this exercise refer to questions in Exercise 19b.

1. Estimate, using linear interpolation, the median of each distribution given in questions 1 to 3.

2. For questions 1 and 3

   (a) construct a cumulative frequency table

   (b) draw a cumulative frequency curve

   (c) estimate the median from the curve.

   (d) Why would parts (a) to (c) give problems if question 2 in Exercise 19b was included? How could you use a cumulative frequency curve to find the median of this distribution?

3. Find the median of the distribution given in question 4. Why does finding the median of this distribution not have the problem associated with finding the mean?

4. Shivani measured, in centimetres correct to 1 dp, the heights of heels of 25 different pairs of women's shoes and recorded the information below.

   0.4, 0.6, 1.0, 1.0, 1.3, 1.2, 1.3, 1.5, 1.3, 2.0, 2.0, 2.5, 2.2,
   2.5, 3.0, 3.3, 3.0, 3.4, 3.8, 4.0, 4.5, 6.0, 5.5, 5.7, 8.0, 8.0

   (a) Make a frequency table using classes 0 up to 2 cm, 2 cm up to 4 cm, and so on.

   (b) Add another row to the table for cumulative frequencies.

   (c) Draw a cumulative frequency curve and use it to estimate the median heel height.

Mean, Median or Mode?

If you need to choose a central value to represent some data, which of mean, median and mode you use will depend to a large extent on context.

Consider this situation. The manager of a shop selling shoes and other related goods needs to provide data on various aspects of the operation of his shop for the owners of the chain of shops.
One set of information would be the number of items sold for each category of goods. Suppose this table shows such information.

Number of items sold in week 34				
Fashion shoes	Sports shoes	Slippers	Tights and socks	Bags
256	350	104	500	75

If the owners want to know which category of goods sold in largest numbers, the manager would use the mode, which in this case is the number of tights and socks.

Another set of information would be weekly turnover, e.g.

Weekly takings (£)					
Week 1	Week 2	Week 3	Week 4	Week 5	Week 6
2805	3657	4741	3378	5681	2923

This data would almost certainly be summarised to give the mean weekly takings, which in this case is £3864. The mean uses all the figures, and averages them out to one figure.

Now consider the figures giving the earnings of each employee for one week.

Week 25: Pay Bill					
A	B	C	D	E	F
£56	£60	£70	£120	£120	£450

If the owners of the chain want figures for the average pay bill per employee per shop, the mean would be appropriate as it takes account of all the figures.

However, for someone looking for work in such a shop the mean pay, £146, is not representative as it is higher than all but one of the figures in the table. The median (£95) gives a better representative figure and would provide a fairer summary for a potential employee. This is because it is not affected by the one large sum of £450.

In order to illustrate some general ideas only a few figures have been used in the tables on the previous page so the detail is easily understandable without any need for a summary. If, on the other hand, similar information was considered about a large supermarket selling several hundred different items and with many staff, it can be imagined that the full figures would contain too much detail to be easily comprehensible and that some form of summary would be needed.

EXERCISE 19d

1. A salesman produces several sets of figures for his own and his employer's information. Extracts are produced below. For each of these, state, with a reason, what would be the best measure of central tendency to use.

(a)

Week Ending 03/02/93			Number of items sold		
Office Tables	Office Chairs	Filing Cabinets	Computer Stands	Desks	Waste Paper Bins
45	96	186	31	84	2500

(b)

Week Ending 03/02/93			Turnover from items sold (£)		
Office Tables	Office Chairs	Filing Cabinets	Computer Stands	Desks	Waste Paper Bins
4500	4800	7998	1953	7890	3750

(c)

Weekly commission earned (£)					
w/e 03/02/93	w/e 10/02/93	w/e 17/02/93	w/e 24/02/93	w/e 03/03/93	w/e 10/03/93
329	275	25	412	315	354

(d)

Units of alcohol consumed					
w/e 03/02/93	w/e 10/02/93	w/e 17/02/93	w/e 24/02/93	w/e 03/03/93	w/e 10/03/93
15	10	50	10	12	14

2. The column graph represents some data.

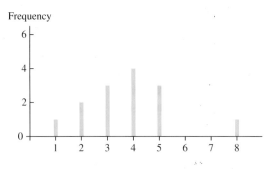

Give a measure of central tendency for the data and justify your choice if the diagram represents

(a) the scores when an ordinary die is thrown

(b) the number of £1 coins in each of several collecting tins

(c) the amount of spending money, in £, given to each of several 14-year-olds.

3. In question 2, part (c) give a circumstance when the measure of central tendency that would be appropriate would be

(a) the mean   (b) the median   (c) the mode.

4.

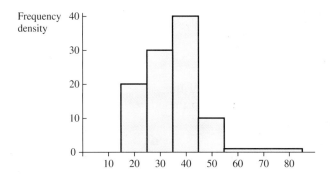

State with reasons, but without working it out, which measure of central tendency it would be appropriate to use to summarise this information if it represents the distribution of

(a) heights, in cm, of one-year-old apple tree seedlings

(b) ages, in years, of students enrolling for a course in word-processing skills

(c) number of pizzas served per day by a fast-food restaurant.

RANGE

Data can also be summarised by giving a figure that represents the spread of the values in a distribution.
The crudest measure of spread is the range, where

**the range of a set of values
is the difference between the highest and lowest values in the set.**

The range can be expressed either by giving the highest and lowest figures or by stating the difference between them.

Quarter	Sales (£)
1990–1	12 000
2	15 000
3	18 000
4	13 000
1991–1	13 000
2	16 000
3	18 000
4	14 000
1992–1	15 000
2	17 000
3	20 000
4	17 000

From this table the quarterly sales range from £12 000 to £20 000 so we say that the range is £8000.

For a grouped frequency distribution, we assume that the lowest value is the lowest class boundary and the highest value is the highest class boundary.

Mass, m grams	Frequency
$85.5 \leqslant m < 87.5$	13
$87.5 \leqslant m < 89.5$	16
$89.5 \leqslant m < 91.5$	14
$91.5 \leqslant m < 93.5$	27
$93.5 \leqslant m < 95.5$	14
$95.5 \leqslant m < 97.5$	11

For this distribution, the masses of the apples go from 85.5 g to 97.5 g so the range is 12 g.

Now consider the distribution of pay considered on page 297.

A	B	C	D	E	F
£56	£60	£70	£120	£120	£450

The range of pay is £394. Without access to the original data, the range gives a misleading impression of the spread of pay, as all the values except the largest are at the lower end of the distribution. (Knowing also that the median was £95 would give an indication that most values were towards the lower end of the range.)

The range of a distribution is easily understood but, as we see from the examples above, on its own it gives no indication about the spread of values within the range.

Even when both the median and the range are known, it tells us very little about the shape of the distribution as we can see from these diagrams.

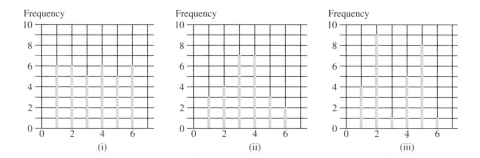

They all have a range of 5 and a median of 3.5; the shape of each is roughly symmetrical but the similarity ends there.
We need more information than the range and median before we can make any sensible speculations about the shape of a distribution.

QUARTILES

The median is the value of the middle ranked item in a distribution, i.e. half of the items in the distribution have a value less than the median and half have a value greater than the median.
We can further divide the number of ranked items in a distribution into quarters. We then define the *lower quartile*, Q_1, as a value such that one-quarter of the items have a value less than or equal to the lower quartile, where,

when *n* items have been arranged in ascending order of size,

the *lower quartile* is the value of the $\dfrac{n+1}{4}$th item.

For a grouped distribution, Q_1 is the $\dfrac{n}{4}$th value.

Similarly, the *upper quartile, Q_3* , is a value such that three-quarters of the items have values less than or equal to it.

When *n* items have been arranged in ascending order of size,

the *upper quartile* is the value of the $\dfrac{3(n+1)}{4}$ th item.

For a grouped distribution, Q_3 is the $\frac{3n}{4}$ th value.

The simplest way to identify the quartiles is from a cumulative frequency curve. The cumulative frequency curve for the distribution of the masses of apples (p. 277) is given overleaf, showing the quartiles.
Reading from the diagram, the lower quartile is just below 89 grams and the upper quartile is between 93.5 grams and 94 grams.

The difference between the upper quartile and the lower quartile
is called the *interquartile range*.

Therefore for the distribution overleaf, the interquartile range is 5 grams to the nearest gram. This means that the weights of the middle half of the masses are within at most 5 grams of each other..

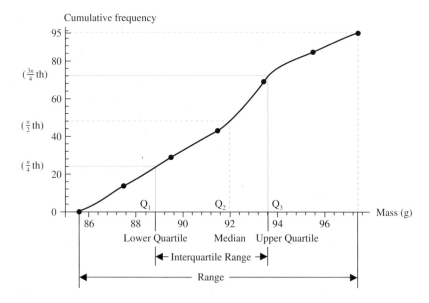

When there are only a few items at the lower and/or higher ends of a distribution, the interquartile range often gives a better indication of spread than the full range because it discounts the effect of these items on the range.

This cumulative frequency polygon illustrates the distribution of the masses of 1000 apples. Explain why the interquartile range might be a better indicator of the spread of the masses than the range.

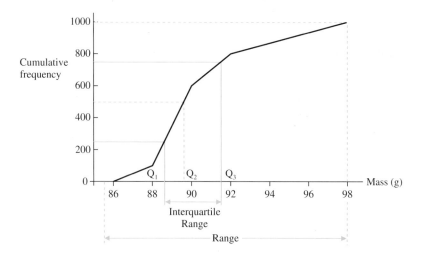

There are 1000 items. Q_1 is the value of the 250 th item and Q_3 is the value of the 750 th item.

From the diagram, the lower quartile is 88.8 g, the median is 89.6 g, and the upper quartile is 91.4 g.
The heaviest quarter of the distribution has a range greater than the remaining three quarters of the distribution. This indicates that there are relatively few apples that are heavier than the bulk of the apples. Using the interquartile range (i.e. the middle half) as a measure of dispersion ignores these few heavier apples and is probably more representative of the variation in mass of the majority of the apples.

For the distributions in questions 1 to 4

(a) draw (i) a histogram (ii) a cumulative frequency polygon

(b) read the values of the median, the range and the interquartile range from your cumulative frequency curve.

1. This table shows the heights, to the nearest cm, of some five-week-old seedlings.

Height (cm)	14	15	16	17	18	19	20	21
Frequency	5	8	15	21	19	14	8	6

2. The results of an examination in science are shown in this table.

Mark	0–9	10–19	20–29	30–39	40–49	50–59	60–69
Frequency	2	3	5	10	25	40	10

3. The lengths, to the nearest millimetre, of some worms are recorded in this table.

Length, mm	0	1	2	3	4	5	6	7
Frequency	8	20	15	10	4	3	1	1

4. Some screws were selected from a box of mixed screws and their lengths were measured. The results are recorded in the table.

Length, correct to nearest cm	1	2	3	4	5	6
Frequency	20	26	23	28	25	22

5. Use the cumulative frequency diagram to find the median and the interquartile range.

(a)

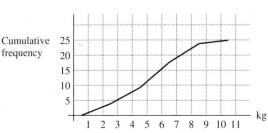

(b)

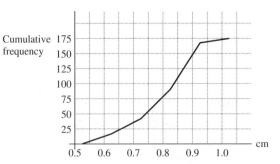

PERCENTILES

Although the interquartile range has the advantage of discounting the 'stretching' effect that a small number of items can have on the range, it also ignores 50% of the items. Therefore it is sometimes useful to consider sections other than quarters of a distribution. This is conventionally done by dividing the number of ranked (arranged in ascending order of size) items in a distribution into hundredths, i.e. percentages.

This diagram shows the cumulative percentage frequency of the masses of the 1000 apples considered in Example 19e.

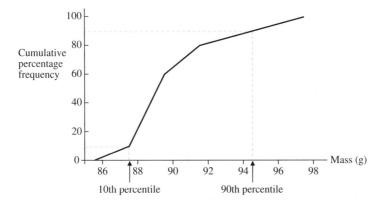

From this diagram we can see that 10% of the items have a value less than or equal to 87.6 g. This value is called the 10th *percentile*. Similarly, 90% of the items have a value less than or equal to 94.4 g, so the 90th percentile is 94.4 g.

For any distribution,

**if n % of the items have a value
less than or equal to the value of a particular item,
that item is the n th percentile**

Using percentiles gives more flexibility in the choice of a section for which to give the range.

From the cumulative percentage frequency diagram above, it is clear that the heaviest 10% of the apples have a large range compared to any other 10% section of the distribution. In this case we could use the range of the central 80% to give a reasonable indication of the spread of the masses, i.e. the range from the 10th to the 90th percentile, which is 6.6 g. This has the advantage over the interquartile range of including more of the distribution while keeping the advantage of excluding the relatively few heavier items that make the full range unrepresentative of the bulk of items in the distribution.

We could also decide to give the range of any other section, e.g. the middle 90%, or even the lower 95%, of the distribution.

However it must be remembered that the 'range of a distribution' is taken to mean the *full* range, so it is important, when *less* than the full range is given, that *the section of the distribution for which the range is given, is stated.*

Percentiles are also used to give an idea of the ranking of a particular item in a distribution. In an intelligence test, for example, a particular score of 130 might be said to be above the 95th percentile; this means that more than 95% of scores obtained from that test are below 130.

EXERCISE 19f

1. Use the cumulative frequency polygons drawn for the following questions from Exercise 19e to find the specified range.

   (a) Question 1; the middle 80%

   (b) Question 2; the lower 90%

   (c) Question 4; the middle 90%

2. For each part of question 1, state whether the range asked for is reasonably representative of the bulk of items in the distribution. If your answer is no, suggest a section whose range is more representative. Give reasons for your answers.

3. This frequency polygon illustrates a distribution of 2000 scores from a verbal reasoning test.

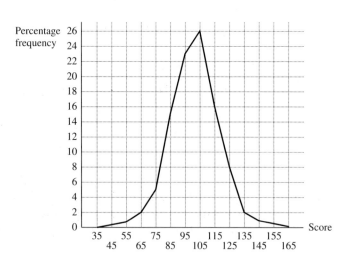

(a) Draw a cumulative percentage frequency curve, using a big enough scale on the vertical axis to enable integer percentages to be read. (Remember that a frequency polygon is drawn by joining the midpoints of the bars in a histogram; use this to find the class widths and the percentage frequency in each class.)

(b) From your curve find approximate values for

(i) the median

(ii) the 80th percentile

(iii) the 99th percentile.

(c) Find, approximately, the percentage of the individuals whose score was

(i) more than 140

(ii) less than 120

(iii) more than 100.

Mean Deviation

As a measure of spread, the range has serious disadvantages; it uses only two values from the distribution and is unduly affected by a few extreme values. Using the range of a central section overcomes the second problem but still takes account of only two values. By finding the deviation (i.e. difference) of each value from some central value (usually the mean) and then finding the mean of these values, averages out the spread.

Now if x_i is a value in the distribution, its deviation from the mean, $\bar{x}$, can be written $|x_i - \bar{x}|$.

Then the mean deviation of n values from the mean is given by

$$\frac{\Sigma |x_i - \bar{x}|}{n}$$

Note that the modulus sign is necessary because some values in the distribution will be less than $\bar{x}$ and, for these values, $x_i - \bar{x}$ is negative whereas the *difference* between two numbers is positive. Sometimes $|x_i - \bar{x}|$ is called the *absolute* value of $x_i - \bar{x}$, and may be denoted by $\text{abs}(x_i - \bar{x})$.

STANDARD DEVIATION

The mean deviation is easily understood, but it is very difficult to work with in expressions and equations because of the modulus sign. Another way of overcoming the problem that $x_i - \bar{x}$ can be either positive or negative is to square $x_i - \bar{x}$ and then find the mean of these squares, i.e. $\dfrac{\Sigma(x_i - \bar{x})^2}{n}$.

This gives the mean square deviation which is called the *variance*.

The square root of this expression is the *root mean square deviation* which is known as the *standard deviation*.

The standard deviation of n values, $x_1, x_2, \ldots, x_n$, is given by

$$\sqrt{\frac{\Sigma(x_i - \bar{x})^2}{n}} \quad \textbf{where } \bar{x} \textbf{ is the mean value.}$$

This formula can be simplified to make the calculations easier.

Now $\quad (x_i - \bar{x})^2 = x_i^2 - 2x_i\bar{x} + \bar{x}^2$

So $\quad \dfrac{\Sigma(x_i - \bar{x})^2}{n} = \dfrac{\Sigma x_i^2}{n} - \dfrac{\Sigma 2 x_i \bar{x}}{n} + \dfrac{\Sigma \bar{x}^2}{n}$

But $\quad \dfrac{\Sigma 2 x_i \bar{x}}{n} = 2 \dfrac{\Sigma x_i}{n} \bar{x} = 2(\bar{x})^2 \quad$ since $\quad \dfrac{\Sigma x_i}{n} = \bar{x}$

and $\quad \dfrac{\Sigma \bar{x}^2}{n} = \dfrac{n\bar{x}^2}{n} = \bar{x}^2 \quad$ (there are n values of x so n brackets contain $\bar{x}$)

Therefore $\quad \dfrac{\Sigma(x_i - \bar{x})^2}{n} = \dfrac{\Sigma x_i^2}{n} - 2\bar{x}^2 + \bar{x}^2 = \dfrac{\Sigma x_i^2}{n} - \bar{x}^2$

Hence $\quad$ **standard deviation** $= \sqrt{\dfrac{\Sigma(x_i - \bar{x})^2}{n}} = \sqrt{\dfrac{\Sigma x_i^2}{n} - \bar{x}^2}$

The second version of the formula for standard deviation (s.d.) is easier to work with and, when $\bar{x}$ is not exact, it has another advantage over the first version. This is because the value used for $\bar{x}$ is rounded, i.e. will contain an error, and the first version magnifies that error more than the second version.

Example 19g

Find the standard deviation of the quarterly sales figures given in this table.

Quarter	Sales (£)
1990–1	12 000
2	15 000
3	18 000
4	13 000
1991–1	13 000
2	16 000
3	18 000
4	14 000
1992–1	15 000
2	17 000
3	20 000
4	17 000

The mean value is calculated in Example 19a

The mean value of these sales is £15 667 (nearest £).

Using the second version of the formula for s.d., we need the square of each value. Adding another column to the table for these values helps.

Quarter	Sales (£)	(Sales)^2
1990–1	12 000	144×10^6
2	15 000	225×10^6
3	18 000	324×10^6
4	13 000	169×10^6
1991–1	13 000	169×10^6
2	16 000	256×10^6
3	18 000	324×10^6
4	14 000	196×10^6
1992–1	15 000	225×10^6
2	17 000	289×10^6
3	20 000	400×10^6
4	17 000	289×10^6
	total:	3010×10^6

Using   s.d. $= \sqrt{\dfrac{\Sigma x_i^2}{n} - \bar{x}^2}$,   gives   s.d. $= £\sqrt{\dfrac{3010 \times 10^6}{12} - (15\,667)^2}$

$$= £2319$$

$$= £2300 \quad (\text{nearest } £100)$$

Note that the value of the standard deviation can, and usually should, be found directly from a calculator, without needing to know the mean value. Use your calculator to check the answer given.

EXERCISE 19g

The following distributions are from Exercise 19a. Find the standard deviation of each one.

1. This is the set (unordered) of class marks in an end-of-session maths test.

    47, 54, 43, 50, 61, 36, 65, 54, 50, 43, 62, 59,
    36, 45, 45, 33, 53, 67, 21, 45, 50, 36, 58

2. The number of packets of fancy napkins made each month by a company is shown in the table.

Jan	Feb	Mar	Apr	May	Jun	Jly	Aug	Sep	Oct	Nov	Dec
430	560	450	550	760	430	525	110	635	450	800	950

3. This list is the sorted values of the lengths, in centimetres correct to 1 dp, of some cucumbers.

    32.5, 40.1, 40.3, 44.5, 49.8, 50.6, 51.2, 53.4, 55.5, 56.1, 56.4, 57.2,
    57.4, 58.0, 58.7, 58.8, 58.9, 59.1, 59.3, 59.4, 60.1, 60.3, 60.5, 62.8

Finding the Standard Deviation from a Frequency Table

Consider again this frequency table, which appears in Chapter 18, showing the frequency of various numbers of heads obtained when four coins were tossed.

Number of heads, x	0	1	2	3	4
Frequency, f	6	18	38	23	5

The mean, $\bar{x}$, is 2.03.

To use $\text{s.d.} = \sqrt{\dfrac{\Sigma x_i^2}{n} - \bar{x}^2}$, we need to evaluate Σx_i^2 and n.

For Σx_i^2 we find (6 lots of 0^2) + (18 lots of 1^2) + ..., which is $\Sigma f x^2$. and n is the total number of tosses, i.e. Σf.

So the formula for s.d. becomes $\sqrt{\dfrac{\Sigma fx^2}{\Sigma f} - \bar{x}^2}$

The statistical functions in a calculator should normally be used to find the standard deviation. The data is entered and the σ button then gives the standard deviation directly; there is no need to find $\bar{x}$ first.

If you have to calculate the standard deviation without using statistical functions, it is easier to keep track of the calculations needed if columns are added to the table, one each for values of x^2 and fx^2 with a row for totals at the end,

i.e.

	x	f	x^2	fx^2
	0	6	0	0
	1	18	1	18
	2	38	4	152
	3	23	9	207
	4	5	16	80
Total		90		457

Then s.d. $= \sqrt{\dfrac{\Sigma fx^2}{\Sigma f} - \bar{x}^2} = \sqrt{\dfrac{457}{90} - (2.03)^2} = 0.978\ldots = 0.98$ (2 sf)

The value of $\bar{x}$ is not exactly 2.03, so the value calculated for s.d. is not exact either.

If we need to find the standard deviation of a grouped frequency distribution, we take the values of x as the mid-class values. In this case the value found for the s.d. is an estimate.

To sum up,

the standard deviation of a frequency distribution is given by

$$\sqrt{\dfrac{\Sigma fx^2}{\Sigma f} - \bar{x}^2}$$

**where, for a grouped distribution, x is the mid-class value
in which case the result is an estimate.**

Example 19h

(a) Estimate the mean and standard deviation of this distribution of the weights, in
     kg to the nearest kg, of the male rowers taking part in an international
     competition in 1991.

Weight (kg)	71–80	81–90	91–100	101–110	111–120	121–130
Frequency	8	33	45	29	15	3

(b) The mean weight of the male rowers taking part in the same competition in
     1990 was 97.1 kg with standard deviation 9.3 kg. Compare the two
     distributions.

(a) Using the statistical functions in a calculator gives $\bar{x} = 96.9$ (3 sf)

     and   s.d. $= 11.6$ (3 sf)

     Therefore the estimated mean weight is 97 kg (nearest kg) and the
     estimated standard deviation is 12 kg (nearest kg).

(b) The mean weights are very close (both 97 kg to the nearest kg). The
     standard deviation weight in 1990 is less than that in 1991, so the dispersion
     is less, i.e. fewer weights in 1990 are much above or below the mean weight.

EXERCISE 19h

Estimate the standard deviation of each of the distributions given in questions 1
to 4 of Exercise 19e using a calculator or computer.

5. The mean and standard deviation weights of potatoes grown in field A is
   100 g and 18 g respectively.
   The mean and standard deviation weights of potatoes grown in other fields are
   given below. Compare each distribution with the distribution from field A.

   (a) $\bar{x} = 120$ g, s.d. $= 10$ g   (b) $\bar{x} = 100$ g, s.d. $= 12$ g
   (c) $\bar{x} = 90$ g, s.d. $= 8$ g   (d) $\bar{x} = 140$ g, s.d. $= 25$ g

CHAPTER 20

PROBABILITY

Basic Definitions

Most statistical investigations use only some of all the relevant information. In a general election, for example, a poll on voting patterns might use information from a thousand people (a *sample*) out of all the people who actually voted. Any conclusions as to whether or not the results from a sample apply to all the possible information from which it is drawn, can only be given in terms of likelihood. This is because the members of a properly drawn sample are chosen at *random,* i.e. all members have an equally likely chance of being chosen. We now develop methods to deal with problems concerned with chance events.

Probability is a measure of the likelihood that something happens. However probability can never tell us the number of times that something actually happens. But being able to quantify the likely occurrence of an event is important because most of the decisions that affect our daily life are based on likelihoods and not on certainties. On a personal level, for example, if it is known that it is likely to rain on two days out of five at a place where your are taking a holiday, it does *not* mean that it *will* rain on four days out of a ten-day holiday, but that it would sensible to equip yourself at least with an umbrella.

An Event

An event is a defined occurrence or situation. For example,

- scoring a six on the throw of an ordinary six-sided die,

- being dealt a hand of four aces from an ordinary pack of playing cards,

- more than ten people queuing at a checkout desk at a given time,

- an apple picked from my Bramley apple tree this year weighing more than 500 grams.

A particular event is denoted by a capital letter, e.g. *A, B,* ...

POSSIBILITY SPACE

In each of the examples above, there is an implied set of circumstances from which there are several possible outcomes, including the event described. This set of possible outcomes is called the *possibility space*.

In the first example, the possible outcomes are any of the scores on the throw of a die, so the possibility space is 1, 2, 3, 4, 5, 6.

In the second case, the possibility space is all the possible combinations of four cards taken from a pack of 52 playing cards.

For the third example, it is possible that, at the given time, the number of people queuing is 0, or 1, or 2, or 3, or 4, or ... So the possibility space is 0, 1, 2, 3, 4, ... (There is no upper limit to the number of people who *might be* queuing.)

In the fourth example, the possibility space is the weights of all the apples on my tree this year.

PROBABILITY THAT AN EVENT OCCURS

The probability that an event A happens, is defined as the number of ways in which A can happen expressed as a fraction of the number of ways in which all *equally likely* events, including A, occur.

The term 'equally likely' is important. For example, if a die has been tampered with so that it is more likely to land showing six than a one, then the events 'scoring six' and 'scoring one' with this die are not equally likely. (Such a die is said to be *biased*. A die where each score is equally likely is called *unbiased* or *fair*.)

The probability that an event A occurs is denoted by $P(A)$. Hence

$$P(A) = \frac{\text{number of ways in which } A \text{ occurs}}{\text{number of ways in which all equally likely events, including } A \text{ occur}}$$

This is the basic definition of probability. All other developments of probability theory are derived from this definition and a large number of problems can be solved directly from it.

The top of the fraction is less than or equal to the denominator, so for any event A,

$$0 \leqslant P(A) \leqslant 1$$

If $P(A) = 1$, the event is certain.

If $P(A) = 0$, the event is impossible.

For example, if a pen is taken from a box containing only red pens,

$$P(\text{pen is red}) = 1 \quad \text{and} \quad P(\text{pen is blue}) = 0$$

Probability that an Event Does Not Happen

If a box contains 150 pens, 50 of which are red, and a pen is chosen at random,

then $P(\text{pen is red}) = \frac{50}{150}$

and $P(\text{pen is not red}) = \dfrac{150 - 50}{150} = 1 - \dfrac{50}{150} = 1 - P(\text{pen is red})$

This is true in all cases. Consider a possibility space in which there are n equally likely outcomes and an event A is the result of r of these outcomes. Then A does not happen on $(n - r)$ of these possible outcomes.

Therefore $P(A \text{ does not happen}) = \dfrac{n - r}{n} = 1 - \dfrac{r}{n} = 1 - P(A)$

The event 'A does not happen' is denoted by $\overline{A}$, or A' (read as 'not A').

$$P(\overline{A}) = 1 - P(A) \quad \text{or} \quad P(A) + P(\overline{A}) = 1$$

This relation is useful in the 'at least' type of problem.

Complementary Events

Two events are *complementary* if together they form all the possible outcomes, i.e. one or other of the two events must happen. For example, when a die is thrown, 'scoring an even number' and 'scoring an odd number' are complementary events as, together, they cover all possibilities.
Hence if events A and B are complementary, it follows that B is the same as 'not A' and A is the same as 'not B', i.e. $P(A) + P(B) = 1$

Examples 20a

1. A single card is chosen at random from an ordinary pack of 52 playing cards. What is the probability that the card is

   (a) the ace of spades   (b) an ace?

   (a) There are 52 different cards in the pack and therefore 52 different equally likely possible outcomes. There is only 1 ace of spades.

   $\therefore \qquad P(\text{ace of spades}) = \frac{1}{52}$

   (b) 'An ace' can be the ace of spades, clubs, hearts or diamonds.

   4 out of the 52 possible outcomes result in an ace,

   $\therefore \qquad P(\text{an ace}) = \frac{4}{52} = \frac{1}{13}$

2. If the first card removed from a pack of 52 playing cards is an ace, what is the probability that a second card, drawn at random from the remaining cards, is also an ace?

As the first card removed is an ace, then the second card is drawn from 51 cards, of which 3 are aces.

$$P(\text{second card is an ace}) = \tfrac{3}{51} = \tfrac{1}{17}$$

3. One integer is chosen at random from the set of integers from 1 to 50 inclusive. What is the probability that it is

(a) not prime

(b) at least 4?

(a) It is easier to list and count the prime numbers than the non-prime numbers.

The prime numbers from 1 to 50 are

2, 3, 5, 7, 11, 13, 17, 19, 23, 29, 31, 37, 41, 47

There are 50 integers and, of these, 14 are prime.

$$\therefore \qquad P(\text{prime}) = \tfrac{14}{50} = \tfrac{7}{25}$$

$$\Rightarrow \qquad P(\text{not prime}) = 1 - \tfrac{7}{25} = \tfrac{18}{25}$$

(b) If a number is at least 4, then it can be any number *except* 1, 2 or 3.

$$P(1, 2 \text{ or } 3 \text{ chosen}) = \tfrac{3}{50} \quad \Rightarrow \quad P(\text{at least } 4) = 1 - \tfrac{3}{50} = \tfrac{47}{50}$$

EXERCISE 20a

1. One integer is chosen at random from 1, 2, 3, 4, 5, 6, 7, 8.
   What is the probability that it is (a) prime   (b) divisible by 3?

2. A card is drawn at random from a pack of 52 playing cards.
   What is the probability that it is

   (a) the seven of hearts

   (b) a 'seven'

   (c) neither a 'seven' nor a 'two'?

3. One child is chosen at random from a class of 30 children. The children are numbered 01, 02, ... 29, 30. Lucy is number 12. What is the probability that

   (a) Lucy is chosen

   (b) Lucy is not chosen

   (c) a child whose number is less than Lucy's is chosen?

4. The masses, in grams to the nearest gram, of ten apples are listed below.

  100, 115, 116, 117, 119, 120, 122, 150, 170, 250

One of these apples is chosen at random.

(a) Find the probability that its mass is
  (i) more than 180 g   (ii) less than 118 g   (iii) at least 110 g
(b) Estimate the probability that its mass is more than 120 g and explain why the answer can only be an estimate.

5. There are 50 identical looking packets in a 'lucky dip' tub. Ten of the packets contain sweets, ten contain marbles and the rest contain crayons.

(a) What is the probability that the first packet removed from the tub contains sweets?
(b) If the first packet removed contains crayons, what is the probability that the second packet removed contains sweets?

6. In question 4, if the first apple chosen weighs less than 130 g, what is the probability that an apple chosen at random from the remaining apples weighs more than 130 g?

Listing Possibility Spaces

In the exercise on the previous page, the possibility spaces were small and obvious, so the possible outcomes were easy to count. Sometimes the possible outcomes may be small in number but not obvious, so it is possible to miss some. For example, suppose that two ordinary dice are rolled together. Assuming the dice to be unbiased, the equally likely outcomes are all the possible scores on one die combined with all the possible scores on the other. Tabulating the scores ensures that no combination is missed, e.g.

		Black die					
		1	2	3	4	5	6
	1	1,1	1,2	1,3	1,4	1,5	1,6
	2	2,1	2,2	2,3	2,4	2,5	2,6
Red die	3	3,1	3,2	3,3	3,4	3,5	3,6
	4	4,1	4,2	4,3	4,4	4,5	4,6
	5	5,1	5,2	5,3	5,4	5,5	5,6
	6	6,1	6,2	6,3	6,4	6,5	6,6

The body of the table shows the 36 possible outcomes when both dice are rolled. If the two numbers are added to give a total score then, for example, we can find the probability of getting a score of at least 7. From the table we see there are 21 outcomes giving a score of at least 7.

Therefore $P(\text{score at least } 7) = \frac{21}{36} = \frac{7}{12}$

EXERCISE 20b

1. Use the possibility space on the previous page to find the probability that

   (a) at least one 6 is rolled

   (b) at least one 6 or a double is rolled

   (c) exactly one 6 is rolled.

2. Two unbiased tetrahedral dice each have faces numbered 1 to 4. The outcome when one of these dice is thrown is the number on the face on which it lands. Tabulate a possibility space for the outcomes when both dice are thrown and use it to find the probability that

   (a) at least one 4 is thrown

   (b) a total score of 5 is thrown.

3. Two unbiased six-sided dice are thrown. One is numbered 1 to 6, the other is numbered 1, 2, 2, 3, 3, 4. Draw up a possibility space for the outcomes and use it to find the probability of

   (a) a total score of 5

   (b) a total score of at least 5

   (c) scoring 3 on either die, but not both.

4. Two normal unbiased dice, each numbered 1 to 6, are tossed simultaneously. What is the probability of getting a total score greater than 7 *and* at least one of the dice scores 5?

5. One student asserted that the probability of getting a score of 12 was the same whether two dice were tossed and the scores added or the score on one die was doubled. State with reasons whether or not the student is right.

6. There are 6 questions in this exercise, including this one. The letter at the end of the first line of question 1 is 't'. List the letters at the end of the first line of the remaining five questions, ignoring numbers. One of these letters (including the 't') is selected at random. The process is then repeated. What is the probability that

   (a) the first letter chosen is 't' and the second letter chosen is 'e'

   (b) both letters chosen are 't'?

Combined Events

We have considered cases concerned with two or more events, such as getting a 5 or a 6 when a die is rolled, or getting a 5 and a 6 when two dice are rolled. The examples so far have been such that we can enumerate all the equally likely outcomes and the outcomes resulting in the combination of events that we are interested in.

It is not always practical or even possible to do this; there are a lot of different ways of getting a total score of more than 10 on four throws of a die but listing them all so that they can be counted is tedious. If a die is rolled repeatedly until a 6 is scored, the number of possible outcomes is infinite as it is possible (although extremely unlikely) that a 6 will never turn up.

We will now develop methods to deal with combined events that do not involve counting all possibilities.

Combined events can be loosely divided into two categories,

- 'either ... or' events such as scoring *either* 5 *or* 6 with one throw of a die

- 'both ... and' events such as scoring *both* a 6 *and* a 5 with two throws of a die.

Some events fall into both categories; scoring at least one 6 when two dice are tossed involves *either* a 6 on just one die *or* a 6 on both dice but at this stage we will consider them separately.

MUTUALLY EXCLUSIVE EVENTS

If a die is thrown once, it is possible to score either a 5 or a 6. It is not possible to score both a 5 and a 6. Such events are called *mutually exclusive*.

> **Two events *A* and *B* are mutually exclusive when
> either *A* or *B* can occur but not both.**

INDEPENDENT EVENTS

Consider a bag that contains 2 blue counters and 3 red counters. Suppose that a counter is removed at random and then put back. If a second counter is then chosen at random, the possibilities for its colour do not depend on the colour of the first counter removed. Such events are called *independent*. However if the first counter is not put back, then the possibilities for the colour of the second counter *do* depend on the colour of the first one removed, i.e. the colour of the first counter and the colour of the second counter are *not* independent.

**Two events, *A* and *B*, are independent
if the occurrence or non-occurrence of either *A* or *B*
has no influence on the occurrence or non-occurrence of the other.**

It is important to be able to distinguish between events that are mutually exclusive, independent or neither.

EXERCISE 20c

State whether the events described are mutually exclusive, independent or neither.

1. An integer is chosen at random from the integers 1 to 9 inclusive.

   (a) The integer is either 5 or 6.

   (b) The integer is either prime or even.

2. Two integers are chosen, with replacement, from 1, 2, 3, 4, 5 or 6.

   (a) Both integers are even.

   (b) Either the first integer is even or the second one is even.

3. Repeat question 2 if the first integer is not replaced before the second integer is selected.

4. Three coins are tossed. The result is

   (a) either two heads and one tail or two tails and one head

   (b) two heads and a tail

   (c) at least one head.

   (Hint: at least one head $\Rightarrow$ either one or two or three heads.)

5. Two discs are removed from a bag holding 2 red, 2 green and 2 blue discs.
   The first disc is red and the second disc is not blue.

SUM AND PRODUCT LAWS

A bag contains 3 red, 2 white and 5 black counters.
Suppose that one counter is selected at random. The probability that the counter is white (W) is given by $P(W) = \frac{2}{10}$, and the probability that the counter is black (B) is given by $P(B) = \frac{5}{10}$

Now W and B are mutually exclusive as a single counter cannot be both white and black. The probability that the counter is either white or black can be found directly; the number of ways of selecting a white or black counter is $2 + 5$ (there are 7, i.e. $2 + 5$, counters that are white or black) and the number of ways of selecting any counter is 10,

$$\therefore \quad P(W \text{ or } B) = \frac{2+5}{10} = \frac{2}{10} + \frac{5}{10} = P(W) + P(B)$$

The result is true for any two events A and B, that are mutually exclusive:

if A occurs in r out of n equally likely ways

and if B occurs in s out of n equally likely ways

then   'A or B' occurs in $(r+s)$ ways out of n equally likely ways

$$\therefore \quad P(A \text{ or } B) = \frac{r+s}{n} = \frac{r}{n} + \frac{s}{n} = P(A) + P(B)$$

'Either A or B' is denoted by $A \cup B$
Hence if A and B are mutually exclusive events,
$$P(A \cup B) = P(A) + P(B)$$

Now suppose that the first counter is replaced and a second counter is selected at random. The possible choices for the colour of the second counter are not influenced by the colour of the first counter removed, so the colours of the two counters selected are independent of each other.

If W_1 is the selection of a white counter first and B_2 is the selection of a black counter second, then $P(W_1) = \frac{2}{10}$ and $P(B_2) = \frac{5}{10}$

The number of ways of selecting a white counter and then a black counter is 2×5.
Therefore the probability of selecting a white counter and then a black counter can be found by dividing 2×5 by the number of ways of selecting any counter first and any counter second, i.e. 10×10.

$$\therefore \quad P(W_1 \text{ and } B_2) = \frac{2 \times 5}{10 \times 10} = \frac{2}{10} \times \frac{5}{10} = P(W_1) \times P(B_2)$$

This argument applies to any two independent events, A and B.

'Both A or B' is denoted by $A \cap B$
Hence if A and B are independent events,
$$P(A \cap B) = P(A) \times P(B)$$

A way of remembering which of the symbols '$\cup$' and '$\cap$' means 'and' is to think of $\cap$ as the letter n which is sometimes used as an abbreviation for 'and', as in 'Pick 'n' Mix'.

Examples 20d

1. One card is selected at random from a set of eight cards numbered 1, 2, 3, 3, 4, 5, 5, 6. What is the probability that the number on the card is at least 5?

A card numbered at least 5 means a card numbered either 5 or 6 and these are mutually exclusive events. There are eight cards, 2 of which are numbered 5 and one of which is numbered 6.

$$P(5) = \tfrac{2}{8} \quad \text{and} \quad P(6) = \tfrac{1}{8}, \qquad \therefore \quad P(5 \cup 6) = \tfrac{2}{8} + \tfrac{1}{8} = \tfrac{3}{8}$$

This can also be solved directly from the basic definition of probability; there are 3 ways of selecting a card numbered 5 or 6 out of 8 equally likely selections

$$\Rightarrow \qquad P(5 \text{ or } 6) = \tfrac{3}{8}.$$

2. Three coins are tossed simultaneously. Two of the coins are fair and one is biased so that a head is twice as likely as a tail. What is the probability that all three coins turn up heads?

There are three coins which we will call a, b and c. We can then denote the events that each coin turns up a head by H_a, H_b, and H_c.

If a and b are the fair coins and c is the biased coin,

$$P(H_a) = \tfrac{1}{2}, \quad P(H_b) = \tfrac{1}{2}, \quad P(H_c) = \tfrac{2}{3}$$

The way that any one coin lands has no influence on the way the other two coins land so H_a, H_b, and H_c are independent events.

$$\therefore \qquad P(H_a \cap H_b \cap H_c) = \tfrac{1}{2} \times \tfrac{1}{2} \times \tfrac{2}{3} = \tfrac{1}{6}$$

3. An ordinary unbiased six-sided die is rolled three times. Find the probability of rolling

   (a) three sixes

   (b) at least one six

   (c) exactly one six.

The outcome of each roll of the die is an independent event. As the die is unbiased, the probability of a six on any one roll is $\tfrac{1}{6}$. We will use the suffixes 1, 2 and 3 to denote the first, second and third roll of the die.

   (a) $P(\text{three sixes}) = P(6_1 \text{ and } 6_2 \text{ and } 6_3)$

   Now $P(6_1 \cap 6_2 \cap 6_3) = P(6_1) \times P(6_2) \times P(6_3)$

   $$= \tfrac{1}{6} \times \tfrac{1}{6} \times \tfrac{1}{6} = \tfrac{1}{216} = 0.0046\ldots$$

(b) $P(\text{at least one six}) = 1 - P(\text{not at least one six, i.e. no sixes})$

Now $P(\text{no sixes}) = P(\text{not } 6_1 \text{ and not } 6_2 \text{ and not } 6_3)$

The probability of not getting a six on any one roll is $\frac{5}{6}$,

$\therefore \qquad P(\bar{6}_1 \cap \bar{6}_2 \cap \bar{6}_3) = \frac{5}{6} \times \frac{5}{6} \times \frac{5}{6} = \frac{125}{216}$

$\Rightarrow \qquad P(\text{at least one six}) = 1 - \frac{125}{216} = \frac{91}{216} = 0.4212\ldots$

(c)   If exactly one six is thrown, then

    either   the first throw is six and the others are not six (A)

    or   the second throw is six and the first and third are not (B)

    or   the third throw is six and the first two throws are not (C)

    These are mutually exclusive events.

Now $\qquad P(A) = P(6_1 \cap \bar{6}_2 \cap \bar{6}_3) = \frac{1}{6} \times \frac{5}{6} \times \frac{5}{6} = \frac{25}{216}$

Similarly $\qquad P(B) = \frac{25}{216} \qquad$ and $\qquad P(C) = \frac{25}{216}$

$\therefore \qquad P(A \text{ or } B \text{ or } C) = P(A) + P(B) + P(C) = \frac{75}{216} = 0.3472\ldots$

4. Pete and Carmen gamble for a box of chocolate by rolling a die. The first person to roll a six wins. If the die is unbiased and Carmen rolls first, find the probability that

(a)   Carmen wins on her second attempt

(b)   Carmen wins.

Each roll of the die is an independent event and, on any one roll, $P(6) = \frac{1}{6}$ and $P(\bar{6}) = \frac{5}{6}$.

Let C_n denote Carmen winning on her nth attempt and P_n denote Pete winning on his nth attempt.

(a)   For Carmen to win on her 2nd attempt, both she and Pete must lose on their first attempt, i.e.

$P(C_2) = P(\bar{C}_1 \text{ and } \bar{P}_1 \text{ and } C_2)$

$\qquad = P(\bar{C}_1) \times P(\bar{P}_1) \times P(C_2) = \frac{5}{6} \times \frac{5}{6} \times \frac{1}{6} = \frac{25}{216}$

(b)   If Carmen wins then   *either* she wins on her first roll

                              or   she wins on her second roll

                              or   she wins on her third roll   and so on.

Now $\qquad P(C_1) = \frac{1}{6}$

$P(C_2) = \frac{5}{6} \times \frac{5}{6} \times \frac{1}{6} = \left(\frac{5}{6}\right)^2 \times \frac{1}{6}$

$P(C_3) = P(\bar{C}_1) \times P(\bar{P}_1) \times P(\bar{C}_2) \times P(\bar{P}_2) \times P(C_3) = \left(\frac{5}{6}\right)^4 \times \frac{1}{6}$

$P(C_4) = P(\bar{C}_1) \times P(\bar{P}_1) \times P(\bar{C}_2) \times P(\bar{P}_2) \times P(\bar{C}_3) \times P(\bar{P}_3) \times P(C_4)$

$\qquad = \left(\frac{5}{6}\right)^6 \times \frac{1}{6}$

and so on.

These are mutually exclusive events, so

$$P(\text{Carmen wins}) = \left\{\tfrac{1}{6}\right\} + \left\{\left(\tfrac{5}{6}\right)^2 \times \tfrac{1}{6}\right\} + \left\{\left(\tfrac{5}{6}\right)^4 \times \tfrac{1}{6}\right\} + \left\{\left(\tfrac{5}{6}\right)^6 \times \tfrac{1}{6}\right\} + \ldots$$

$$= \tfrac{1}{6} + \tfrac{1}{6}\left(\tfrac{5}{6}\right)^2 + \tfrac{1}{6}\left(\tfrac{5}{6}\right)^4 + \tfrac{1}{6}\left(\tfrac{5}{6}\right)^6 + \ldots$$

This is a G.P. with first term $\tfrac{1}{6}$ and common ratio $\left(\tfrac{5}{6}\right)^2$

The sum to infinity is $\dfrac{\tfrac{1}{6}}{1 - \left(\tfrac{5}{6}\right)^2} = \tfrac{6}{11} = 0.545$

Therefore $P(\text{Carmen wins}) = 0.55 \quad (2 \text{ dp})$

Carmen has a slightly better then even chance of winning, so going first gives her a slight advantage.

5. In a game of darts, the probability that Steve aims at and hits a double is 0.4. How many throws are necessary for the probability that Steve hits at least one double to exceed 0.95?

The probability of hitting a double, $P(D)$, is 0.4, so the probability of not hitting a double, $P(\overline{D})$ is 0.6, $P(\overline{D}$ in two throws$) = (0.6)^2, \ldots$

$$P(D \text{ at least once in } n \text{ throws}) = 1 - P(\overline{D} \text{ on all } n \text{ throws})$$

$$= 1 - (0.6)^n$$

$\therefore$ if $P(D$ at least once in n throws$) > 0.95$, then

$$1 - (0.6)^n > 0.95$$

$\Rightarrow$ $(0.6)^n < 0.05$

$\Rightarrow$ $n \log 0.6 < \log 0.05$

$\Rightarrow$ $n > \dfrac{\log 0.05}{\log 0.6} = 5.8\ldots$

In the line above, the inequality sign is reversed because log 0.6 is negative.

Therefore 6 throws are necessary.

6. Two nails are taken from a box holding a large number of nails, 10% of which are bent. What is the probability that exactly one of the nails removed is bent?

We will use suffixes to denote the first and second nails. If one of the nails is bent, it is either the first nail or the second and these are mutually exclusive events.

$$P(\text{exactly 1 nail bent}) = P(B_1 \text{ and } \overline{B}_2 \text{ or } \overline{B}_1 \text{ and } B_2)$$

$$= P(B_1 \text{ and } \overline{B}_2) + P(\overline{B}_1 \text{ and } B_2)$$

We now need to make an assumption: we will assume that the number of nails in the box is large enough for the first nail selected to have no effect on the possible selection of the second nail, i.e. that the two events are independent.

$\therefore$ $P(\text{exactly one nail bent}) = P(B_1) \times P(\overline{B}_2) + P(\overline{B}_1) \times P(B_2)$

$$= 0.1 \times 0.9 + 0.9 \times 0.1 = 0.18$$

Give answers correct to 2 dp where necessary.

1. Three unbiased coins are tossed. Find the probability of

   (a) three heads   (b) at least one head   (c) exactly one head.

2. The probability that an archer hits the gold (i.e. the centre of the target) with any one shot is $\frac{1}{5}$. Find the probability that he hits the gold

   (a) with his second shot

   (b) exactly once with his first three shots

   (c) at least once in his first four shots.

3. In a multiple choice test each question has five possible answers, only one of which is correct. A candidate guesses his answers at random. What is the probability that in a test of ten such questions, he gets none right?

4. Two coins are tossed. One coin is fair but the other is biased so that a head is three times as likely as a tail. Find the probability that

   (a) on one toss of both coins, they both land tails up

   (b) on two tosses of both coins, they both land tails up both times

   (c) on three tosses of both coins, at least one head is obtained.

5. One card is selected at random from a set of cards numbered 1, 2, 3, 4, and 5. The card is replaced before another selection is made. Find the probability that

   (a) on one selection, card number 4 is selected

   (b) on two selections, the total of the numbers taken is 2

   (c) on three selections, the total of the numbers taken is at least 4.

6. Two people, A and B gamble for 10 p by tossing a fair coin in turn. The first to toss a head wins. A tosses first. Find the probability that

   (a) A wins on her second throw

   (b) B wins on her second throw.

7. In question 6, what is the probability that A wins?

8. Three children, A, B and C, decide which of them goes in first to the school medical by rolling an ordinary fair die. The first to roll a six is the 'winner'. They toss the die in the order A, B then C. What is the probability that B 'wins'?

9. A sack contains a large number of crocus bulbs and 5% or these bulbs are unsound. A customer helps himself to 2 bulbs. His selection can be treated as random and the number of bulbs in the sack is large enough for the selection of each bulb to be considered as an independent event. What is the probability that the 2 bulbs selected

(a) are both sound

(b) contain exactly one unsound bulb

(c) are both unsound?

10. In a quality control exercise, a delivery of eggs was checked for damaged eggs. Four eggs were chosen at random and three were found to be damaged, so the delivery was returned to the supplier. When the supplier checked the whole consignment, it was found that 2% of the eggs were damaged. The consignment was returned to the customer. What is the probability that another random sample of four eggs will contain three damaged eggs? (The number of eggs is large enough for the selection of each egg to be treated as an independent event.)

CONDITIONAL PROBABILITY

Consider again the bag containing 3 red, 2 white and 3 black counters. If one counter is removed at random, there are 8 possible outcomes. If the counter is not replaced, the possible outcomes when a second counter is removed are reduced by one, and the events are *not* independent. The possible outcomes for the colour of the second counter *depend* on which colour was removed first, i.e. the number of possibilities for a red second counter (R_2) is 3 if the first counter was white or black but 2 if the first counter was red. Therefore

$$P(\text{2nd counter is red, given that 1st counter is white}) = \tfrac{3}{7}$$

This is called the *conditional probability that R_2 occurs when W_1 has happened.* It is written $P(R_2|W_1)$ and read as 'probability that R_2 occurs given W_1 has occurred',

i.e. $P(R_2|W_1) = \tfrac{3}{7}$

If a red counter was removed first then $P(R_2|R_1) = \tfrac{2}{7}$

For any two events A and B,

$P(B|A)$ means the probability that B occurs
given that A has occurred.

PROBABILITY THAT TWO EVENTS BOTH HAPPEN

Consider the probability that *both* a red counter is removed first *and* a red counter is removed second, i.e. $P(R_1$ and $R_2)$. The number of ways that a red counter first and a red counter second can occur is 3×2, and the number of ways of removing any counter first and removing any counter second is 8×7.

Therefore
$$P(R_1 \cap R_2) = \frac{3 \times 2}{8 \times 7} = \frac{3}{8} \times \frac{2}{7} = P(R_1) \times P(R_2|R_1)$$

If A and B are any two events then

the probability that both A and B occur is given by

$$P(A \cap B) = P(A) \times P(B|A)$$

If A and B are independent, then the fact that A has occurred has no influence on the probability that B occurs, i.e.

for independent events $P(B|A) = P(B)$

PROBABILITY THAT EITHER OF TWO EVENTS HAPPENS

We continue to consider the bag containing 3 red, 2 white and 3 black counters.

If, as before, a counter is removed, not replaced, and then a second counter is removed, consider the probability of removing *either* a red counter first *or* a red counter second, i.e. $P(R_1 \cup R_2)$.
Now R_1 and R_2 are *not* mutually exclusive because both events can occur.

The number of ways that it is possible for the first of the two counters to be red is 3×7 as there are 3 choices for the first, and the second can be any of the remaining 7 counters. This number, 3×7, includes the case when both counters are red.
The same argument applies to the choice of the second counter, i.e. there are 7×3 ways it can be red, including the cases when both are red.

If we add 3×7 to 7×3, we get the number of times that either the first or the second is red, but this includes the number of times that the counters are both red *twice over*.
Therefore the number of ways that *either* R_1 *or* R_2 occur is the number of ways the each occurs minus the number of ways that both occur,

i.e. $3 \times 7 + 7 \times 3 - 3 \times 2$

As before, there are 8×7 ways of selecting any two counters,

therefore $\qquad P(R_1 \cup R_2) = \dfrac{3 \times 7 + 7 \times 3 - 3 \times 2}{8 \times 7}$

$$= P(R_1) + P(R_2) - P(R_1 \cap R_2)$$

The result is true for any two events A and B, as we now show.

Suppose that there are n equally likely outcomes,

of which r is the number of ways that A can happen,

$\qquad\qquad s$ is the number of ways that B can happen,

and $\qquad t$ is the number of ways that A and B can happen,

This information can be illustrated in a diagram, called a *Venn diagram*, where overlapping circles represent the numbers of ways that each event can happen.

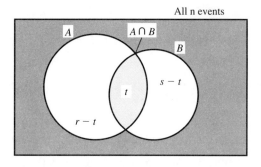

Now we can see that the r ways that A happens includes the t ways that A and B happen. Similarly the s ways that B can happen includes the t ways that both A and B happen.

$$\therefore \qquad P(A \cup B) = \frac{r + s - t}{n} = \frac{r}{n} + \frac{s}{n} - \frac{t}{n} = P(A) + P(B) - P(A \cap B)$$

If events A and B are not mutually exclusive then
$$P(A \cup B) = P(A) + P(B) - P(A \cap B)$$

If A and B *are* mutually exclusive, A *and* B cannot happen so $P(A \cap B) = 0$

From the diagram, there is clearly no difference between the number of ways that 'A and B' happen and that 'B and A' can happen, so it is obvious (but worth remembering) that

$$P(A \cup B) = P(B \cup A)$$

Similarly

$$P(A \cap B) = P(B \cap A) \quad \Rightarrow \quad P(A)P(B|A) = P(B)P(A|B)$$

Note that it does *not* follow that $P(B|A)$ and $P(A|B)$ are equal; in general they are not.

The following worked examples deal with a variety of probability problems but do not attempt to cover all possible types. When analysing any problem, it is important to make sure that you understand exactly what combination of events is involved. This is particularly important with 'either ... or' situations.

Examples 20e

1. In a group of students, 10% are left-handed, 8% are short-sighted and 2% are left-handed and short-sighted. Find the probability that

   (a) a left-handed student is short-sighted

   (b) a short-sighted student is left-handed.

Let $\quad P(L) = P(\text{student is left-handed})$

and $\quad P(S) = P(\text{student is short-sighted})$

then $\quad P(L) = 0.1, \; P(S) = 0.08 \quad \text{and} \quad P(L \cap S) = 0.02$

(a) $P(\text{left-handed student is short-sighted})$ means $P(\text{student is short-sighted given that the student is left-handed})$, i.e. $P(S|L)$

   Using $\quad P(L \cap S) = P(L) \times P(S|L) \quad$ gives

   $$0.02 = 0.1 \times P(S|L) \quad \Rightarrow \quad P(S|L) = 0.2$$

(b) For this part we want $P(L|S)$.

   Using $\quad P(S \cap L) = P(S) \times P(L|S) \quad$ we have

   $$0.02 = 0.08 \times P(L|S) \quad \Rightarrow \quad P(L|S) = 0.25$$

   This example can also be done very simply from the basic definition of probability, i.e. for (a)

   $P(S|L) =$ number that are short-sighted as a fraction of number that are left-handed

   $\quad\quad\quad = 2\% \text{ over } 10\% = 0.2$

   and for (b)

   $P(L|S) =$ number that are left-handed as a fraction of number that are short-sighted

   $\quad\quad\quad = 2\% \text{ over } 8\% = 0.25$

2. Two unbiased dodecahedral (twelve faces) dice with their faces numbered 1 to 12, are thrown. The score is taken from the face on which a die lands.

Find, correct to 2 sf, the probability that

(a)   at least one 12 is thrown

(b)   given that a 12 is thrown on the first die, a total score of at least 20 is obtained.

(a) $P(\text{at least one } 12) = 1 - P(\overline{12}) = 1 - \frac{121}{144} = 0.16$ (2 sf)

   This can also be done using

   $P(\text{at least one } 12) = P(12 \cup 12) = P(12) + P(12) - P(12 \cap 12)$

   $\qquad\qquad\qquad\qquad\quad = \frac{1}{12} + \frac{1}{12} - \frac{1}{144} = 0.16$ (2 sf)

   but the first method is more direct.

(b)   If 12 is scored on the first die, a score of 8, 9, 10, 11 or 12 is needed on the second die. The score obtained on the second die is independent of that obtained on the first die, i.e.

   $P(8 \text{ or more on 2nd die} \,|\, 12 \text{ on first}) = P(8 \text{ or more on 2nd die})$

   $\qquad\qquad\qquad\qquad\qquad\qquad\qquad\quad = \frac{5}{12} = 0.416\ldots = 0.42$ (2 sf)

3. A and B are two events such that $P(A) = \frac{1}{3}$, $P(B) = \frac{2}{9}$ and $P(A|B) = \frac{1}{2}$.
Find

(a) $P(A \cap B)$ (b) $P(A \cup B)$ (c) $P(B|\overline{A})$

(a)   We are given $P(A|B)$ so we will use $P(B \cap A)$

   $P(A \cap B) = P(B \cap A) = P(B)P(A|B) = \frac{2}{9} \times \frac{1}{2} = \frac{1}{9}$

(b) $P(A \cup B) = P(A) + P(B) - P(A \cap B)$

   $\qquad\qquad = \frac{1}{3} + \frac{2}{9} - \frac{1}{9} = \frac{4}{9}$

(c)   Using $P(\overline{A} \cap B) = P(\overline{A}) \times P(B|\overline{A})$ to find $P(B|\overline{A})$

   We have $P(\overline{A}) = 1 - P(A) = \frac{2}{3}$

   and from the diagram

   $P(\overline{A} \cap B) = P(B) - P(A \cap B) = \frac{2}{9} - \frac{1}{9} = \frac{1}{9}$

   $\therefore \qquad \frac{1}{9} = \frac{2}{3} \times P(B|\overline{A}) \quad \Rightarrow \quad P(B|\overline{A}) = \frac{1}{6}$

1. The diagram illustrates the percentages of students studying one or more of the subjects physics, chemistry and biology in the sixth form.

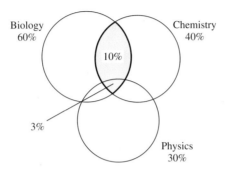

(a) What is the probability that a student chosen at random is studying biology?

(b) Given that a student is studying biology, what is the probability that the student is also studying chemistry?

(c) Find the probability that the student who is studying biology and chemistry is also studying physics.

2. Three cards are drawn, without replacement, from a well shuffled pack of 52 playing cards. Find, correct to 2 significant figures, the probability that

(a) all three cards are aces

(b) the first two cards are not aces but the third card is

(c) the third card is an ace. (Hint: this is an 'either ... or' situation, i.e. either AAA or $\overline{A}AA$ or $\overline{A}\,\overline{A}A$ or $\overline{A}\,\overline{A}A$)

3. A and B are two events such that $P(A) = \frac{1}{4}$, $P(B) = \frac{1}{3}$ and $P(A \cup B) = \frac{1}{2}$. Find $P(A \cap B)$.

4. The probabilities for two events, A and B are such that $P(A) = 0.3$, $P(B) = 0.4$ and $P(A \cup B) = 0.5$
Show that A and B are neither independent nor mutually exclusive.

5. A and B are two events such that $P(A) = \frac{2}{7}$, and $P(A \cap B) = \frac{1}{5}$. If A and B are independent find $P(B)$ and $P(A \cup B)$.

6. A and B are two events such that $P(A) = \frac{2}{5}$, $P(A|B) = \frac{1}{3}$ and $P(B|A) = \frac{1}{2}$. Find $P(A \cup B)$ and say, with reasons, whether A and B are mutually exclusive.

7. A school offers children a first-aid course followed by a test. Experience shows that the probability that a child will pass the test first time is 0.6. If a child fails the first test, it can be taken again one month later and the probability of success with the second test is 0.75.
   Find the probability that a randomly chosen child will

   (a) not pass at the first attempt

   (b) will pass on the first or second attempt

   (c) will not pass either test.

8. The probability that a child selected at random likes cola is 0.8. The probability that a child likes lemonade is 0.6 and the probability that a child likes both is 0.5.

   (a) Given that a child likes cola, what is the probability that the child likes both?

   (b) What is the probability that a child likes either lemonade or cola but not both?

9. The table gives information about proportions of students with various attributes.

Male	Females Studying Maths	Studying Maths
50%	20%	45%

   If a student is chosen at random what is the probability that

   (a) the student is studying maths

   (b) a student is studying maths given that he is male

   (c) a student is male given that he/she is studying maths?

10. Two cards are drawn at random from a pack of 7 cards numbered 1, 2, 3, 4, 5, 6, 7. Find the probability that

    (a) the product of the numbers is even

    (b) a prime number is included

    (c) the product of the numbers is even given that exactly one of the numbers is prime.

TREE DIAGRAMS

The number of ways in which a compound event can occur is not always very obvious. This is particularly true of compound events which can involve three or more separate events, so we now introduce a method which is helpful in such problems.

Consider, for example, tossing three coins. To find the probability that a toss gives exactly two heads, we need first to determine the ways that this can happen. There are three ways, because only one of the coins shows a tail,

i.e. if the coins are numbered 1, 2, 3 for identification, they can land

$$H_1 H_2 T_3 \quad \text{or} \quad H_1 T_2 H_3 \quad \text{or} \quad T_1 H_2 H_3$$

Using the laws from the last section we have

$$P(2H \& T) = P(\text{either } H_1, H_2, T_3 \quad \text{or} \quad H_1, T_2, H_3 \quad \text{or} \quad T_1, H_2, H_3)$$

$$= P(H_1)P(H_2)P(T_3) + P(H_1)P(T_2)P(H_3) + P(T_1)P(H_2)P(H_3)$$

Note that *if* the coins are unbiased we can use

$$P(2H \& T) = \frac{\text{number of ways coins can land showing two heads}}{\text{number of ways coins can land}}$$

$$= \frac{3}{2 \times 2 \times 2}$$

because the events are *equally* likely.

An alternative approach is to draw a diagram that shows the outcomes. The table approach shown earlier can cope with the outcomes of only two events. The method shown here can show the outcomes of many more events.

Consider again tossing the three coins, numbered 1, 2 and 3. The possible outcomes when coin 1 is tossed are a head and a tail.
Starting at the left-hand side of the page, we draw two branches, each branch being one of the possible outcomes.

(i)

When coin 2 is tossed there are two possible outcomes (head or tail) either of which may follow either of the possible outcomes of the toss of coin 1. To illustrate this we draw two branches at the end of each of the first two branches in diagram (i).

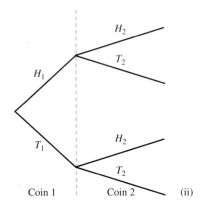

Following the branches from left to right shows there are four possible paths corresponding to the four outcomes, $H_1 \cap H_2$, $H_1 \cap T_2$, $T_1 \cap H_2$, $T_1 \cap T_2$. Each of these is followed by two further possible outcomes when coin 3 is tossed. Branching again from diagram (ii) gives this diagram.

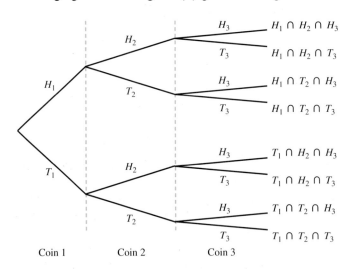

Following any one path from left to right gives one of the possible outcomes:: each one is a combination of three separate events of the form 'A and B and C' so the probability of the combined event is found by multiplying together the probabilities for each event along the path.

As this type of diagram shows all the possible outcomes, the events at the ends of the branches are mutually exclusive. Therefore the probability of either one or other of these outcomes is found by adding the probabilities at the ends of the respective branches.

To illustrate this, suppose that two of the coins are biased, so that

$$P(H_1) = 0.4, \ \ P(H_2) = 0.6 \ \ \text{and} \ \ P(H_3) = 0.5$$

These probabilities can be written on the appropriate branches of the tree diagram and then the probabilities of the combined events at the ends of the branches can also be shown, i.e.

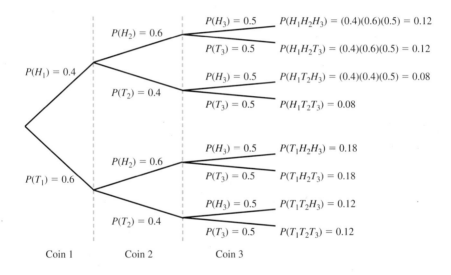

Now the outcome 'two heads and a tail' appears at the ends of the second, third and fifth branches, so $P(2H \ \text{and} \ T) = 0.12 + 0.08 + 0.18 = 0.38$

Tree diagrams are a powerful method for finding the probability of a combination of events in a wide variety of situations because they give a visual way of sorting out the different events.

In many cases a more direct method may be available, but not everyone can spot a shorter approach.

The following worked examples illustrate some cases where tree diagrams may be useful.

Examples 20f

1. Three dice, each numbered 1 to 6 are thrown. Two of the dice are biased so that a 6 on either of them is twice as likely as any other score. The third die is fair. Find the probability that

(a)   exactly one 6 is thrown

(b)   when one 6 is thrown, it turns up on the fair die.

On a biased die, as a 6 is twice as likely as any other score, there are seven equally likely events, two of which are scoring 6, i.e. $P(6) = \frac{2}{7}$

Using a tree diagram to show the probabilities of 6 and $\bar{6}$, and starting with the fair die gives

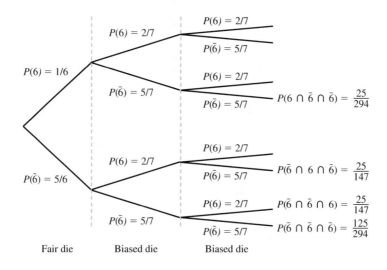

(a)   Exactly one six can occur in three ways,

$$\therefore \qquad P(\text{exactly one six}) = \tfrac{25}{294} + \tfrac{25}{147} + \tfrac{25}{147} = \tfrac{125}{294} = 0.425 \quad (3\text{ sf})$$

(b) $P(\text{when exactly one 6 is thrown it is on the fair die})$

$= P(\text{die is fair} \mid \text{one 6})$

Now $P(\text{one 6 } and \text{ die is fair}) = P(\text{one 6}) \times P(\text{die is fair} \mid \text{one 6})$

From the diagram, $P(\text{one 6 } and \text{ die is fair}) = \tfrac{25}{294}$

and $\qquad\qquad\qquad P(\text{one 6}) = \tfrac{125}{294}$

$$\therefore \qquad P(\text{die is fair} \mid 6) = \tfrac{25}{294} \div \tfrac{125}{294} = 0.2$$

2. Three coins are such that one of them is biased and the other two are fair. The probability that the biased coin lands head up is 0.7. If one of the coins is tossed twice and lands heads up on both occasions, what is the probability that it is the biased coin?

The events here are choosing a coin and tossing a coin twice. The only outcome of tossing that we are concerned with is two heads, so we will only draw these routes on the tree diagram. Starting with the choice of coin gives this tree diagram.

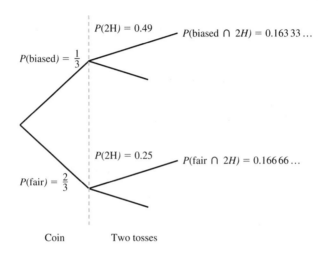

$P(\text{biased} \,|\, 2H)$ can be found from

$P(2H \cap \text{biased}) = P(2H) \times P(\text{biased} \,|\, 2H)$

$P(2H \cap \text{biased}) = 0.163\,33\ldots$

and $P(2H) = 0.163\,33\ldots + 0.166\,66\ldots = 0.33$

$\therefore \quad P(\text{biased} \,|\, 2H) = 0.163\,33\ldots \div 0.33 = 0.495 \quad (3\text{ sf})$

3. A student is allowed three attempts at passing an examination. The probability that a randomly chosen student passes at the first attempt is 0.8. If a student fails the first time, the probability of success at the second attempt is 0.5 and for the third attempt it is 0.3. Find the probability that a student chosen at random,

(a) will pass the examination

(b) from those that pass, does so at the first attempt.

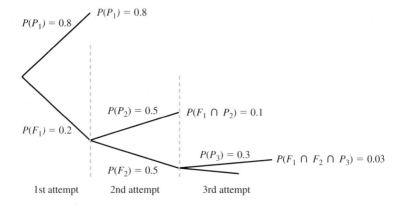

$$\text{1st attempt} \qquad \text{2nd attempt} \qquad \text{3rd attempt}$$

(a) $P(\text{pass}) = P(P_1 \quad \text{or} \quad F_1 \cap P_2 \quad \text{or} \quad F_1 \cap F_2 \cap P_3)$

$$= 0.8 + 0.1 + 0.03 = 0.93$$

(b) $P(P_1 \text{ given the student passes}) = P(P_1 | \text{pass})$

Using $P(P_1 \cap \text{passes}) = P(\text{pass}) \times P(P_1 | \text{pass})$ we have

$(P_1 \cap \text{passes})$ means 'a student passes at the first attempt₁ and passes' which is the same as 'a student passes at the first attempt, i.e.

$$P(P_1 \cap \text{passes}) = P(P_1) = 0.8 \quad \text{and} \quad P(\text{pass}) = 0.93$$

$$P(P_1 | \text{pass}) = \tfrac{0.8}{0.93} = 0.860 \quad (3 \text{ sf})$$

Note that $P(P_1 | \text{pass})$ means the probability that a student who passes does so at the first attempt, i.e.

$P(P_1 | \text{pass}) = $ number who pass first time as a fraction of those who pass

$$= 80\% \text{ over } 93\% = 0.860$$

Accuracy of Answers

It is usually reasonable to give final answers for probabilities corrected to 3 decimal places, unless another degree of accuracy is asked for. There are some contexts where it is clearly not reasonable. It would, for example, be silly to give 3 decimal places for the probability of a thunderstorm occurring as it is probably impossible to predict such an event as accurately as that. On the other hand, the probability of picking the six winning numbers in a draw for the national lottery is $\approx 7.15 \times 10^{-8}$ and, while it might be reasonable to say that this is practically zero, it is nonsense to state that it is 0.000 (to 3 dp).

Use any method for these questions. Give answers correct to an appropriate degree of accuracy.

1. One box contains 50 red pens and 25 blue pens. A second box contains 35 red pens and 40 blue pens. One box is chosen at random and then one pen is chosen at random from that box; if it is red, what is the probability that it came from the first box?

2. Three coins are tossed. Two of the coins are fair and one is biased so that a head is three times as likely as a tail. What is the probability that

   (a) exactly one head shows

   (b) when only one coin lands head up, it is the biased coin?

3. There are 50 straws in a pot. The hidden end of one straw is painted red and if that straw is chosen, a prize is won. Competitors are each allowed to remove up to 4 straws, one at a time and without replacement. What is the probability that a competitor will win a prize?

4. A company offers a repair service for washing machines. On any one day, the probability that no repairs are requested is 0.1, the probability that one repair is requested is 0.3 and the probability that two repairs are requested is 0.5.

   (a) On any one day, what is the probability that more than two repairs are requested?

   (b) What is the probability that there are no requests for repairs on three consecutive days?

5. Two dice are thrown. One die is fair and the other is biased so that the probability of scoring a six on it is 0.12.

   (a) What is the probability of scoring two sixes?

   (b) If one six is scored, what is the probability that it is on the biased die?

6. A survey by a school catering company asks schools to give a questionnaire to each child in the school whose birthday is on the 10th day of any month.

   (a) What is the probability that a randomly chosen child will be asked to take part in the survey? What assumptions do you need to make?

   (b) A sixth form college has 200 students in the first year. What is the probability that none of these students will be asked to take part?

7. A company buys paper from two suppliers A and B and places twice as many orders with A as with B. The probability that supplier A will deliver an order within one week is 0.8 and the probability that supplier B will deliver an order within one week is 0.7.

(a) What is the probability that a randomly selected order takes more than a week to be delivered?

(b) Given that an order has taken more than a week to be delivered, what is the probability that it was placed with supplier B?

8. A box holds a large number of green, red and yellow sweets in the proportions green : red : yellow $= 1:3:2$. Three sweets are selected at random. (The number of sweets is large enough for the selection of a sweet to be treated as independent of the previous selection.)

(a) What is the probability that one sweet of each colour is chosen?

(b) What is the probability that at least one red sweet is chosen?

(c) Given that the sweets are all the same colour, what is the probability that they are all red?

9. Two children, Michael and Sophie, play a game with two coins and tokens. They start with one token each and take it in turns to toss the coins. If the coins land showing 2 heads, Michael loses and has to give Sophie his token. If the coins land showing 2 tails, Sophie loses and has to give Michael her token. The game ends when one of the children loses and has no token to give.

(a) What is the probability that Michael has no tokens after two tosses?

(b) What is the probability that the game has to end after three tosses?

SUMMARY

Variables

A *variable* is any quantity that can have different values.

A *quantitative* variable has numerical values. These values are either *discrete*, i.e. distinct, or *continuous*, i.e. anywhere within a range of values.

A variable that does not have numerical values, e.g. colour, opinion, is called *qualitative*.

Data

The *population* is all the relevant information in an investigation.

A *sample* is just part of the population.

A *sample frame* is a list from which a sample is chosen.

A *census* is an investigation that uses the whole population.

A *survey* is an investigation that uses a sample.

A *frequency table* lists each different item together with the number of times it occurs.

A *grouped frequency table* lists *groups* or *classes* of items together with the number of times the items in each group occur. The values at the ends of a class are called the *class boundaries*.

Cumulative frequency gives the number of items from the lowest in the distribution to the upper end of a class.

DATA REPRESENTATION

A *histogram* is a bar chart drawn such that the area of each bar represents the frequency of items in the class. The height of a bar is the *frequency density*, which is calculated by dividing frequency by class width.

When frequency densities are plotted at the midpoints of the class intervals and the points joined by straight lines, the diagram is called a *frequency polygon*.

MEASURES OF CENTRAL TENDENCY

The *mean* of a set of values is $\dfrac{\text{the sum of all the values}}{\text{the number of values}}$,

i.e. $\bar{x} = \dfrac{\Sigma x}{n}$ for ungrouped data, $\bar{x} \approx \dfrac{\Sigma fx}{\Sigma f}$ for grouped data.

The *median* is the value of the middle item when a set of n values has been arranged in order of size, i.e. the value of the $\dfrac{n+1}{2}$ th item. For grouped data, the median can be estimated by *linear interpolation*, i.e.

$$\text{median} \approx (\text{l.c.b.}) + \dfrac{k}{m} \times (\text{class width})$$

where the median is the kth item of m items in the class.

The *mode* is the value with the greatest frequency. In a grouped frequency distribution, the *modal class* is the class with the greatest frequency.

DISPERSION

The *range* of a set of values is the difference between the greatest and least value in the set.

When n items have been arranged in ascending order of size, the *lower quartile* is the value of the $\dfrac{n+1}{4}$ th item and the *upper quartile* is the value of the $\dfrac{3(n+1)}{4}$ th item.

The difference between the upper and lower quartile is called *the interquartile range*.

When $n\%$ of all the values are less than or equal to a particular item, that item is the nth *percentile*.

The *standard deviation* (or *root mean square deviation*) of n values is

$$\sqrt{\dfrac{\Sigma(x-\bar{x})^2}{n}} = \sqrt{\dfrac{\Sigma x^2}{n} - \bar{x}^2}$$

For a grouped distribution,   s.d. $\approx \sqrt{\dfrac{\Sigma fx^2}{\Sigma f} - \bar{x}^2}$

Probability

BASIC DEFINITIONS

An *event* or *outcome* is what can happen as a result of a given situation.
The *possibility space* is the set of all possible outcomes of a given situation.
The *probability that an event A occurs* is denoted by $P(A)$ and is defined as

$$P(A) = \frac{\text{number of ways that } A \text{ occurs}}{\text{number of ways that all likely events, including } A, \text{ occur}}$$

and it follows that

$$0 \leqslant P(A) \leqslant 1$$

$$P(\overline{A}) = 1 - P(A)$$

where $\overline{A}$ means 'not the event A' or 'the complement of A',
i.e. $\overline{A}$ is all the possible outcomes except A.

Two events, A and B, are *mutually exclusive* if 'either A or B' can occur but 'both A and B' cannot occur.

Two events, A and B, are *independent* if the occurrence of A has no influence on the possible occurrence of B, and vice-versa.

PROBABILITY LAWS

If A and B are mutually exclusive events, then

$$P(A \cup B) = P(A) + P(B)$$

where $A \cup B$ means A or B

If A and B are independent events, then

$$P(A \cap B) = P(A) \times P(B)$$

where $A \cap B$ means A and B
(Note that if A and B are mutually exclusive then $P(A \cap B) = 0$)

If A and B are not independent then

$$P(A \cap B) = P(A) \times P(B|A)$$

where $P(B|A)$ means the probability that B occurs given that A has occurred and it is called the conditional probability that B occurs given that A has occurred.
It follows that if A and B are independent, $P(B|A) = P(B)$

If A and B are not mutually exclusive then

$$P(A \cup B) = P(A) + P(B) - P(A \cap B)$$

Type I

1. This frequency curve illustrates the distribution of heights of a group of adults.

The fraction of group represented by the shaded area is approximately

  A $\frac{1}{2}$ **B** $\frac{3}{4}$ **C** $\frac{1}{10}$ **D** $\frac{1}{3}$ **E** $\frac{9}{10}$

2. The mean of the set of numbers 0, 3, 5, 12 is

  A 4   **B** 12   **C** 20   **D** 6.7   **E** 5

3. A couple is picked at random from a large group of couples with two children. Assuming that the birth of a girl or a boy is equally likely, the probability that the couple have an older girl and a younger boy is

  A $\frac{1}{2}$ **B** 1   **C** $\frac{1}{4}$ **D** 0   **E** $\frac{3}{4}$

4. A and B are independent events. $P(A) = 1/4$, $P(B) = 1/4$, $P(\bar{A} \cap \bar{B}) =$

  A $\frac{1}{16}$ **B** $\frac{9}{16}$ **C** $\frac{1}{2}$ **D** $\frac{1}{4}$ **E** $\frac{3}{4}$

5. The median of this frequency distribution

x	1	2	3
f	2	5	3

 is

  A 3   **B** 2.1   **C** 2   **D** 2.3   **E** 5

6. The mean of the distribution given in question 5 is

  A 3   **B** 2.1   **C** 2   **D** 2.3   **E** 1

7. The shaded area in the diagram represents

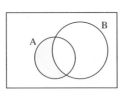

  A A   **B** B   **C** $A \cap \bar{B}$ **D** $A \cup \bar{B}$ **E** $A \mid B$

8. One letter is chosen at random from the letters of the word CHOICE. The probability that it is C is

  A $\frac{1}{6}$ **B** $\frac{1}{5}$ **C** $\frac{1}{3}$ **D** $\frac{2}{3}$ **E** 1

Type II

9. The mean score in a test was 6, so half the candidates got a score less than 6.

10. The probability of getting a score of 5 or more with a biased die is $1/3$.

11. Two cards are selected at random from four cards numbered 1, 2, 3, 4. The probability that one of the cards selected is numbered 4 is $1/2$.

12. The table gives the number of houses sold by an agent in various price ranges.

Price range	0–£100 000	£100 000–£200 000	over £200 000
Houses sold	95	47	22

The median house price is less than £100 000

13. A and B are two events and $P(A) = 1/2$, $P(B) = 2/5$,
$P(A \cap B) = 0 \Rightarrow P(A|B) = 2/5$

14. The mean value of a set of values is 6, the median value is 3 and the standard deviation is 2. Therefore the sum of the squares of the values is 40 times the number of values in the set.

15. A and B are mutually exclusive events, so it follows that
$$P(A \cup B) = P(A) + P(B)$$

MISCELLANEOUS EXERCISE E

1. Newborn babies are routinely screened for a serious disease which affects only 2 per 1000 babies. The result of screening can be positive or negative. A positive result suggests that the baby has the disease, but the test is not perfect. If a baby has the disease, the probability that the result will be negative is 0.01. If the baby does not have the disease, the probability that the result will be positive is 0.02.

    (a) Find the probability that a baby has the disease, given that the result of the test is positive.

    (b) Comment on the value you obtain. (ULEAC)$_s$

2. In an engineering experiment, a student took 10 measurements of the time for a trolley to run down an inclined plane. The following times in seconds were obtained:

    | 8.2 | 8.7 | 8.4 | 9.0 | 8.4 |
    | 8.5 | 8.3 | 8.7 | 8.4 | 8.8 |

    Calculate the mean, the variance and the standard deviation of these times.
    (ULEAC)$_s$

3. Two events A and B are such that $P(A) = 0.4$, $P(B) = 0.7$, $P(A \text{ or } B) = 0.8$, Calculate

   (a) $P(A \text{ and } B)$;

   (b) the conditional probability $P(A|B)$.   (AEB)$_s$

4. (a) Explain briefly

      (i) why it is often desirable to take samples,

      (ii) what you understand by a sampling frame.

   (b) Give an example of a sampling frame suitable for use in a survey of attitudes of pupils in a school to a proposal to start the school day 15 minutes earlier.   (ULEAC)$_s$

5. An urn contains 3 red, 4 white and 5 blue discs. Three discs are selected at random from the urn.

   Find that probability that

   (a) all three discs are the same colour, if the selection is with replacement,

   (b) all three discs are of different colours, if the selection is without replacement.   (ULEAC)

6.

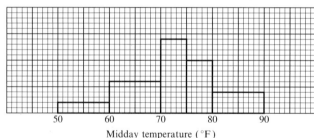

Midday temperature (°F)

The histogram shows the distribution of midday temperatures in degrees Fahrenheit during the 120 days of the months from November to February on a Pacific island.

   (i) How many small squares in the figure represent one reading?

   (ii) Make a frequency table for the data.

   (iii) Find an approximate value for the mean midday temperature over the period.

   Why is the answer to (iii) only approximate and in what way has the original distribution been modelled?   (OCSEB)$_s$

7. In an investigation of delays at a roadworks, the times spent, by a sample of commuters, waiting to pass through the roadworks were recorded to the nearest minute. Shown below is part of a cumulative frequency table resulting from the investigation.

Upper class boundary	2.5	4.5	7.5	8.5	9.5	10.5	12.5	15.5	20.5
Cumulative number of commuters	0	6	21	48	97	149	178	191	200

(a) For how many of the commuters was the time recorded as 11 minutes or 12 minutes?

(b) Estimate   (i) the lower quartile,

(ii) the 81st percentile, of these waiting times   (ULEAC)

8. The medical test for a certain infection is not completely reliable: if an individual has the infection there is a probability of 0.95 that the test will prove positive, and if an individual does not have the infection there is a probability of 0.1 that the test will prove positive. In a certain population, the probability that an individual chosen at random will have the infection is p. This information is shown on the tree diagram below.

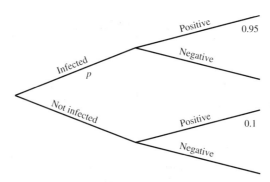

(i) An individual is chosen at random and tested. Write down the probability that the individual is infected and the test is positive.

(ii) An individual is chosen at random and tested. Show that the probability of the test being positive is $0.1 + 0.85p$.

(iii) Express in terms of p the conditional probability that an individual whose test is positive has the infection. Given that this probability is 0.6, find the conditional probability that an individual whose test is negative does not have the infection.   (UCLES)$_s$

9. The following information appears in the Annual Report 1992 of Eurotunnel PLC.

Size of shareholding		Number of shareholders
1 –	99	133 853
100 –	499	347 495
500 –	999	79 087
1 000 –	1 499	31 638
1 500 –	2 499	27 547
2 500 –	4 999	10 655
5 000 –	9 999	3 842
10 000 –	49 999	2 188
50 000 –	99 999	283
100 000 –	249 999	216
250 000 –	499 999	82
500 000 –	999 999	54
1 000 000 and over		50
Total		636 990

(i) Explain briefly what difficulties would arise in attempting to

   (a) represent the data by means of an accurate and easily comprehensible diagram,

   (b) find the mean size of shareholding per shareholder.

(ii) Estimate the median size of shareholding. (UCLES)$_s$

10. A toy factory produces plastic elephants on a machine. It has been found by experience that 5% of the output is defective.

(a) Find the probability that, in a random sample of 10 elephants selected for inspection,

   (i) none is defective,   (ii) more than one is defective.

(b) To meet the pre-Christmas rush the factory brings two old machines back into operation. The usual machine (known as machine A) produces 50% of the output, machine B produces 30% and machine C produces 20%. The percentages of defective items in the output of B and C are 10% and 15% respectively.

Draw a tree diagram to represent this information, marking the corresponding probabilities on each branch. Hence, show that the probability that an elephant chosen at random from the whole output is defective is 0.085. (OCSEB)$_s$

11. Summarised below are the values of the orders (to the nearest £) taken by a sales representative for a wholesale firm during a particular year.

Value of order (£)	Number of orders
Less than 10	3
10–19	9
20–29	15
30–39	27
40–49	29
50–59	34
60–69	19
70–99	10
100 or more	4

(a) Using interpolation, estimate the median and the semi-interquartile range for these data. [The semi-interquartile range is $\frac{1}{2}(Q_3 - Q_1)$]

(b) Explain why the median and semi-interquartile range might be more appropriate summary measures for these data than the mean and standard deviation.   (ULEAC)

12. A player has two dice, which are indistinguishable in appearance. One die is fair, so that the probability of getting a six on any throw is $\frac{1}{6}$, and one is biased in such a way that the probability of getting a six on any throw is $\frac{1}{3}$.

(i) The player chooses one of the dice at random and throws it once.

(a) Find the probability that a six is thrown.

(b) Show that the conditional probability that the die is the biased one, given that a six is thrown, is $\frac{2}{3}$.

(ii) The player chooses one of the dice at random and throws it twice.

(a) Show that the probability that two sixes are thrown is $\frac{5}{72}$.

(b) Find the conditional probability that the die is the biased one, given that two sixes are thrown.

(iii) The player chooses one of the dice at random and throws it n times. Show that the conditional probability that the die is the biased one, given that n sixes are thrown is $\dfrac{2^n}{2^n + 1}$.   (UCLES)$_s$

ANSWERS

CHAPTER 1

Exercise 1a – p. 2

1. $\dfrac{b-a}{ab}$

2. $\dfrac{8}{15x}$

3. $\dfrac{q-p}{pq}$

4. $\dfrac{11}{10x}$

5. $\dfrac{x^2+1}{x}$

6. $\dfrac{x^2-y^2}{xy}$

7. $\dfrac{2p^2-1}{p}$

8. $\dfrac{7x+3}{12}$

9. $\dfrac{5x-1}{6}$

10. $\dfrac{11-7x}{15}$

11. $\dfrac{\sin B + \sin A}{\sin A \sin B}$

12. $\dfrac{\sin A + \cos A}{\cos A \sin A}$

13. $\dfrac{12x^2+1}{4x}$

14. $\dfrac{2x^2+x-2}{2x+1}$

15. $\dfrac{x^2+2x+2}{x+1}$

16. $\dfrac{2x+3}{2x}$

17. $\dfrac{1+x-x^2}{x}$

18. $\dfrac{n+1}{n^2}$

19. $\dfrac{x(b^2+a^2)}{a^2b^2}$

20. $\dfrac{a^2+3a+1}{a+1}$

21. $\dfrac{e^x(2+x)}{2x}$

22. $\dfrac{\sin A - \cos A}{(\cos A + 1)(\sin A + 1)}$

Exercise 1b – p. 3

1. $\dfrac{2x}{(x+1)(x-1)}$

2. $\dfrac{2x-1}{(x+1)(x-2)}$

3. $\dfrac{7x+18}{(x+2)(x+3)}$

4. $\dfrac{x}{(x-1)(x+1)}$

5. $\dfrac{-1-3a}{(a-1)(a+1)} = \dfrac{1+3a}{(1-a)(1+a)}$

6. $\dfrac{x+2}{(x+1)^2}$

7. $\dfrac{1-4x}{(2x+1)^2}$

8. $\dfrac{-3x-10}{(x+1)(x+4)} = -\dfrac{3x+10}{(x+1)(x+4)}$

9. $\dfrac{2x+6}{(x+1)^2}$

10. $\dfrac{8-x-x^2}{(x+2)^2(x+4)}$

11. $\dfrac{7x+8}{6(x-1)(x+4)}$

12. $\dfrac{8-3x}{5(x+2)(x+4)}$

13. $\dfrac{15x-58}{6(x+1)(3x-5)}$

14. $\dfrac{5x^2-9x-32}{(x+1)(x-2)(x+3)}$

15. $\dfrac{2x^2+6x+6}{(x+1)(x+2)(x+3)}$

$= \dfrac{2(x^2+3x+3)}{(x+1)(x+2)(x+3)}$

16. $-\dfrac{1}{x(x+1)^2}$

17. $\dfrac{7t+3}{(t+1)^2}$

18. $\dfrac{-t^4+2t^3-2t^2-2t-1}{(t^2+1)(t^2-1)}$

19. $\dfrac{1+3y-3x}{(y-x)(y+x)}$

20. $\dfrac{n^3+6n^2+8n+2}{n(n+1)(n+2)}$

21. $\dfrac{e^3-1}{(e^2-1)(e+1)} = \dfrac{e^2+e+1}{(e+1)^2}$

22. $\dfrac{\cos A - \sin A}{(\sin A - 1)(\cos A - 1)}$

Exercise 1c – p. 5

1. (a) Q: $x + 1$, R: $-6x + 3$
 (b) Q: $x^3 - x^2 - 4x + 4$, R: -2
 (c) Q: $2x - 4$, R: $5x - 5$
 (d) Q: $3x^2 + 6x + 12$, R: 19
 (e) Q: x^2, R: $-6x^2 + 1$
 (f) Q: $2x^2 + x + 9$, R: 29
 (g) Q: $x - 10$, R: 32
 (h) Q: $5x - 1$, R: $5x$
 (i) Q: 3, R: -10
 (j) Q: $2x^2 + x - 4$, R: -3

2. (a) $1 + \dfrac{3}{x+1}$

 (b) $2 + \dfrac{4}{x-2}$

 (c) $1 + \dfrac{4}{x^2 - 1}$

 (d) $x + 2 + \dfrac{4}{x-2}$

 (e) $x + 7 + \dfrac{28}{x-4}$

 (f) $1 - \dfrac{x+4}{x(x+1)}$

Exercise 1d – p. 8

1. (a) 3   (b) 18   (c) 47
 (d) $\frac{35}{16}$ (e) $-\frac{16}{27}$
 (f) $a^3 - 2a^2 + 6$ (g) $c^2 - ac + b$
 (h) $\dfrac{1}{a^4} - \dfrac{2}{a} + 1$

2. (a) yes   (b) no   (c) no
 (d) yes   (e) no   (f) yes

3. (a) $(x-1)(x+2)(x+1)$
 (b) $(x-2)(x^2+x+1)$
 (c) $(x-1)(x+1)(x^2+1)$
 (d) $(x+2)(x^2+x+1)$
 (e) $(2x-1)(x^2+1)$
 (f) $(3x-1)(9x^2+3x+1)$
 (g) $(x+a)(x^2-ax+a^2)$
 (h) $(x-y)(x^2+xy+y^2)$

4. -7
5. 5
6. -3
7. 1

Exercise 1e – p. 10

1. $(x-1)(2x-1)(x+1)$; $-1, \frac{1}{2}, 1$
2. $f(x) = (x-2)(x^2+x+1) \Rightarrow f(x) = 0$
 only when $x = 2$

3. 8
4. $x = -2.56, 1, 1.56$
5. $(x+1)(x-2)(x-3)$

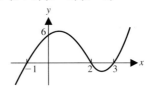

6. (a) $p = 11$, $q = -5$ (b) 3
7. (a) $8 + 4p + q = 0$, $-64 + 16p + q = 0$;
 $p = -6$, $q = 32$
 (b) $x = 2$, ± 4

Exercise 1f – p. 16

1. 0.2033
2. 0.2153
3. 0.1610
4. 0.9043
5. 0.4698
6. 1.24
7. (a) 7   (b) 6.70
8. (a) 2   (b) 1.53

Mixed Exercise 1 – p. 17

1. (a) $\dfrac{2p^2 - 3r}{rp}$ (b) $\dfrac{4x^2 - x + 3}{4x - 1}$

2. $(x-2)(x^2+3x+1)$
3. -1.3

4. $\dfrac{5x^2 + x - 1}{x(x+1)(2x-1)}$

5. $x^2 - x + 2 + \dfrac{1}{x-1}$

6. 2.6
7. -2

8. (a) $\dfrac{x(x^2+x-4)}{(x-1)^2(x+1)}$

 (b) $\dfrac{(2x-1)(x+1)}{2x(x+1)}$

 (c) $\dfrac{bc + ac + ab}{abc}$

9. 0.45
10. (a) $(x-1)(x+1)(x^2+1)$
 (b) $x(3x-5)(3x+5)(9x^2+25)$

11. $\frac{11}{3}$
12. $(a+2b)(a-b)(2a+3b)$
13. 3.39
14. (b) 0.511
15. (b) 2 positive roots in the intervals
 $[0.5, 1]$ and $[1, 1.5]$

 (c) -1.62, $x = -\sqrt{\left(2 - \dfrac{1}{x}\right)}$

CHAPTER 2

Exercise 2a – p. 20

1. (a) $\left(\frac{5}{2}, 4\right)$ (b) $\left(\frac{5}{2}, \frac{1}{2}\right)$

(c) $\left(3, \frac{7}{2}\right)$ (d) $\left(\dfrac{a+b}{2}, \dfrac{a+b}{2}\right)$

2. (a) $\left(\frac{1}{2}, 1\right)$ (b) $\left(-\frac{1}{2}, -1\right)$

(c) $(-2, -3)$ (d) $\left(\dfrac{t^2+1}{2t}, \dfrac{t^2+2t}{2}\right)$

3. $(2, -4)$

4. (b) $\left(-\frac{7}{2}, -\frac{1}{2}\right)$ (c) $\frac{35}{2}$

5. (a) $\sqrt{5}(2+\sqrt{2})$ (b) $\left(0, \frac{9}{2}\right)$ (c) $2\frac{1}{2}$

6. $(-5, -3)$

7. $(0, -11)$

8. $(-2, 8)$

Exercise 2b – p. 24

1. (a) Parallel   (b) Neither

    (c) Perpendicular (d) Neither

    (e) Parallel

2. $x - 2y + 9 = 0$

3. (a) $2x - y + 5 = 0$ (b) $x + 2y - 5 = 0$

4. $10x - 26y - 1 = 0$

5. $2sx + 4ty - 3s^2 - 8t^2 = 0$

6. $4x - 3y - 5 = 0$

7. $6x + 9y - 5 = 0$

8. $(-8, -4)$, $\sqrt{130}$

10. $(q-4)(q+3) + (p-1)(p-5) = 0$

Exercise 2c – p. 27

1. (a) $A(0, 2, 6)$, $B(5, 2, 6)$, $C(5, 2, 0)$,

    $D(0, 2, 0)$, $E(0, 0, 6)$, $F(5, 0, 6)$,

    $G(5, 0, 0)$

    (b) $A(0, 2, 1)$, $C(3, 2, 0)$, $D(0, 2, 0)$,

    $E(0, 0, 1)$, $F(3, 0, 1)$, $G(3, 0, 0)$

2. (a) $A(4, 0, 0)$, $B(4, 0, 2)$, $C(-4, 0, 2)$,

    $D(-4, 0, 0)$, $E(4, 3, 0)$, $F(4, 3, 2)$,

    $G(-4, 3, 2)$, $H(-4, 3, 0)$

    (b) $A(5, 0, 0)$, $B(5, 0, 4)$, $C(-4, 0, 4)$,

    $D(-4, 0, 0)$, $E(5, 6, 0)$, $H(-4, 6, 0)$

3. (a) $A(3, 0, 2)$, $C(0, 5, 2)$, $D(0, 0, 2)$,

    $E(3, 0, 0)$, $F(3, 5, 0)$, $G(0, 5, 0)$

    (b) (i) $\sqrt{13}$ (ii) $\sqrt{38}$ (iii) $\sqrt{29}$

Exercise 2d – p. 28

1. (a) (i) $(3, 0, 0)$ (ii) $(3, 7, 0)$

    (iii) $(0, 7, 4)$

    (b) (i) $\left(0, \frac{7}{2}, 0\right)$ (ii) $\left(\frac{3}{2}, \frac{7}{2}, 4\right)$

    (c) (i) 5   (ii) $\sqrt{74}$ (iii) $\sqrt{74}$

2. (b) 6

3. (b) $\sqrt{29}$

4. (a) $31.0°$ (b) 70.0 sq units

5. (a) $A\left(\frac{9}{14}\pi, \frac{3}{14}\pi, \sin\frac{27}{196}\pi\right)$,

    $B\left(\frac{2}{7}\pi, \frac{9}{14}\pi, \sin\frac{9}{49}\pi\right)$,

    $C\left(\frac{5}{7}\pi, \frac{5}{7}\pi, \sin\frac{25}{49}\pi\right)$

    (b) 1.76

    (c) $z = \sin \pi y$

CHAPTER 3

Exercise 3a – p. 33

Answers are correct to 3 sf

1. 11.1

2. 13.7

3. 10.2

4. 8.83

5. 156

6. 113

7. 7.01

8. 89.1

9. 581

10. 141

11. 16.3

12. 28.1

13. 51.3

14. no; an angle and the side opposite to it are not known

Exercise 3b – p. 38

Answers are correct to the nearest degree.

1. $18°$

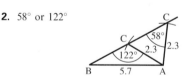

2. $58°$ or $122°$

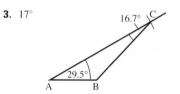

3. $17°$

4. 35°

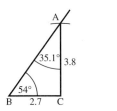

5. 57° or 123°

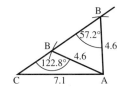

6. 30°

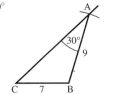

Exercise 3c – p. 41
1. 5.29 cm
2. 12.9 cm
3. 53.9 cm
4. 4.04 cm
5. 101 cm
6. 12.0 cm
7. 64.0 cm
8. 31.8 cm

Exercise 3d – p. 43
1. 38°
2. 55°
3. 45°
4. 94°
5. (a) 18°　　　(b) 126°
6. 29°
7. 114 cm, 68°

Exercise 3e – p. 44
1. $\angle A = 50°$; $a = 68$ cm; $b = 87.4$ cm
2. $\angle A = 22°$; $\angle B = 46°$; $c = 23.8$ cm
3. $\angle C = 70°$; $b = 17.5$ cm; $c = 17.6$ cm
4. $\angle B = 81°$; $a = 112$ cm; $\angle C = 41°$
5. $\angle A = 37°$; $a = 164$ cm; $c = 272$ cm
6. $\angle B = 34°$; $a = 37.0$ cm; $b = 22.8$ cm
7. $\angle C = 43°$; $b = 19.4$ cm; $c = 13.5$ cm
8. $\angle A = 52°$; $a = 33.2$ cm; $c = 41.5$ cm
9. $\angle B = 43°$; $\angle C = 60°$; $a = 27.1$ cm
10. $\angle A = 22°$; $\angle C = 33°$; $b = 30.3$ cm

11. 14.1 m
12. 40°, 53°, 87°

Mixed Exercise 3 – p. 45
1. (a) 116°　　　　　(b) 86°
2. $-\frac{24}{25}$
3. (a) $\dfrac{5}{\sqrt{39}}$　　　(b) $\dfrac{-5}{\sqrt{39}}$
4. (a) $\dfrac{1}{\sqrt{2}}, \dfrac{-1}{\sqrt{2}}$　　　(b) $\dfrac{3}{\sqrt{13}}, \dfrac{-2}{\sqrt{13}}$
5. $-\frac{5}{13}$
6. 9.05 cm
7. 4.82
8. 83°
9. 54° or 126°
10. 108°, 50°, 22°

CHAPTER 4

Exercise 4a – p. 47
1. 12 300 cm²
2. 2190 cm²
3. 1680 cm²
4. 453 square units
5. 42.9 square units
6. 51°, 21.0 cm²
7. 10.6 cm, 59.8 cm²
8. 52°, 151 cm (or 150 cm)
9. 5.25 cm

Exercise 4b – p. 50
1. 11.0 cm, 67.7 cm² (or 67.8 cm²)
2. 58.5 km
3. 477 m
4. $\angle$BAO = 74°, $\angle$CAO = 52°; 22 cm²
5. (a) 60.8 cm　　　　(b) 35°
　　(c) 42.8 cm　　　　(d) 2140 cm²
　　(e) 427 000 cm³

Exercise 4c – p. 57
1. (a) $4\sqrt{2}$ cm　　(b) $2\sqrt{29}$ cm
　　(c) $2\sqrt{29}$ cm　　(d) $2\sqrt{33}$ cm
2. (a) 6 cm, $6\sqrt{2}$ cm, $2\sqrt{34}$ cm
　　(b) 53°　　(c) 43°　　(d) 53°
3. (a) $5\sqrt{2}$ cm　　(b) $5\sqrt{5}$ cm
　　(c) $\sqrt{109}$ cm　　(d) 21°　　(e) 34°
4. $\frac{1}{3}$
5. P = 55.4°, Q = 31.2°, R = 93.4°
6. 71°
7. (a) 6 m　　(b) 9 m　　(c) 48°
　　(d) 240 m²

8. (a) 15.3 m   (b) 2.29 cm
    (c) 15°   (d) 30°
9. 420 m; 31°
10. 46°
11. (a) 55.6°   (b) 25.5 sq. units

Mixed Exercise 4 – p. 61
1. 75.8 cm^2
2. $\frac{1}{2}$; yes, $\angle A$ can be 30° or 150°
3. (a) 98°   (b) 19.8 cm^2
    (c) 3.96 cm
4. PQ = 8.29 cm, QR = 6 cm,
    RP = 3.46 cm; 9 cm^2
5. (a) 45°   (b) 35°
6. 35°
7. $\dfrac{a\sqrt{3}}{49}$; $\dfrac{2a}{7\sqrt{7}}$
8. $a\sqrt{\frac{37}{39}} = 0.974a$
11. (a) 1050 m   (b) 2720 m
    (c) 3470 m

Multiple Choice Exercise A – p. 66
1. C
2. E
3. A
4. D
5. A
6. E
7. A
8. D
9. T
10. F
11. T
12. T

Miscellaneous Exercise A – p. 67
1. $\theta = 41.4°$, $\angle ACB = 55.8°$, $\angle ACB > \theta$
2. $6y - 5x + 2 = 0$
3. $a = 1$, $b = -4$, $x + 1$
4. 1.87
5. $\cos R = \dfrac{3 + k^2}{4k}$; $1.5 < k < 2$
6. $x < 2$, $3 < x < 4$
7. (a) $a = 7$, $b = 3$ (b) 1.28
8. (a) $(5, 0)$
9. (b) 837 m   (c) 556 m
10. $a = -13$, $b = 12$, $x + 4$
11. (a) 62.5°   (b) 7.10 m
    (c) 45.2°   (d) 55.0°
12. $x^2 + x + 2$

13. (a) 7260 km   (b) 6020 km
14. (b) $x = -4, -1, 3$
15. $a = -1$, $b = 2$
16. 2.218, $x = 2(1 + e^{-x})$
17. (a) 53.1°   (c) 60°   (d) 43.9°
18. 1.613, 1.663

CHAPTER 5
Exercise 5a – p. 74
1. $\log_{10} 1000 = 3$
2. $\log_2 16 = 4$
3. $\log_{10} 10\,000 = 4$
4. $\log_3 9 = 2$
5. $\log_4 16 = 2$
6. $\log_5 25 = 2$
7. $\log_{10} 0.01 = -2$
8. $\log_9 3 = \frac{1}{2}$
9. $\log_5 1 = 0$
10. $\log_4 2 = \frac{1}{2}$
11. $\log_{12} 1 = 0$
12. $\log_8 2 = \frac{1}{3}$
13. $\log_q p = 2$
14. $\log_x 2 = y$
15. $\log_p r = q$
16. $10^5 = 100\,000$
17. $4^3 = 64$
18. $10^1 = 10$
19. $2^2 = 4$
20. $2^5 = 32$
21. $10^3 = 1000$
22. $5^0 = 1$
23. $3^2 = 9$
24. $4^2 = 16$
25. $3^3 = 27$
26. $36^{1/2} = 6$
27. $a^0 = 1$
28. $x^z = y$
29. $a^b = 5$
30. $p^r = q$

Exercise 5b – p. 75
1. 2
2. 6
3. 6
4. 4
5. 2
6. 3
7. $\frac{1}{2}$
8. -2
9. -1

10. $\frac{1}{2}$
11. 0
12. 1
13. $\frac{1}{3}$
14. 0
15. $\frac{1}{3}$
16. 3

Exercise 5c – p. 76

1. $\log p + \log q$
2. $\log p + \log q + \log r$
3. $\log p + \log q$
4. $\log p + \log q - \log r$
5. $\log p - \log q - \log r$
6. $2 \log p + \log q$
7. $\log q - 2 \log r$
8. $\log p + \frac{1}{2} \log q$
9. $2 \log p + 3 \log q - \log r$
10. $\frac{1}{2} \log q - \frac{1}{2} \log r$
11. $n \log q$
12. $n \log p + m \log q$
13. $\log 2 + \log p + \log q$
14. $\log p + \log q - \log 2$
15. $\log 2 + \log p + 2 \log q$
16. $\log p - \log q - \log 2$
17. $\log pq$
18. $\log p^2 q$
19. $\log q/r$
20. $\log q^3 p^4$
21. $\log p^n/q$
22. $\log pq^2/r^3$
23. $\log p/2$
24. $\log p^2/2^p$
25. $\log \dfrac{pq}{3}$
26. $\log (p + 2)/(q - 2)$

Exercise 5d – p. 78

1. (a) 0.602   (b) 1.61
   (c) −0.699   (d) −1.61
   (e) 0.531   (f) 0.587

2. (a) $\dfrac{\log 8}{\log 3}$ (b) $\dfrac{1}{\log 5}$ (c) $\dfrac{\log 5}{2}$

3. (a) $\dfrac{\ln 3}{\ln 4}$ (b) $\dfrac{\ln 2}{\ln 7}$ (c) $\dfrac{\ln 8.2}{\ln 0.1}$

4. (a) $\dfrac{\log_a y}{\log_a x}$ (b) $\dfrac{1}{\log_a x}$ (c) $\dfrac{\log_a 8}{\log_a y}$

5. (a) 1.19   (b) 2.86   (c) 2.46
   (d) −0.415   (e) −1.69   (f) 0.846

Exercise 5e – p. 79

1. 1.63
2. 0.861
3. 10.9
4. 3.20
5. 1.16
6. 2.77
7. 1.42
8. 1.14
9. −0.104
10. 1.22
11. 0.195
12. 0.536

Exercise 5f – p. 80

1. −1.1
2. 2
3. 3
4. 2
5. 3
6. $\sqrt{3}$ or $1/\sqrt{3}$
7. (a) $x = 1.05$ (3 sf)
   (b) $x = 1$
   (c) $x = -2$
   (d) $x = 2^{1/\sqrt{2}}$
   (e) $x = 2^{11} \times 3^{\frac{1}{3}} = 2954$ (4 sf)
   (f) $x = 1.09$ (3 sf)
8. $x = \frac{1}{2}$, $y = 1$
9. $x = 2$, $y = 4$
10. $x = 1$, $y = 0$
11. $x = 1$, $y = 0$
12. (a) $\log x/\log e$
   (b) one-way stretch parallel to Oy by

     factor $\dfrac{1}{\log e}$

   (c)

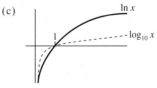

   (d) $y = 3^x$
   (e)

13. (a) increases slowly; when $a > 1$, powers of a increase when x increases

(b) for $x > 1$, $\log_a x < 0$ and decreases as x increases; powers of a number less than 1 need to be negative to give a result greater than 1.

CHAPTER 6

Exercise 6a – p. 85

1. (a) $60°$, $300°$ (b) $59.0°$, $239.0°$
(c) $41.8°$, $138.2°$

2. (a) $-140.2°$, $39.8°$
(b) $-131.8°$, $131.8°$
(c) $-150°$, $-30°$

3. $-\frac{1}{2}\pi$, $\frac{1}{2}\pi$

4. (a) 1 (b) $-\sqrt{2}$ (c) -2

5.

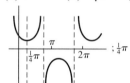 ; $\frac{1}{4}\pi$

6.

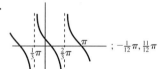 ; $-\frac{1}{12}\pi$, $\frac{11}{12}\pi$

8. $\cot\theta = \tan(\frac{1}{2}\pi - \theta)$

Exercise 6b – p. 89

	$\sin\theta$	$\cos\theta$	$\tan\theta$
1. (a)	$-\frac{12}{13}$	$-\frac{5}{13}$	$\frac{12}{5}$
(b)	$\frac{3}{5}$	$-\frac{4}{5}$	$-\frac{3}{4}$
(c)	$\frac{7}{25}$	$\frac{24}{25}$	$\frac{7}{24}$
(d)	0	±1	0

2. $\tan^4 A$
3. 1
4. $\sec\theta\,\text{cosec}\,\theta$
5. $\sec^2\theta$
6. $\tan\theta$
7. $\sin^3\theta$
8. $x^2 - y^2 = 16$
9. $b^2x^2 - a^2y^2 = a^2b^2$
10. $y^2(4 + x^2) = 36$
11. $(1-x)^2 + (y-1)^2 = 1$
12. $y^2(x^2 - 4x + 5) = 4$
13. $x^2(b^2 - y^2) = a^2b^2$

Exercise 6c – p. 92

1. $30°$, $150°$, $210°$, $330°$
2. $70.5°$, $109.5°$, $250.5°$, $289.5°$
3. $36.9°$, $143.1°$, $216.9°$, $323.1°$
4. $45°$, $225°$
5. $270°$
6. 0, $180°$, $360°$
7. $57.7°$, $122.3°$, $237.7°$, $302.3°$
8. $190.1°$, $349.9°$
9. $38.2°$, $141.8°$
10. $30°$, $150°$
11. $30°$, $150°$
12. $0°$, $131.8°$, $228.2°$, $360°$
13. ±0.723 rad
14. -0.314 rad, -2.83 rad
15. $-\frac{3}{4}\pi$, -0.245 rad, $\frac{1}{4}\pi$, 2.90 rad
16. $-\pi$, $-\frac{1}{3}\pi$, $\frac{1}{3}\pi$, π
17. $-\pi$, $-\frac{2}{3}\pi$, 0, $\frac{2}{3}\pi$, π
18. $-\frac{1}{2}\pi$, $\frac{1}{6}\pi$, $\frac{1}{2}\pi$, $\frac{5}{6}\pi$
19. $-\pi$, $-\frac{1}{6}\pi$, 0, $\frac{1}{6}\pi$, π
20. $-\frac{1}{2}\pi$, $\frac{1}{2}\pi$

Exercise 6d – p. 94

1. $22.5°$, $112.5°$, $202.5°$, $292.5°$
2. $40°$, $80°$, $160°$, $200°$, $280°$, $320°$
3. no values in the given range
4. $12°$, $60°$, $84°$, $132°$, $156°$, $204°$, $228°$, $276°$, $300°$, $348°$
5. no values in the given range
6. $25.5°$, $154.5°$, $205.5°$, $334.5°$
7. $135°$, $315°$
8. $240°$
9. $60°$, $240°$
10. 0, $120°$, $360°$
11. $-\frac{7}{12}\pi$, $\frac{1}{12}\pi$
12. $-\frac{11}{12}\pi$, $\frac{1}{12}\pi$
13. 0, $\frac{2}{3}\pi$
14. $-\pi$, $-\frac{1}{3}\pi$, π
15. $149.5°$, $59.5°$, $30.5°$, $120.5°$
16. $-105.2°$, $14.8°$, $134.8°$, $-74.8°$, $45.2°$, $165.2°$
17. $\pm63.6°$
18. $\frac{1}{6}\pi$, $\frac{5}{12}\pi$, $\frac{2}{3}\pi$, $\frac{11}{12}\pi$, $\frac{7}{6}\pi$, $\frac{17}{12}\pi$, $\frac{5}{3}\pi$, $\frac{23}{24}\pi$
19. $\frac{1}{15}\pi$, $\frac{1}{3}\pi$, $\frac{7}{15}\pi$, $\frac{11}{15}\pi$, $\frac{13}{15}\pi$, $\frac{17}{15}\pi$, $\frac{19}{15}\pi$, $\frac{23}{15}\pi$, $\frac{5}{3}\pi$, $\frac{29}{15}\pi$
20. $\frac{3}{2}\pi$

Mixed Exercise 6 – p. 94

1. $x^2 + \dfrac{1}{y^2} = 1$

2. $\sin \beta = \pm \dfrac{\sqrt{3}}{2}$, $\tan \beta = \pm \sqrt{3}$

3. $\dfrac{2}{\sin^2 \theta}$; $\frac{1}{4}\pi$, $\frac{3}{4}\pi$, $\frac{5}{4}\pi$, $\frac{7}{4}\pi$

4. $60°$, $109.5°$

5. (a)

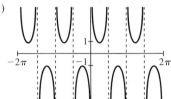

(b)

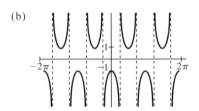

6. 1

8. $-\frac{29}{36}\pi$, $-\frac{17}{36}\pi$, $-\frac{5}{36}\pi$, $\frac{7}{36}\pi$, $\frac{19}{36}\pi$, $\frac{31}{36}\pi$

9. (a) $(x-2)^2 + (y+1)^2 = 1$
(b) $(x+3)^2 = 1 + (2-y)^2$

10. $\frac{1}{8}\pi$, $\frac{3}{8}\pi$, $\frac{5}{8}\pi$, $\frac{7}{8}\pi$

12. $\sin^2 A$

13. 0, $180°$, $360°$, $70.5°$, $289.5°$

14. $\sec^2\theta \tan^2\theta$

CHAPTER 7

Exercise 7b – p. 100

1. 0

2. $\frac{1}{2}$

3. $\frac{1}{4}(\sqrt{6} - \sqrt{2})$

4. $-(2 + \sqrt{3})$

5. $\frac{1}{4}(\sqrt{6} - \sqrt{2})$

6. $\frac{1}{4}(\sqrt{6} + \sqrt{2})$

7. $\sin 3\theta$

8. 0

9. $\tan 3A$

10. $\tan \beta$

11. (a) $\frac{3}{5}$ (b) $-\frac{4}{5}$ (c) $-\frac{3}{4}$

12. (a) 1, $115°$ (b) 1, $30°$
(c) 1, $310°$ (d) 1, $330°$

18. $67.5°$, $247.5°$

19. $7.4°$, $187.4°$

20. $37.9°$, $217.9°$

21. $15°$, $195°$

Exercise 7c – p. 104

1. $\frac{1}{2}$

2. $\dfrac{1}{\sqrt{2}}$

3. $\frac{1}{2}\sin 2\theta$

4. $\cos 8\theta$

5. $-\dfrac{1}{\sqrt{3}}$

6. $\tan 6\theta$

7. $-\dfrac{1}{\sqrt{2}}$

8. $\dfrac{1}{\sqrt{2}}$

9. (a) $\frac{24}{25}$, $-\frac{7}{25}$ (b) $\frac{336}{625}$, $\frac{527}{625}$
(c) $\frac{120}{169}$, $-\frac{119}{169}$

10. (a) $-\frac{336}{527}$ (b) $\frac{527}{625}$
(c) $-\frac{336}{625}$ (d) $\frac{164\,833}{390\,625}$

11. (a) $x(1 - y^2) = 2y$
(b) $x = 2y^2 - 1$
(c) $x = 1 - \dfrac{2}{y^2}$
(d) $2x^2 y + 1 = y$

12. (a) $-\cos 2x$
(b) $3 - \cos 2x$
(c) $\frac{1}{2}(\cos 2x + 3)$
(d) $\frac{1}{2}(\cos 2x + 1)(3 + \cos 2x)$
(e) $\frac{1}{4}(1 + \cos 2x)^2$
(f) $\frac{1}{4}(1 - \cos 2x)^2$

14. (a) $\frac{3}{2}\pi$, $\frac{1}{6}\pi$, $\frac{5}{6}\pi$
(b) $\frac{1}{2}\pi$, $\frac{7}{6}\pi$, $\frac{11}{6}\pi$, $\frac{3}{2}\pi$
(c) 0, $\frac{2}{3}\pi$, $\frac{4}{3}\pi$, 2π
(d) $\frac{1}{6}\pi$, $\frac{5}{6}\pi$, $\frac{1}{2}\pi$, $\frac{3}{2}\pi$
(e) $+\frac{1}{3}\pi$, $\frac{5}{3}\pi$
(f) $\frac{1}{4}\pi$, $\frac{1}{2}\pi$, $\frac{5}{4}\pi$, $\frac{3}{2}\pi$

15. (a) (i) $\dfrac{1 - 2\tan \theta - \tan^2 \theta}{1 + \tan^2 \theta}$
(ii) $\dfrac{1}{\tan \theta}$
(b) (i) $135°$ (ii) $45°$

Mixed Exercise 7 – p. 105

1. $y = 1 - 2x^2$

4. $\frac{56}{65}, -\frac{16}{65}$

5. $x = 2y - 1$

6. $-155.7°, 24.3°, -114.3°, 65.7°$

7. $-\pi, 0, \pi$

9. $\cot^2 x$

10. $90°, 270°$

11. (a) $2 - \cos 2\theta$ (b) $2 + 2\cos 4\theta$

12. (e) $y = c\sin(x + \alpha)$

Multiple Choice Exercise B – p. 108

1. A

2. E

3. D

4. B

5. C

6. C

7. A

8. B

9. T

10. T

11. F

12. F

13. F

14. T

Miscellaneous Exercise B – p. 109

1. $194°, 346°$

2. 0.774

3. $0, 60°, 120°, 180°$

4. 3

5. $45.6°, 91.9°, 268.1°, 314.4°$

6. $x < -7.21$

7.

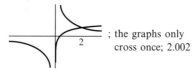

; the graphs only cross once; 2.002

8. $75°, 195°, 255°$

9. $-108°, 108°$

10. (a) $\frac{1}{3}, -\frac{1}{2}$

(b) $70.5°, 120°, 240°, 289.5°$

11. (i) translation by 1 unit left, one-way stretch parallel to y-axis, scale factor 2

(ii) $\ln 3/\ln 2$, one-way stretch parallel to x-axis, scale factor $1/a$

12. (a) $60°, -60°$ (b) $37.5°, -22.5°$

13. (b) $(x - 1)(x + 2)(3x + 2)$

(c) $90°, 228°, 318°$

14. 6

15. $(y^3 + 12)/6$

CHAPTER 8

Exercise 8a – p. 114

1. $2; -\frac{1}{2}$

2. $-\frac{1}{3}; 3$

3. $\frac{1}{4}; -4$

4. $6; -\frac{1}{6}$

5. $1; -1$

6. $5; -\frac{1}{5}$

7. $11; -\frac{1}{11}$

8. $-11; \frac{1}{11}$

9. $4; -\frac{1}{4}$

10. $\frac{4}{27}; -\frac{27}{4}$

11. $\frac{5}{4}; -\frac{4}{5}$

12. $2; -\frac{1}{2}$

13. $(2, 2)$ and $(-2, 4)$

14. $(1, 0)$ and $\left(-\frac{1}{3}, \frac{4}{27}\right)$

15. $(3, 0)$ and $(-3, 18)$

16. $(-1, -2)$ and $(1, 2)$

17. $(1, -16)$

18. $\left(-1, \frac{1}{4}\right)$

19. $(0, -5)$

20. $(1, -2)$ and $(-1, 2)$

Exercise 8b – p. 116

1. $6(3x + 1)$

2. $4(x - 3)^3$

3. $20(4x + 5)^4$

4. $21(2 + 3x)^6$

5. $18(6x - 2)^2$

6. $-10(4 - 2x)^4$

7. $-10(1 - 5x)$

8. $-6(3 - 2x)^2$

9. $-12(4 - 3x)^3$

10. $-3(3x + 1)^{-2}$

11. $-8(2x - 5)^{-5}$

12. $10(1 - 2x)^{-6}$

13. $(2x + 3)^{-1/2}$

14. $-(8 - 3x)^{-2/3}$

15. $\dfrac{-2}{\sqrt{(1 - 4x)}}$

16. $\dfrac{7}{(2 - 7x)^2}$

17. $\dfrac{1}{2(3 - x)^{3/2}}$

18. $\dfrac{10}{(1 - 5x)^3}$

Exercise 8c – p. 117

1. $\dfrac{3}{x}$

2. $2x\,e^{x^2+1}$

3. $\dfrac{1}{x}$

4. $\frac{1}{2}e^{\frac{1}{2}(x+1)}$

5. $-\dfrac{2}{x}$

6. $\dfrac{2x}{x^2-3}$

7. $\dfrac{3}{3x-4}$

8. $-2\,e^{-2x}$

9. $\dfrac{1}{2x}$

10. $\dfrac{6}{2x-3}$

11. $\dfrac{1}{x}-\dfrac{1}{x+1}$

12. $\dfrac{1}{2(x-1)}$

13. (a) e^{2x}, e^{x^2} (b) $2e^{2x}$, $2x\,e^{x^2}$

Exercise 8d – p. 121

1. $(0, 0)$, point of inflexion

2. $(-1, -8)$, minimum; $(1, 8)$, maximum

3. $(-1, 2)$, maximum; $(1, -2)$, minimum

4. $(\frac{1}{2}, \frac{1}{4})$, maximum

5. $(-2, -4)$, maximum; $(2, 4)$, minimum

6. $(-1, 2)$ and $(1, 2)$, both minimum

7. $(-\sqrt{\frac{3}{2}}, 3\sqrt{\frac{3}{2}})$, maximum; $(0, 0)$, point of inflexion; $(\sqrt{\frac{3}{2}}, -3\sqrt{\frac{3}{2}})$, minimum

8. $(0, 0)$, minimum

9. $(1, \frac{3}{2})$, minimum

10. (a) y decreases to zero, then increases, i.e. y has a minimum value at O.

(b) $\dfrac{dy}{dx}$ increases from $-$ve values through 0 at O to +ve value

(c) $\dfrac{dy}{dx} = 3x^3$

(d) 0; $\dfrac{dy}{dx}$ has a stationary value at O.

11. Curves become less steep for $-1 < x < 1$ and steeper for $x < -1$ and $x > 1$

Exercise 8e – p. 124

1. $3(x-1)(x-3)$

2. $\dfrac{3\sqrt{x}}{2}-\dfrac{3}{\sqrt{x}}$

3. $(x-2)^4(6x+8)$

4. $(2x+3)^2(8x+3)$

5. $(x-1)^3(5x+3)$

6. $\dfrac{1}{2\sqrt{x}}(x-3)^3+3\sqrt{x}(x-3)^2$

$= \dfrac{(x-3)^2(7x-3)}{2\sqrt{x}}$

7. $\dfrac{(x+5)^3(3x-17)}{(x-3)^2}$

8. $\dfrac{2}{(3x+2)^2}$

9. $\dfrac{(2x-7)^{1/2}(x+7)}{x^2}$

10. $\dfrac{x^2(7x-6)}{2\sqrt{(x-1)}}$

11. $3(x+3)^{-2}$

12. $2x(2x-3)(4x-3)$

13. $x\,e^x$

14. $2e^{2x}+2x^2\,e^{2x}$

15. $1+\ln x$

16. $\dfrac{2x}{2x-1}+\ln(2x-1)$

17. $e^{3x-1}+3x\,e^{3x-1}$

18. $\dfrac{x^2-1}{2x}+2x\ln\sqrt{x}$

Exercise 8f – p. 126

1. $\dfrac{2x(x-3)-(x-3)^2}{x^2}=\dfrac{(x-3)(x+3)}{x^2}$

2. $\dfrac{(x+3)(2x)-x^2}{(x+3)^2}=\dfrac{x(x+6)}{(x+3)^2}$

3. $\dfrac{-x^2-2x(4-x)}{x^4}=\dfrac{x-8}{x^3}$

4. $\dfrac{2x^3(x+1)-(x+1)^2(3x^2)}{x^6}$

$= \dfrac{-(x+1)(x+3)}{x^4}$

5. $\dfrac{4(1-x)^3+12x(1-x)^2}{(1-x)^6}=\dfrac{4(1+2x)}{(1-x)^4}$

6. $\dfrac{(x-2)(4x)-2x^2}{(x-2)^2}=\dfrac{2x(x-4)}{(x-2)^2}$

7. $\dfrac{\frac{5}{3}x^{2/3}(3x-2)-3x^{5/3}}{(3x-2)^2}=\dfrac{2x^{2/3}(3x-5)}{3(3x-2)^2}$

8. $\dfrac{-3(1-2x)^2}{x^4}$

9. $\dfrac{(3x-2)(x+1)^{3/2}}{2x^2}$

10. $\dfrac{-e^x}{(e^x-1)^2}$

11. $\dfrac{2x+1-2x\ln x}{x(2x+1)^2}$

12. $\dfrac{(x+1)\ln(x+1)-x\ln x}{x(x+1)[\ln(x+1)]^2}$

Mixed Exercise 8 – p. 127

1. $\dfrac{3x+2}{2\sqrt{(x+1)}}$

2. $6x(x^2-8)^2$

3. $\dfrac{1-x^2}{(x^2+1)^2}$

4. $\dfrac{-4x^3}{3(2-x^4)^{2/3}}$

5. $\dfrac{2x}{(x^2+2)^2}$

6. $\frac{1}{2}x(5\sqrt{x}-8)$

7. $6x(x^2-2)^2$

8. $\dfrac{1-2x}{2\sqrt{(x-x^2)}}$

9. $\dfrac{\sqrt{x}+2}{2(\sqrt{x}+1)^2}$

10. $\dfrac{x(5x-8)}{2\sqrt{(x-2)}}$

11. $\dfrac{-(3x+4)}{2x^3\sqrt{(x+1)}}$

12. $6x^5(x^2+1)^2(2x^2+1)$

13. $\dfrac{x}{\sqrt{(x^2-8)}}$

14. $x^2(5x^2-18)$

15. $6x(x^2-6)^2$

16. $\dfrac{-(x^2+6)}{(x^2-6)^2}$

17. $-8x^3(x^4+3)^{-3}$

18. $\dfrac{(2-x)^2(2-7x)}{2\sqrt{x}}$

19. $\dfrac{2+5x}{2\sqrt{x}(2-x)^4}$

20. $(x-2)(3x-4)$

21. $30x^2(2x^3+4)^4$

22. $e^x(x+1)$

23. $x(2\ln x+1)$

24. $e^x(x^3+3x^2-2)$

25. $12x\ln(x-2)+\dfrac{6x^2}{(x-2)}$

26. xe^x

27. $x\ln x+\dfrac{(x^2+4)}{2x}$

28. $\dfrac{4+3x}{2\sqrt{(2+x)}}$

29. $\frac{1}{2}\ln(x-5)+\dfrac{x}{2(x-5)}$

30. $(x^2+2x-2)e^x$

31. $\dfrac{1-x}{e^x}$

32. $\dfrac{e^x(x-2)}{x^3}$

33. $\dfrac{1-3\ln x}{x^4}$

34. $\dfrac{x\ln x-2(x+1)}{2x\sqrt{(x+1)}(\ln x)^2}$

35. $\dfrac{e^x(x^2-2x-1)}{(x^2-1)^2}$

36. $\dfrac{-2}{(e^x-e^{-x})^2}$

37. $4e^{4x}$

38. $\dfrac{2x}{x^2-1}$

39. $2xe^{x^2}$

40. $-6e^{(1-x)}$ or $-6e(e^{-x})$

41. $2xe^{(x^2+1)}$

42. $\dfrac{1}{2(x+2)}$

43. $\dfrac{2\ln x}{x}$

44. $\dfrac{-1}{x(\ln x)^2}$

45. $\frac{1}{2}\sqrt{(e^x)}$

46. $1+\ln x$

47. $\frac{8}{3}(4x-1)^{-1/3}$

48. $\dfrac{e^x(x-2)}{(x-1)^2}$

49. $3(\ln 10)10^{3x}$

50. $\dfrac{2x}{(1+x^2)^2}$

51. $\dfrac{2}{x^2e^{2/x}}$

52. $\dfrac{-e^x}{1-e^x}$ or $\dfrac{e^x}{e^x-1}$

53. $3x^2e^{3x}(x+1)$

54. $\dfrac{e^{x/2}(x-10)}{2x^6}$

55. $\dfrac{5x+9}{x(x+3)}$

56. $\dfrac{4}{x}(\ln x)^3$

57. $\dfrac{(x+3)^2(x^2-6x+6)}{(x^2+2)^2}$

58. $\dfrac{8x}{x^2+1}$

59. $\dfrac{dy}{dx}=\dfrac{1}{x(x+1)}$; $\dfrac{d^2y}{dx^2}=\dfrac{(2x+1)}{x^2(x+1)^2}$

60. $\dfrac{dy}{dx}=\dfrac{-4e^x}{(e^x-4)^2}$; $\dfrac{d^2y}{dx^2}=\dfrac{4e^x(e^x+4)}{(e^x-4)^3}$

61. (a) $(-1,-2e)$, $(3,6e^{-3})$
   (b) minimum when $x=-1$,
     maximum when $x=3$
   (c) (i) $y\to 0$ (ii) $y\to\infty$
   (d)

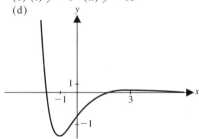

   (e) 2; the curve crosses the x-axis twice.
   (f) $\pm\sqrt{3}$

62. (a) $\simeq 1.5$; $y=6.7(15^{x/14})$
   (b) (i) 37   (ii) 79
   (c) ≈ 5.5 per day; larger than the actual
     increase

CHAPTER 9

Exercise 9a – p. 134

1. (a) $1+36x+594x^2+5940x^3$
   (b) $1-18x+144x^2-672x^3$
   (c) $1+35x+525x^2+4375x^3$
   (d) $1-\frac{20}{3}x+\frac{190}{9}x^2-\frac{380}{9}x^3$
   (e) $1-4x+\frac{20}{3}x^2-\frac{160}{27}x^3$
   (f) $1+12x+\frac{342}{5}x^2+\frac{6156}{25}x^3$

2. (a) $1024+5120x+11\,520x^2$
   (b) $128-672x+1512x^2$
   (c) $\left(\frac{3}{2}\right)^9+\frac{3^{10}}{2^7}x+\frac{3^9}{2^3}x^2$

3. (a) $336x^2$ (b) $-10x$
   (c) $1760x^3$ (d) $9x^2$
   (e) $-455(2)^{12}x^3$ (f) $7920x^4$
   (g) $63x^5$ (h) $3360p^6q^4$
   (i) $56a^3b^5$ (j) $16a^7b$

4. (a) $1-8x+27x^2$
   (b) $1+19x+160x^2$
   (c) $2-19x+85x^2$
   (d) $1-68x+2136x^2$

5. (a) $64x^6-192x^5+240x^4-160x^3$
       $+60x^2-12x+1$
   (b) $243x^5-810x^4+1080x^3-720x^2$
       $+240x-32$
   (c) $x^9-36x^8+576x^7-5376x^6$
       $+32\,256x^5-129\,024x^4$
       $+344\,064x^3-589\,824x^2$
       $+589\,824x-262\,144$

6. $1+4x-11x^2-70x^3$

7. $5^5(25-140x+301x^2)$

8. $X=x(2-x)$, $1+40x+740x^2$

Exercise 9b – p. 137

1. (a) $1-20x$ (b) $1+\frac{6}{5}x$
   (c) $2^7+2^6(7x)$

4. (a) $1-50x$ (b) $256-1024x$
   (c) $1-19x$

5. $a=114$, $b=16$, $c=1$

6. $y=1-4x$

7. (a) $a=115$, $b=-28$, $c=3$
   (b) 0.0314 (4 sf)
   (c) $|x|\leqslant 0.068$

8. (a) $1+100x+\dfrac{(100)(99)}{2}x^2$
      $+\dfrac{(100)(99)(98)}{3!}x^3$
      $+\dfrac{(100)(99)(98)(97)}{4!}x^4$
      $+\dfrac{(100)(99)(98)(97)(96)}{5!}x^5$
   (b) $u_{n+1}=\dfrac{(100-n)u_n}{n+1}$

CHAPTER 10

Exercise 10a – p. 141

1. $\frac{40}{3}$

2. $\frac{4}{3}$

3. 4.55

4. 4.8

5. 60; no, outside range of t.

6. 120, 80

7. π

8. $v \propto l^3$

9. 0.3 m above bottom; no, because $h \neq 0$ for any value of t but the ball will reach the bottom, also near the bottom, the size of the ball is large compared with h.

Exercise 10b – p. 146

1. (a) $\left(\dfrac{1}{y}\right) = a(x) + b$

 (b) $(xy) = (x) + a$

 (c) $(y^2) = a(x) + b$

 (d) $(xy/y) = a(x^2/y) - b$

2. $a = 2$, $b = -4$ (exactly)

3. $a = 0.5$, $b = -2$

4. $a = 2$, $b = \frac{1}{2}$

Exercise 10c – p. 150

1. (a) $(y) = a(e^x/y) + b$

 (b) $(\ln y) = (n+2)(\ln x) + \ln a$

 (c) $(x) = a(\ln y) - k$

 There are other alternatives.

2. $a = 6$, $b = 4$

3. $a = 30$, $b = 2$

4. $k = 4$, $n = 0.07$

5. $k = 5500$, $n = 1.5$

6. (a) 10 000, 1.1

 (b) $\ln(y - 2000) = \ln a - x \ln b$

 (c) gradient and intercept different so values of constants different.

 (d) $y - 500 = 11\,500\,(1.1)^{-x}$

7. (a) $\frac{1}{30}$

 (d) $s = 730$, $s = \frac{1}{69}\,u^2 + 20$

CHAPTER 11

Exercise 11a – p. 153

1. Approximately equal

2. No

3. (c) $y = -\sin a$

 (d) No; gives different and much smaller values for gradients

Exercise 11b – p. 155

1. (a) $3\cos x$

 (b) $1 + \sin x$

 (c) $\cos x - 2\sin x$

 (d) $8x - 3\cos x$

2. (a) $\cos x + \sin x$ (b) $\cos \theta$

 (c) $-3\sin \theta$ (d) $5\cos \theta$

 (e) $3\cos \theta - 2\sin \theta$

 (f) $4\cos x + 6\sin x$

3. (a) -1 (b) 1 (c) -1

 (d) 1 (e) $2(\pi - 1)$ (f) 4

4. (a) $\frac{1}{6}\pi$ (b) $\frac{1}{6}\pi$ (c) $\frac{1}{4}\pi$

 (d) π

5. (a) $\left(\frac{1}{3}\pi, \sqrt{3} - \frac{1}{3}\pi\right)$, max;

 $\left(\frac{5}{3}\pi, -\sqrt{3} - \frac{5}{3}\pi\right)$, min

 (b) $\left(\frac{1}{6}\pi, \frac{1}{6}\pi + \sqrt{3}\right)$, max;

 $\left(\frac{5}{6}\pi, \frac{5}{6}\pi - \sqrt{3}\right)$, min

6. $y + \theta = 3 + \frac{1}{2}\pi$

7. $2\pi y + x = 2\pi^3 - \pi$

8. $(0, 1)$

Exercise 11c – p. 158

1. $4\cos 4x$

2. $2\sin(\pi - 2x)$ or $2\sin 2x$

3. $\frac{1}{2}\cos\left(\frac{1}{2}x + \pi\right)$ or $-\frac{1}{2}\cos\frac{1}{2}x$

4. $\dfrac{x\cos x - \sin x}{x}$

5. $-\dfrac{(\cos x + \sin x)}{e^x}$

6. $\dfrac{\cos x}{2\sqrt{(\sin x)}}$

7. $2\sin x\cos x$ or $\sin 2x$

8. $\cos^2 x - \sin^2 x$ or $\cos 2x$

9. $\cos x\,e^{\sin x}$

10. $-\tan x$

11. $e^x(\cos x - \sin x)$

12. $x^2\cos x + 2x\sin x$

13. $2x\cos x^2$

14. $-\sin x\,e^{\cos x}$

15. $3\cot x$

16. $\dfrac{\sin x}{\cos^2 x} = \sec x\tan x$

17. $\sec^2 x$

18. $\dfrac{-\cos x}{\sin^2 x} = -\csc x\cot x$

19. $\dfrac{-1}{\sin^2 x} = -\csc^2 x$

Mixed Exercise 11 – p. 158

1. (a) $-4\cos 4\theta$ (b) $1 + \sin \theta$

 (c) $3\sin^2 \theta\cos \theta + 3\cos 3\theta$

2. (a) $3x^2 + e^x$ (b) $2e^{(2x+3)}$
   (c) $e^x(\sin x + \cos x)$

3. (a) $-\dfrac{3}{x}$ (b) $-\dfrac{2}{x}$ (c) $\dfrac{1}{2x}$

4. (a) $3\cos x + e^{-x}$ (b) $\dfrac{1}{2x} + \dfrac{1}{2}\sin x$

   (c) $4x^3 + 4e^x - \dfrac{1}{x}$

   (d) $-\dfrac{1}{2}(e^{-x} + x^{-3/2}) - \dfrac{1}{x}$

5. $1 + \dfrac{1}{x} + \ln x$

6. $3\sin 6x$

7. $\frac{8}{3}(4x-1)^{-1/3}$

8. $9 - 18\sqrt{x} + 8x$

9. $(x^4 + 4x^3 + 3)/(x+1)^4$

10. $\dfrac{(x-1)\ln(x-1) - x\ln x}{x(x-1)\{\ln(x-1)\}^2}$

11. $-1/\sin x \cos x$ or $-2\operatorname{cosec} 2x$

12. $2x\sin x + x^2\cos x$

13. $e^x(x-2)/(x-1)^2$

14. $2\cos x/(1 - \sin x)^2$

15. $x(5x-4)/2\sqrt{(x-1)}$

16. $-2(1-x)^2(2x+1)$

17. $\dfrac{3}{2(x+3)} - \dfrac{x}{x^2+2}$

18. $\cos^2 x(4\cos^2 x - 3)$

19. $-\sin 2x\, e^{\cos^2 x}$

20. (a) $x = \ln 3$ (b) $x = 1\,(not\,-1)$
   (c) $x = \frac{1}{4}$

21. (a) 1 (b) $y - x = 1 - \frac{1}{2}\pi$
   (b) $y + x = 1 + \frac{1}{2}\pi$

22. (a) $1 + e$ (b) $y = x(1+e)$
   (c) $y(1+e) + x = (1+e)^2 + 1$

23. (a) 2 (b) $y = 2x + 1$
   (c) $2y + x = 2$

24. (a) -1 (b) $x + y = 3$
   (c) $x - y + 1 = 0$

25. (a) $\left(\frac{1}{2}\pi, 0\right)$, min; $\left(\frac{3}{2}\pi, 2\right)$, max
   (b) $\left(\frac{1}{6}\pi, \left\{\frac{1}{12}\pi + \frac{1}{2}\sqrt{3}\right\}\right)$, max;
     $\left(\frac{5}{6}\pi, \left\{\frac{5}{12}\pi - \frac{1}{2}\sqrt{3}\right\}\right)$, min
   (c) $(\ln 3, \{3 - 3\ln 3\}$, min; only one
     turning point

26. (a) $\left(\frac{1}{6}\pi, \left\{\frac{1}{2}\pi - \sqrt{3}\right\}\right)$
   (b) $(1, -1)$

27. (b) $\dfrac{\sin(x+\delta x) - \sin x}{\delta x}$
   (c) $\cos x$
   (d) δx is small

CHAPTER 12

Exercise 12a – p. 164

1. $2x + 2y\dfrac{dy}{dx} = 0$

2. $2x + y + (x + 2y)\dfrac{dy}{dx} = 0$

3. $2x + x\dfrac{dy}{dx} + y = 2y\dfrac{dy}{dx}$

4. $-\dfrac{1}{x^2} - \dfrac{1}{y^2}\dfrac{dy}{dx} = e^y\dfrac{dy}{dx}$

5. $-\dfrac{2}{x^3} - \dfrac{2}{y^3}\dfrac{dy}{dx} = 0$

6. $\dfrac{x}{2} - \dfrac{2y}{9}\dfrac{dy}{dx} = 0$

7. $\cos x + \cos y\dfrac{dy}{dx} = 0$

8. $\cos x\cos y - \sin x\sin y\dfrac{dy}{dx} = 0$

9. $e^y + xe^y\dfrac{dy}{dx} = 1$

10. $(1 + x)\dfrac{dy}{dx} = -y$

11. $\dfrac{dy}{dx} = \pm\dfrac{1}{\sqrt{(2x+1)}}$

12. $\dfrac{d^2y}{dx^2} = \pm\dfrac{x}{\sqrt{(2-x^2)^3}}$

13. $\pm\frac{1}{4}\sqrt{2}$

15. (a) $xx_1 - 3yy_1 = 2(y + y_1)$
   (b) $x(2x_1 + y_1) + y(2y_1 + x_1) = 6$

17. $3x + 12y - 7 = 0$

Exercise 12b – p. 167

1. $3^x\ln 3$

2. $2(1.5)^x\ln 1.5$

3. $(2\ln 3)3^{2x}$

4. $3^x(1 + x\ln 3)$

5. $-3^x\ln 3$

6. $5a^x\ln a$

7. $\dfrac{dy}{dx} = 2x^{2x}(1 + \ln x)$

8. $\dfrac{dy}{dx}\left(\dfrac{1}{y+1} - \ln x\right) = \dfrac{y}{x}$

9. $\dfrac{1}{y}\dfrac{dy}{dx} = \ln(x + x^2) + \dfrac{1 + 2x}{1 + x}$

10. $4 + 8\ln 2$

11. (a) $x = \ln 2 - 1$
     (b) $(1, 0)$
     (c)

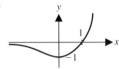

Exercise 12c – p. 171

1. (a) $\dfrac{1}{4t}$ (b) $-\tan\theta$ (c) $-\dfrac{4}{t^2}$

2. (a)

     (b)

     (c)

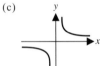

3. $\dfrac{dy}{dx} = 2t - t^2; \frac{3}{4}$

4. (a) $\frac{3}{2}t$ (b) $\frac{3}{2}\sqrt{x}$
     (c) $x = t^2 \Rightarrow t = \sqrt{x}$

5. (a) $x = 2y^2; \dfrac{dy}{dx} = \dfrac{1}{4y} = \dfrac{1}{4t}$

     (b) $x^2 + y^2 = 1; \dfrac{dy}{dx} = -\dfrac{x}{y} = -\tan\theta$

     (c) $xy = 4; \dfrac{dy}{dx} = -\dfrac{4}{x^2} = -\dfrac{4}{t^2}$

6. $\left(-\frac{1}{3}\sqrt{3}, \frac{2}{9}\sqrt{3}\right)$, max;

     $\left(\frac{1}{3}\sqrt{3}, -\frac{2}{9}\sqrt{3}\right)$, min

7. $\frac{1}{2}\pi$

8. $2x + y + 2 = 0$

9. $y = x; \left(-\frac{1}{2}\sqrt{2}, -\frac{1}{2}\sqrt{2}\right)$

Mixed Exercise 12 – p. 172

1. (a) $4y^3 \dfrac{dy}{dx}$ (b) $y^2 + 2xy \dfrac{dy}{dx}$

     (c) $-\dfrac{1}{y^2} \dfrac{dy}{dx}$ (d) $\ln y + \dfrac{x}{y} \dfrac{dy}{dx}$

     (e) $\cos y \dfrac{dy}{dx}$ (f) $e^y \dfrac{dy}{dx}$

     (g) $\dfrac{dy}{dx} \cos x - y \sin x$

     (h) $(\cos y - y \sin y) \dfrac{dy}{dx}$

2. $\dfrac{x}{2y}$

3. $-\dfrac{y^2}{x^2}$

4. $-\dfrac{2y}{3x}$

5. $3(1.1)^x \ln(1.1)$

6. $-\dfrac{(x-1)^3(3x+13)}{(x+3)^3}$

7. $-\dfrac{y(y+1)(3x+2)}{x(x+1)(y+2)}$

8. $3t/2$

9. $\dfrac{t}{t+1}$

10. $-\frac{3}{2}\cos\theta$

11. $-\dfrac{1}{t^2}$

12. $\dfrac{2y+7}{2y-2x+3}$

13. $2t - t^2$

Multiple Choice Exercise C – p. 175

1. A
2. D
3. D
4. C
5. C
6. C
7. D
8. B
9. F
10. F
11. T
12. T
13. F
14. F
15. F
16. T

Miscellaneous Exercise C – p. 177

1. -380
2. $1/2,\ -2$
3. $k = 0.039,\ a = 0.90,\ -0.3$
4. $a = 60.3,\ b = 6.05$
5. $1 + 8ax + 28a^2x^2;\ a = 1,\ b = -8$
6. (a)

   (b) $-1/t^2$ (d) $-1/8$
7. $33\,000,\ -1.4$
8. $73.5\,\text{m},\ 94.5\,\text{m}$
9. $1 + 5x + 10x^2 + 10x^3 + 5x^4 + x^5;\ 30$
10. (i) $x = 2(t_1 + t_2),\ y = 2(1/t_1 + 1/t_2)$
    (iii) $(2\sqrt{2},\ 4\sqrt{2}),\ (-2\sqrt{2},\ -4\sqrt{2})$
11. (ii) $(1,\ 3),\ (-1,\ -3),\ (3,\ 1),\ (-3,\ -1)$
    (iii) $(\sqrt{8},\ 0),\ (-\sqrt{8},\ 0),\ (0,\ \sqrt{8}),$
    $(0,\ -\sqrt{8});\ 3,\ 1/3$
12. (i) $1.65,\ 65\%$ (ii) 0.5 (iv) $1/2$
13. 3.3 to 3.5
14. 240
15. (i) (a) $2/x$
    (b) $2x\sin 3x + 3x^2\sin 3x$
    (ii) $1/4$
16. (a) $k = 3/2,\ p = 63,\ q = 189$
    (b) 126
17. $8x + 8x^3$

CHAPTER 13

All answers to indefinite integrals in this chapter require the term $+K$.

Exercise 13a – p. 183

1. $\frac{1}{16}(4x + 1)^4$
2. $\frac{1}{24}(2 + 3x)^8$
3. $\frac{1}{30}(5x - 4)^6$
4. $\frac{1}{24}(6x - 2)^4$
5. $-\frac{1}{3}(4 - x)^3$
6. $-\frac{1}{8}(3 - 2x)^4$
7. $-\frac{1}{3}(3x + 1)^{-1}$
8. $-\frac{1}{6}(2x - 5)^{-3}$
9. $\frac{1}{3}(4 - 3x)^{-1}$
10. $\frac{1}{3}(2x + 3)^{3/2}$
11. $\frac{1}{4}(3x - 1)^{4/3}$
12. $(2x - 1)^{1/2}$
13. $-\frac{2}{3}(8 - 3x)^{1/2}$

14. $-\frac{1}{6}(1 - 4x)^{3/2}$
15. $-\frac{1}{5}(3 - 2x)^{5/2}$
16. $\dfrac{1}{7(2 - 7x)}$
17. $-2\sqrt{(3 - x)}$
18. $-\frac{3}{10}(1 - 5x)^{2/3}$
19. $\frac{1}{4}e^{4x}$
20. $-4e^{-x}$
21. $\frac{1}{3}e^{(3x-2)}$
22. $-\frac{2}{5}e^{(1-5x)}$
23. $-3e^{-2x}$
24. $5e^{(x-3)}$
25. $2e^{(x/2+2)}$
26. $\dfrac{2^x}{\ln 2}$
27. $\dfrac{4^{(2+x)}}{\ln 4}$
28. $\dfrac{1}{2}e^{2x} - \dfrac{1}{2e^{2x}}$
29. $\dfrac{a^{(1-2x)}}{-2\ln a}$
30. $\dfrac{2^x}{\ln 2} + \dfrac{1}{3}x^3$
31. $\frac{1}{2}\{e^4 - 1\}$
32. $2\{e^2 - 1\}$
33. $1 - \dfrac{1}{e}$
34. $1 - e^2$
35. $12\frac{2}{3}$
36. -10
37. $\frac{1}{2}$
38. 6

Exercise 13b – p. 187

Answers can be expressed in other forms also. Where A is used, $A > 0$.

1. (a) $\frac{1}{2}\ln|x| + K$
   (b) $\frac{1}{2}\ln\{A|x|\}$
2. (a) $4\ln|x| + K$
   (b) $4\ln\{A|x|\}$ or $\ln(Ax^4)$
3. (a) $\frac{1}{3}\ln|3x + 1| + K$
   (b) $\frac{1}{3}\ln A|3x + 1|$ or $\ln A|\sqrt[3]{(3x + 1)}|$
4. (a) $-\frac{3}{2}\ln|1 - 2x| + K$
   (b) $-\frac{3}{2}\ln A|1 - 2x|$
   or $\ln A|(1 - 2x)^{-3/2}|$

5. (a) $2\ln|2+3x|+K$
(b) $\ln A(2+3x)^2$

6. (a) $-\frac{3}{2}\ln|4-2x|+K$
(b) $\ln A|(4-2x)^{-3/2}|$

or $\ln \dfrac{A}{|(4-2x)^{3/2}|}$

(or each of these results using $\ln|2-x|$)

7. (a) $-4\ln|1-x|+K$
(b) $\dfrac{A}{(1-x)^4}$

8. (a) $-\frac{5}{7}\ln|6-7x|+K$
(b) $\ln \dfrac{A}{|(6-7x)^{5/7}|}$

9. $3\ln 2$
10. $\frac{1}{2}\ln 2 = \ln\sqrt{2}$
11. $2\ln 2 = \ln 4$
12. $\ln 2$

Exercise 13c – p. 188

1. $-\frac{1}{2}\cos 2x$
2. $\frac{1}{7}\sin 7x$
3. $\frac{1}{4}\tan 4x$
4. $-\cos\left(\frac{1}{4}\pi + x\right)$
5. $\frac{3}{4}\sin\left(4x-\frac{1}{2}\pi\right)$
6. $\frac{1}{2}\tan\left(\frac{1}{3}\pi + 2x\right)$
7. $-\frac{2}{3}\cos(3x-\alpha)$
8. $-10\sin\left(\alpha-\frac{1}{2}x\right)$
9. $\frac{1}{3}\sin 3x - \sin x$
10. $\frac{1}{3}$
11. $-\frac{1}{4}$
12. 0
13. $\frac{1}{2}$

Exercise 13d – p. 188

1. $\frac{1}{2}\cos\left(\frac{1}{2}\pi - 2x\right)$
2. $\frac{1}{4}e^{(4x-1)}$
3. $\frac{1}{7}\tan 7x$
4. $\frac{1}{2}\ln|2x-3|$
5. $\sqrt{(2x-3)}$
6. $-1/\{3(3x-2)\}$
7. $5^x/\ln 5$
8. $\frac{3}{4}\ln(4x-1)$
9. $\frac{1}{9}(3x-5)^3$
10. $\frac{1}{4}e^{(4x-5)}$

11. $\frac{1}{6}(4x-5)^{3/2}$
12. $-\frac{1}{4}\cos\left(4x-\frac{1}{3}\pi\right)$
13. $-\frac{3}{2}\ln|1-x|$
14. $10^{(x+1)}/\ln 10$
15. $\frac{1}{3}\sin\left(3x-\frac{1}{3}\pi\right)$
16. $\frac{1}{9}(3x-4)^6$
17. $\frac{2}{3}\sqrt{2}$
18. $2e(e-1)$
19. $-\frac{1}{2}$
20. $5\ln 2$

Exercise 13e – p. 193

1. e^{x^4}
2. $-e^{\cos x}$
3. $e^{\tan x}$
4. e^{x^2+x}
5. $-e^{(1-\tan x)}$
6. $e^{(x+\sin x)}$
7. $e^{(1+x^2)}$
8. $e^{(x^3-2)}$
9. $\frac{1}{10}(x^2-3)^5$
10. $-\frac{1}{3}(1-x^2)^{3/2}$
11. $\frac{1}{6}(\sin 2x+3)^3$
12. $-\frac{1}{6}(1-x^3)^2$
13. $\frac{2}{3}(1+e^x)^{3/2}$
14. $\frac{1}{5}\sin^5 x$
15. $\frac{1}{4}\tan^4 x$
16. $\dfrac{1}{3(n+1)}(1+x^{n+1})^3$
17. $-\frac{1}{4}\cos^4 x$
18. $\frac{4}{9}(1+x^{3/2})^{3/2}$
19. $\frac{1}{12}(x^4+4)^3$
20. $-\frac{1}{4}(1-e^x)^4$
21. $\frac{2}{3}(1-\cos\theta)^{3/2}$
22. $\frac{1}{3}(x^2+2x+3)^{3/2}$
23. $\frac{1}{2}e^{(x^2+1)}$
24. $\frac{1}{2}(1+\tan x)^2$

Exercise 13f – p. 195

1. $\frac{1}{2}(e-1)$
2. $\frac{1}{5}$
3. $\frac{1}{2}(\ln 2)^2$
4. $\frac{1}{15}7^5$
5. $e-1$

6. 9

7. $\frac{1}{2}(e^3 - 1)$

8. $\frac{13}{24}$

9. $\frac{1}{3}(\ln 3)^3$

10. $\frac{7}{3}$

11. $e - 1$

12. $\frac{19}{9}$

Exercise 13g – p. 200

1. $x \sin x + \cos x$

2. $e^x(x^2 - 2x + 2)$

3. $\frac{1}{16} x^4 (4 \ln |3x| - 1)$

4. $-e^{-x}(x + 1)$

5. $3(\sin x - x \cos x)$

6. $\frac{1}{5} e^x(\sin 2x - 2\cos 2x)$

7. $\frac{1}{5} e^{2x}(\sin x - 2\cos x)$

8. $\frac{1}{32} e^{4x}(8x^2 - 4x + 1)$

9. $-\frac{1}{2} e^{-x}(\cos x + \sin x)$

10. $x(\ln |2x| - 1)$

11. $x e^x$

12. $\frac{1}{72}(8x - 1)(x + 1)^8$

13. $\sin\left(x + \frac{1}{6}\pi\right) - x\cos\left(x + \frac{1}{6}\pi\right)$

14. $\frac{1}{n^2}(\cos nx + nx \sin nx)$

15. $\frac{x^{n+1}}{(n+1)^2}[(n+1)\ln |x| - 1]$

16. $\frac{3}{4}(2x \sin 2x + \cos 2x)$

17. $\frac{1}{5} e^x(\sin 2x - 2\cos 2x)$

18. $(2 - x^2)\cos x + 2x \sin x$

19. $\frac{e^{ax}}{a^2 + b^2}(a \sin bx - b \cos bx)$

20. $\frac{1}{3} \sin\theta(3\cos^2\theta + 2\sin^2\theta)$

21. $\frac{1}{2} e^{x^2 - 2x + 4}$

22. $(x^2 + 1)e^x$

23. $-\frac{1}{4}(4 + \cos x)^4$

24. $e^{\sin x}$

25. $\frac{2}{15}\sqrt{(1 + x^5)^3}$

26. $\frac{1}{5}(e^x + 2)^5$

27. $\frac{1}{4} e^{2x-1}(2x - 1)$

28. $-\frac{1}{20}(1 - x^2)^{10}$

29. $\frac{1}{6} \sin^6 x$

Exercise 13h – p. 201

1. 1

2. $\frac{32}{3} \ln 2 - \frac{7}{4}$

3. e

4. $-\frac{1}{2}(e^\pi + 1)$

5. $\frac{16}{15}$

6. $\frac{1}{4}\pi^2 - 2$

7. $\frac{1}{4}(1 + e^2)$

8. $e - 2$

9. $-\frac{1}{2}$

Mixed Exercise 13 – p. 201

1. $\frac{1}{4} e^{2x}(2x^2 - 2x + 1)$

2. e^{x^2}

3. $\frac{1}{6}(3 \tan x - 4)^2$

4. $\frac{1}{4}(x + 1)^2\{2\ln(x + 1) - 1\}$

5. $\frac{1}{4} \tan^4 x$

6. $(x^2 - 2)\sin x + 2x \cos x$

7. $-e^{\cos x}$

8. $\frac{1}{288}(2x + 3)^8(16x - 3)$

9. $-\frac{1}{2} e^{(1-x)^2}$

10. $\frac{1}{4} e^{(2x-1)}(2x - 1)$

11. $\frac{1}{6} \sin^6 x$

12. $-\frac{1}{4}(4 + \cos x)^4$

13. $\frac{1}{2} e^{(x^2 - 2x + 3)}$

14. $-\frac{1}{30}(1 - x^3)^{10}$

15. $\frac{1}{4} x \sin 4x + \frac{1}{16} \cos 4x$

16. e^{x^2}

17. $\frac{3}{8}(x^2 - 1)^4$

18. $\frac{3}{20}(4x + 1)(x - 1)^4$

19. $\frac{1}{3}(e^9 - e^3)$

20. $\frac{1}{4}$

21. $\ln \frac{1}{2}$

22. 1

23. $\frac{1}{9}(e^3 - 1)$

24. $\frac{1}{8}(\pi - 2)$

25. $\frac{1}{2}(\ln 2)^2$

26. $\frac{1}{3} \ln 2^8 - \frac{7}{9}$

CHAPTER 14

Exercise 14a – p. 204

1. $\dfrac{3}{2(x + 1)} - \dfrac{1}{2(x - 1)}$

2. $\dfrac{13}{6(x - 7)} - \dfrac{1}{6(x - 1)}$

3. $\dfrac{4}{5(x - 2)} - \dfrac{4}{5(x + 3)}$

4. $\dfrac{7}{9(2x-1)} + \dfrac{28}{9(x+4)}$

5. $\dfrac{1}{x-2} - \dfrac{1}{x}$

6. $\dfrac{3}{x-2} - \dfrac{1}{x-1}$

7. $\dfrac{1}{2(x-3)} - \dfrac{1}{2(x+3)}$

8. $\dfrac{7}{3x} - \dfrac{1}{3(x+1)}$

9. $\dfrac{9}{x} - \dfrac{18}{2x+1}$

10. $\dfrac{2}{5(x-1)} - \dfrac{1}{5(3x+2)}$

Exercise 14b – p. 206

1. $\dfrac{9}{8(x-5)} - \dfrac{1}{8(x+3)}$

2. $\dfrac{4}{(x-1)} - \dfrac{5}{(2x-1)}$

3. $\dfrac{1}{5(x-2)} + \dfrac{6}{5(4x-3)}$

4. $\dfrac{5}{3(2x-1)} - \dfrac{4}{3(x+1)}$

5. $\dfrac{3}{x} - \dfrac{6}{2x-1}$

6. $\dfrac{4}{9(x-8)} - \dfrac{4}{9(x+1)}$

7. $\dfrac{1}{2x-3} + \dfrac{1}{2x+3}$

8. $\dfrac{5}{x+2} - \dfrac{1}{x}$

9. $\dfrac{1}{(x-2)} + \dfrac{1}{2(x+1)}$

10. $\dfrac{2}{(x-2)} - \dfrac{1}{2x}$

Exercise 14c – p. 209

1. $\dfrac{1}{x-1} - \dfrac{x+1}{x^2+1}$

2. $\dfrac{1}{x} - \dfrac{x}{2x^2+1}$

3. $\dfrac{3}{2x} - \dfrac{x}{2(x^2+2)}$

4. $\dfrac{22}{19(x-3)} + \dfrac{1-6x}{19(2x^2+1)}$

5. $\dfrac{3}{5(x+2)} - \dfrac{3}{5(2x+1)} + \dfrac{x-1}{5(x^2+1)}$

6. $\dfrac{1}{x} - \dfrac{6x+3}{2x^2-1} + \dfrac{2}{x-1}$

7. $\dfrac{1}{x-1} - \dfrac{1}{x-2} + \dfrac{2}{(x-2)^2}$

8. $\dfrac{2}{x} - \dfrac{1}{x^2} - \dfrac{3}{2x+1}$

9. $\dfrac{3}{x} - \dfrac{9}{3x-1} + \dfrac{9}{(3x-1)^2}$

10. $1 + \dfrac{1}{2(x-1)} - \dfrac{1}{2(x+1)}$

11. $1 - \dfrac{7}{4(x+3)} - \dfrac{1}{4(x-1)}$

12. $x + \dfrac{2}{x-1} - \dfrac{1}{x+1}$

13. $\dfrac{1}{x-1} - \dfrac{1}{x+1}$

14. $\dfrac{1}{x-2} - \dfrac{1}{x+1}$

15. $\dfrac{1}{3(x-3)} - \dfrac{1}{3x}$

16. $\dfrac{1}{2(x-1)} - \dfrac{1}{2(x+1)}$

17. (a) $\dfrac{2}{x-1} - \dfrac{2}{x} - \dfrac{2}{x^2}$

(b) $\dfrac{1}{x} - \dfrac{x}{x^2+1}$

(c) $1 + \dfrac{1}{2(x-1)} - \dfrac{1}{2(x+1)}$

(d) $\dfrac{4}{9(x-1)} + \dfrac{1}{3(x-1)^2} - \dfrac{8}{9(2x+1)}$

18. $y = \dfrac{1}{x-1} - \dfrac{1}{x}, \ \dfrac{dy}{dx} = \dfrac{1}{x^2} - \dfrac{1}{(x-1)^2}$

19. $\mathrm{f}(x) = \dfrac{1}{x-1} + \dfrac{1}{x+1}$

$\mathrm{f}'(x) = -\dfrac{1}{(x-1)^2} - \dfrac{1}{(x+1)^2}$

$= \dfrac{-2(x^2+1)}{(x-1)^2(x+1)^2}$

and there is no value of x for which $x^2+1=0$

CHAPTER 15

All indefinite integrals in this chapter require the addition of a constant of integration.

Exercise 15a – p. 213

1. $\ln(4+\sin x)$

2. $\frac{1}{3}\ln|3e^x-1|$

3. $\dfrac{1}{4(1-x^2)^2}$

4. $\dfrac{1}{2\cos^2 x}$

5. $\frac{1}{4}\ln(1+x^4)$

6. $\ln|x^2+3x-4|$

7. $\frac{2}{3}\sqrt{(2+x^3)}$

8. $\dfrac{-1}{\sin x - 2}$

9. $\ln|\ln x|$

10. $\dfrac{-1}{5\sin^5 x}$

11. $-2\sqrt{(1-e^x)}$

12. $\frac{1}{6}\ln|3x^2-6x+1|$

13. $\dfrac{-1}{(n-1)\sin^{n-1}x}\qquad (n\neq 1)$

14. $\dfrac{1}{(n-1)\cos^{(n-1)}x}\qquad (n\neq 1)$

15. $\dfrac{1}{(3+\cos x)}$

16. $\ln 3$

17. $\ln\sqrt{2}$

18. $\frac{1}{18}-\frac{1}{128}$

20. $\dfrac{e-1}{2(e+1)}$

20. 0

21. $\dfrac{1}{\ln 4}$

Exercise 15b – p. 215

1. $2\ln\left|\dfrac{x}{x+1}\right|$

2. $\ln\left|\dfrac{x-2}{x+2}\right|$

3. $\frac{1}{2}\ln|x^2-1|$

4. $\frac{1}{2}\ln\left|\dfrac{(x+2)^3}{x}\right|$

5. $\ln\dfrac{(x-3)^2}{|x-2|}$

6. $\frac{1}{2}\ln\dfrac{|x^2-1|}{x^2}$

7. $x-\ln|x+1|$

8. $x+4\ln|x|$

9. $x-4\ln|x+4|$

10. $\ln\dfrac{|1-x|}{x^4}$

11. $x-\frac{1}{2}\ln\left|\dfrac{x+1}{x-1}\right|$

12. $x+\ln\dfrac{|x+1|}{(x+2)^4}$

13. $\frac{1}{2}\ln|x^2-1|$

14. $\dfrac{-1}{x^2-1}$

15. $\ln\left|\dfrac{x-1}{x+1}\right|$

16. $\ln|x^2-5x+6|$

17. $\ln\dfrac{(x-3)^6}{(x-2)^4}$

18. $\ln\left|\dfrac{(x-3)^3}{x-2}\right|$

19. $4+\ln 5$

20. $\ln\frac{1}{6}$

21. $\frac{1}{2}\ln\frac{12}{5}$

22. $\ln\frac{5}{3}$

23. $\frac{5}{36}$

24. $1-\frac{3}{2}\ln\frac{7}{5}$

Exercise 15c – p. 217

1. $\dfrac{2}{x-2}-\dfrac{2}{x-1}$; $\dfrac{-2}{(x-2)^2}+\dfrac{2}{(x-1)^2}$;

$\dfrac{4}{(x-2)^3}-\dfrac{4}{(x-1)^3}$

2. $\dfrac{\frac{9}{5}}{x-3}-\dfrac{\frac{3}{5}}{2x-1}$;

$\dfrac{6}{5(2x-1)^2}-\dfrac{9}{5(x-3)^2}$;

$\dfrac{-24}{5(2x-1)^3}+\dfrac{18}{5(x-3)^3}$

3. $\dfrac{\frac{2}{3}}{x-4}-\dfrac{\frac{1}{3}}{x+2}$;

$\dfrac{-1}{3(x+2)^2}-\dfrac{2}{3(x-4)^2}$;

$\dfrac{2}{3(x+2)^3}+\dfrac{4}{3(x-4)^3}$

4. $\dfrac{1}{x-3}-\dfrac{1}{x+2}$; $\dfrac{1}{(x+2)^2}-\dfrac{1}{(x-3)^2}$;

$\dfrac{-2}{(x+2)^3}+\dfrac{2}{(x-3)^3}$

5. $\dfrac{3}{2x+3} - \dfrac{1}{x+1}$;

$\dfrac{-6}{(2x+3)^2} + \dfrac{1}{(x+1)^2}$;

$\dfrac{24}{(2x+3)^3} - \dfrac{2}{(x+1)^3}$

6. $\dfrac{\frac{3}{2}}{x-1} - \dfrac{\frac{9}{2}}{3x-1}$;

$\dfrac{27}{2(3x-1)^2} - \dfrac{3}{2(x-1)^2}$;

$\dfrac{-81}{(3x-1)^3} + \dfrac{3}{(x-1)^3}$

Exercise 15d – p. 219

1. $\frac{1}{4}(2x+\sin 2x)$

2. $\sin x - \frac{1}{3}\sin^3 x$

3. $-\frac{1}{15}\cos x(15-10\cos^2 x+3\cos^4 x)$

4. $\tan x - x$

5. $\frac{1}{32}\{12x-8\sin 2x+\sin 4x\}$

6. $\frac{1}{2}\tan^2 x - \ln|\sec x|$

7. $\frac{1}{32}\{12x+8\sin 2x+\sin 4x\}$

8. $\frac{1}{3}\cos x(\cos^2 x-3)$

9. $\frac{1}{15}\sin^3\theta(5-3\sin^2\theta)$

10. $(\sin^{11}\theta)\left(\frac{1}{11}-\frac{1}{13}\sin^2\theta\right)$

11. $(\sin^{(n+1)}\theta)\left(\dfrac{1}{n+1}-\dfrac{1}{n+3}\sin^2\theta\right)$

$(n\neq -1 \text{ or } -3)$

12. $\frac{1}{32}(4\theta-\sin 4\theta)$

13. $\tan\theta-\theta$

14. $\theta-\tan\theta+\frac{1}{2}\tan^2\theta$

Exercise 15e – p. 223

1. $\frac{1}{21}(x+3)^6(3x+2)$

2. $-\frac{2}{3}(x+6)\sqrt{(3-x)}$

3. $\frac{2}{15}(3x-2)(x+1)^{3/2}$

4. $\dfrac{1-5x}{10(x-3)^5}$

5. $\frac{4}{135}(9x+8)(3x-4)^{3/2}$

6. $-\frac{1}{36}(8x+1)(1-x)^8$

7. $\dfrac{5+4x}{12(4-x)^4}$

Mixed Exercise 15 – p. 223

1. $\frac{1}{3}x^3 - \frac{1}{2}x^2$

2. $\frac{1}{4}x^4 - \frac{2}{3}x^3$

3. $\frac{1}{2}e^{2x+3}$

4. $\frac{1}{6}(2x^2-5)^{3/2}$

5. $xe^x - x$

6. $x\ln x$

7. $\frac{1}{12}(6x-\sin 6x)$

8. $-\frac{1}{2}e^{-x^2}$

9. $\frac{1}{3}\sin^3 x$

10. $\frac{1}{110}(10u-7)(u+7)^{10}$

11. $-\dfrac{1}{12(x^3+9)^4}$

12. $\frac{1}{2}\ln|1-\cos 2y|$ $(y\neq n\pi)$

13. $\frac{1}{2}\ln|2x+7|$

14. $-\frac{2}{9}(1+\cos 3x)^{3/2}$

15. $\frac{1}{16}(\sin 4x - 4x\cos 4x)$

16. $\frac{1}{2}\ln|x^2+4x-5|$

17. $\frac{1}{2}\ln|x^2+4x+5|$

18. $-\frac{1}{4}(x^2+4x-5)^{-2}$

19. $-(9-y^2)^{3/2}$

20. $\frac{1}{13}e^{2x}(2\cos 3x+3\sin 3x)$

21. $x(\ln|5x|-1)$

22. $\frac{1}{6}\sin 2x(3-\sin^2 2x)$

23. $e^{\sin x}$

24. $-2\sqrt{(7+\cos y)}$

25. $e^x(x^2-2x+2)$

26. $\frac{1}{2}\ln|x^2-4|$

27. $x+\ln\left|\dfrac{x-2}{x+2}\right|$

28. $\frac{1}{4}\ln\left|\dfrac{x-2}{x+2}\right|$

29. $\frac{1}{4}x^2(2\ln|x|-1)$

30. $\frac{1}{15}\cos^3 u(3\cos^2 u-5)$

31. $\tan\theta-\theta$

32. $x-\ln|1-x|$

33. $-\ln|1-\tan x|$

34. $\frac{1}{3}(7+x^2)^{3/2}$

35. $\frac{1}{10}(1+x^2)^5$

36. $-\frac{1}{9}e^{-3x}(1+3x)$

37. $x+\ln(x+2)$

38. $\ln(x-4)-\ln(x-1)$

39. $\ln x - \frac{1}{2}\ln(2x+1)$

40. $-2\sqrt{(\cos x)}$

41. $-\frac{1}{5}\cos(5\theta - \frac{1}{4}\pi)$

42. $e^{\tan u}$

CHAPTER 16

Exercise 16a – p. 227

1. $y^2 = A - 2\cos x$

2. $\frac{1}{y} - \frac{1}{x} = A$

3. $2y^3 = 3(x^2 + 4y + A)$

4. $\ln x = A - \cos y$

5. $(A - x)y = 1$

6. $y = \ln \dfrac{A}{\sqrt{(1 - x^2)}}$

7. $y = A(x - 3)$

8. $\sin y = 4x + A$

9. $u^2 = v^2 + 4v + A$

10. $16y^3 = 12x^4 \ln|x| - 3x^4 + A$

11. $y^2 + 2(x+1)e^{-x} = A$

12. $\sin x = A - e^{-y}$

13. $2r^2 = 2\theta - \sin 2\theta + A$

14. $u + 2 = A(v+1)$

15. $y^2 = A + (\ln|x|)^2$

16. $y^2 = Ax(x+2)$

17. $4v^3 = 3(2+t)^4 + A$

18. $1 + y^2 = Ax^2$

19. $Ar = e^{\tan\theta}$

20. $y^2 = A - \operatorname{cosec}^2 x$

21. $v^2 + A = 2u - 2\ln|u|$

22. $e^{-x} = e^{1-y} + A$

23. $A - \dfrac{1}{y} = 2\ln|\tan x|$

24. $y - 1 = A(y+1)(x^2+1)$

Exercise 16b – p. 230

1. $y^3 = x^3 + 3x - 13$

2. $e^t(5 - 2\sqrt{s}) = 1$

3. $3(y^2 - 1) = 8(x^2 - 1)$

4. $y = x^2 - x$

5. $y = e^x - 2$

6. $y = 5 - \dfrac{3}{x}$

7. $y = \pm\sqrt{(4 + e^{-3} - e)}$
   $= \pm 1.154$

8. $(y+1)^2(x+1) = 2(x-1)$

9. $2y = x^2 + 6x$

10. $4y^2 = (y+1)^2(x^2+1)$

11. $x^3 y = y - 1$

12. $y = \tan\{\frac{1}{2}(x^2 - 4)\}$

13. $y^2 = 2x$

Exercise 16c – p. 233

1. $s\dfrac{ds}{dt} = k$

2. $\dfrac{dh}{dt} = k\ln|H - h|$

3. (a) $\dfrac{dn}{dt} = 0$ (b) $\dfrac{dn}{dt} = k\sqrt{n}$

4. (a) $\dfrac{dn}{dt} = k_1 n$ (b) $\dfrac{dn}{dt} = \dfrac{k_2}{n}$

   (c) $\dfrac{dn}{dt} = -k_3$

5. $\dfrac{dp}{dt} = kp(s - p)$

Exercise 16d – p. 238

1. $t = 16T/3$

2. $dy/dx = k\sqrt{x}; y = 0.4x^{3/2} + 1.6; 1.2$

3. (a) $dn/dt = kn$
   (b) $t = T\ln 2/\ln 1.5 = 1.71T$
   (c) e.g. if the colony becomes too great
        for the amount of liquid to support

4. 1000 years (3 sf)

5. (a) $-dm/dt = km; m = 50e^{-kt}$
        where $k = 0.002554\ldots$
   (b) 26.8 g (3 sf)

6. (c) $k = 0.357$ (3 sf)
   (d) $44°$ (nearest degree)

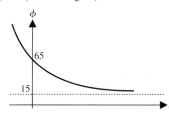

   (e) It may be; $15 + 50e^{-kt} \to 15$ as t
        gets very large, i.e. the water cools to
        room temperature. However during 24
        hours the room temperature may not
        have been constant making the model
        inappropriate.

7. About 4 to $4\frac{1}{4}$ hours before discovery
    assuming Newton's Law of Cooling and
    constant ambient temperature of $10°$C.

CHAPTER 17

Exercise 17a – p. 243
1. $0.099\,\text{cm/s}$
2. $30\,\text{cm}^3/\text{s}$
3. (a) $8\,\text{m}^2/\text{s}$ (b) too small
4. Decreasing at $0.126\,\text{cm/s}$
5. -2
6. $4a\,\text{cm/minute}$

Exercise 17b – p. 247
1. 9
2. (a) $\frac{8}{3}$

   (b) $\frac{16}{3}$

3. $e - 2$
4. $\frac{1}{6}$
5. $\frac{1}{2}\sqrt{3} - \frac{1}{6}\pi$
6. (a) 1   (b) $2 - \ln 3$
7. $8(\ln 2)$
8. 36
9. $\frac{4}{3}$
10. (a) $\frac{4}{3}$ (b) $\frac{1}{3}$
11. $4\sqrt{3}$

Exercise 17c – p. 254
1. $\frac{512}{15}\pi$
2. $\frac{1}{2}\pi(e^6 - 1)$
3. $\frac{1}{2}\pi$
4. $\frac{64}{5}\pi$
5. 2π
6. 8π
7. 8π
8. $\frac{3}{5}\pi(\sqrt[3]{32} - 1)$
9. $\frac{1}{2}\pi(e^2 - 1)$
10. $\frac{16}{15}\pi$
11. $\frac{16}{15}\pi$
12. $\frac{1}{2}\pi^2$
13. $\frac{5}{2}\pi$
14. $\frac{3}{10}\pi$

Mixed Exercise 17 – p. 255
1. $0.15\,\text{cm}^3/\text{s}$
2. $e^4 - 1$

3. (a) $\frac{16}{3}$ (b) 8π
4. (a) $4\sqrt{3}$
   (b) (i) $\dfrac{48\pi\sqrt{3}}{5}$ (ii) $\dfrac{9\pi}{2}$
5. (a) $e - \dfrac{1}{e}$ (b) $\dfrac{1}{2}\left(e^2 - \dfrac{1}{e^2}\right)$
6. (a) $\frac{32}{3}$ (b) 8π
7. $2;\ \frac{1}{2}\pi^2$
8. $\frac{1}{3}$
9. (a) 3092   (b) Slight underestimate

Multiple Choice Exercise D – p. 259
1. A
2. A
3. B
4. B
5. A
6. C
7. B
8. E
9. T
10. T
11. F
12. T
13. F
14. F
15. F
16. F
17. T
18. F
19. T

Miscellaneous Exercise D – p. 262
1. (a) $3x^2/[2\sqrt{(1 + x^3)}]$
   (b) $4/3$
2. (i) $8(3 + 2x)^8$ (ii) 68
3. (i) $1/3y + 1/3(3 - y)$
   (ii) $(1/3)\ln y - (1/3)\ln(3 - y) + K$
   (iii) $y = 3x^3/(4 + x^3)$
4. $\frac{1}{2}\pi(1 - 1/e^4)$
5. (a) $x\sin x + \cos x + c$
   (b) $\frac{1}{2}y + \frac{1}{4}\sin 2y + k$;
        $\frac{1}{2}y + \frac{1}{4}\sin 2y = x\sin x + \cos x + K$
6. (a) $A = 1,\ B = -1,\ C = 2$
7. $\frac{1}{2}x^2\ln x - \frac{1}{4}x^2 + c$
8. $A = 1,\ B = -2,\ C = 2$
9. (i) 1   (ii) $\frac{1}{4}\pi^2$

10. (i) 30   (ii) 50

11. (a) $(1, 0)$, 1   (b) $(e^{-1/2}, -\frac{1}{4}e^{-1})$, 2

12. $A = 1$, $B = -2$, $C = 2$

13. (i) $\dfrac{1 - x^2}{(1 + x)^2}$ (ii) $(1, \frac{1}{2})$, $(-1, -\frac{1}{2})$

     (iii) $\frac{1}{2}\ln(x^2 + 1) + k$

     (iv) $\sqrt{(e^{20} - 1)} = 22\,026$

14. (a) $(0, 1)$ (b) $(e, 1/e)$

     (c) $-1/e^3$ (d) $\frac{1}{2}(\ln x)^2 + c$, $\frac{1}{2}e^{-1}$

15. (a)

     $(0, 1)$, $(0, 3)$, $x = 0$ and $x = 2$

CHAPTER 18

Exercise 18a – p. 270

1. QA = qualitative, QD = qualitative and discrete, QC = quantitative and continuous

     (a) QA   (b) QD   (c) QC

     (d) QC   (e) QA   (f) QD

     (g) QA   (h) QC   (i) QC

2. (a) People on the electoral register

     (b) Finite   (b) Qualitative

3. (a) 0, 1, 2, or 3

     (b) All possible tosses of the three coins

     (c) Infinite

     (d) (ii) is slightly easier to use for most people, but is a personal choice.

4. (a) It is possible to ask all the present people involved but not all future parents and children.

     (b) School register and feeder schools registers.

5. (a) Schools' registers for that age group. (If they are beyond compulsory school age, there is not one.)

     (b) Professional register

     (c) Electoral register

     (d) All schools' rolls.

     (e) There probably isn't one, but one can be made by tagging and numbering each mouse.

6. All road vehicles includes lorries, vans, etc. as well as cars. Also assumes no cars are removed from use – most unlikely.

7. The table shows that a lower number of claims come from 17–20 year-olds; the likely reason for this is because the number of drivers in this age group is a small proportion of all drivers.

8. The sample is biased; the people asked are likely to have an interest in not having a charge made.

9. The sample is biased; there are many people without telephones on the electoral register and these people are likely to have interests different from those who do have telephones.

11. (a) Possible problems with the first question: 'Games machine' is ambiguous, no space allocated to tick 'no'. Second question is not specific enough; for 12 year-olds, it should state whether it is given for personal spending only or if it includes fare and/or lunch money.

     (b) First question: Do you own a games machine (e.g. Game boy, Super Nintendo)? Yes ... No ...

     Second question: How much pocket money do you get each week? (Do not include fare or lunch money.)

12. (a) Government population census

     (b) Opinion poll survey

Exercise 18b – p. 274

1. (a)

Number of flaws/metre	0	1	2	3	4	5
Frequency	15	19	10	4	1	1

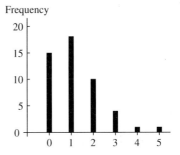

     (b) 1

2. (a)

	TB	German measles	Yellow-fever	Flu	Diphtheria	Polio	Whooping cough	Measles	Anti-tetanus
	55	60	40	180	220	200	100	150	75

(b) 1080

(c) The numbers are all multiples of 5 so they may have been rounded in some way.

Exercise 18c – p. 278

1. e.g.

h mm	5.45–5.65	5.65–5.85	5.85–6.05	6.05–6.25	6.25–6.45	6.45–6.65	6.65–6.85	6.85–7.05
f	5	7	8	7	7	3	4	1

2. e.g.

l cm	97.5–117.5	117.5–137.5	137.5–157.5	157.5–177.5	177.5–197.5	197.5–217.5
f	10	8	14	13	10	1

3. e.g.

t sec	0.5–3.5	3.5–6.5	6.5–9.5	9.5–12.5	12.5–15.5	15.5–35.5
f	3	10	15	15	4	3

4. e.g.

no of goals	0–1	2–3	4–5	6–7	8–9	10–12	13–18
f	14	12	7	4	2	2	2

5. e.g.

no of goals	0–4	5–9	10–14	15–20
f	30	9	2	2

Advantages: few equal-size groups. Disadvantages: loses too much detail.

no. of goals	0	1	2	3	4–6	7–9	10–20
f	6	8	6	6	9	4	4

Advantages: keeps detail for low scores. Disadvantages: groups of widely different widths.
Table given for question 5: Advantages: keeps a reasonable amount of detail except for last group. Disadvantages: unequal sized groups.

Exercise 18d – p. 283

The shape of these histograms depends on the classes chosen.

1.
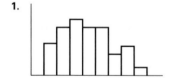

2.
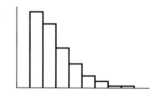

3. (a) 12   (b) 36

(c)

no. of goals	0	1	2	3–4	5–7	8–10
f	7	14	10	10	9	6

4. (a)
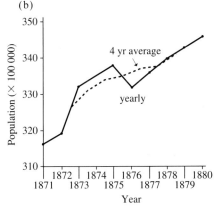

   (b) No, because the numbers of matches played by the two teams are not the same.

5. The class widths vary from £50 to £50 000 and so are very difficult to represent. Also the last class is open ended so it is not known how wide it is.

CHAPTER 19

Exercise 19a – p. 287

(answers given corrected to 3 sf)

1. Mean 48.4, median 50, modes 36, 45, 50

2. Mean 554, median 538, modes 430, 450

3. Mean 54.2, median 57.3, no mode

4. 543

5. (a) 33 360 000
   (b)

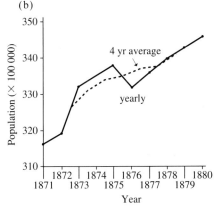

   Using a moving average evens out the odd fluctuations and gives a better idea of longer term trend.

   (c) No, because the figures given include the population of all Ireland. The UK does not now include Eire.

6. Increases the mean because $(6.8 \times n + 9)/(n + 1)$ is greater than 6.8. Moves the median half a score up the ranking order – this may increase the value of the median, it depends if there is another score of 6 above the old median. Does not change the mode – there is only one score of 9.

Exercise 19b – p. 291

1. 6.80 mm

2. 634

3. (a) 98   (b) 66.3 g

4. (a) The last class is open ended, so the midclass value cannot be found.
   (b) Assuming the last class is 61–90 gives a mean age $\simeq$ 47.0 years.

Exercise 19c – p. 296

1. q.1. 6.83   q.2. 770   q.3. 65.3

2. q.1.

< 5.85	5
< 6.25	14
< 6.65	29
< 7.05	48
< 7.45	64
< 7.85	72
< 8.25	74

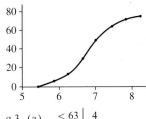

q.3. (a)

< 63	4
< 64	12
< 65	26
< 66	50
< 67	70
< 69	86
< 72	98

(b)

(c) 66

(d) The last class is very wide so the full cumulative frequency curve would have most of the information in half its width. By drawing only the section up to 1199.5.

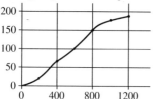

3. 44 years (nearest year)

The middle ranked age is in a class with known boundaries. The age of people in the top class does not affect the median.

Class	f	cum f
$0.05 \leqslant h < 1.95$	9	9
$1.95 \leqslant h < 3.95$	10	19
$3.95 \leqslant h < 5.95$	4	23
$5.95 \leqslant h < 7.95$	1	24
$7.95 \leqslant h < 9.95$	2	26

median ≈ 2.5 cm

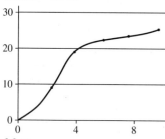

4. 2.9 cm

Exercise 19d – p. 298

1. (a) Mode, if the category that sells most items is required.

(b) Mean, to give the mean turnover/ category.

(c) Median, if the one low figure is exceptional or mean if the low figure is part of expected weekly variation in weekly commission.

(d) Median, to ignore the one large number or mean if a true average weekly consumption is required.

2. (a) Mean 3.46 ignoring the score of 8 as this must be a mistake.

(b) Mean 3.79 including the 8 coins to give the mean number of coins per tin.

(c) Median £4 in case £8 is a mistake.

3. (a) A survey into average spending money of 14-year-olds.

(b) If the £8 is considered exceptional for some reason, e.g. possibly the question was misunderstood and fare money to school was included, so it needs to be partly discounted.

(c) When arguing that 'most of my friends get £4 a week and I only get £2'.

4. (a) Median, to discount the effect on the mean that the very few heights over 55 cm have, as they may be considered unsuitable for use.

(b) Mean, as all ages are relevant.

(c) Median, if the purpose is to cope only with the average demand. Mode, if the purpose is to cope with the demand that occurs most often.

Exercise 19e – p. 303

The answers given here are calculated. Your answers should be close to these.

1. (a)
(i)

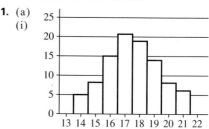

(ii)

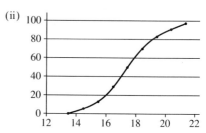

(b) Median 17,
range 8 cm,
interquartile range 3

2. (a) (i)

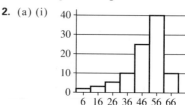

(ii)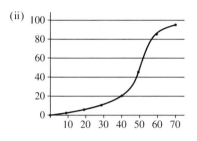

(b) Median 50,
   range 70,
   interquartile range 14

3. (a) (i)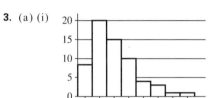

(ii)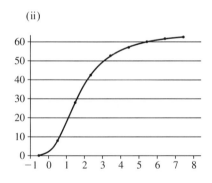

(b) Median 2,
   range 8,
   interquartile range 2

4. (a) (i)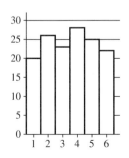

(ii)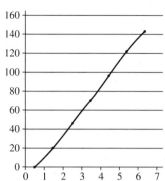

(b) Median 3.5,
   range 6,
   interquartile range 2.4
5. (a) Median 5.5, interquartile range 3.2
   (b) Median 0.81, interquartile range 0.13

Exercise 19f – p. 306
1. (a) 5   (b) 58   (c) 5.5
2. (a) Yes, it excludes the relatively few
      heights at each end.
   (b) No, it includes the few very low
      marks. It would be better to cut these
      out and use all the higher marks, i.e.
      use the top 90%.
   (c) No, the values are fairly evenly
      distributed so all of them should be
      included. 100%
3. (a)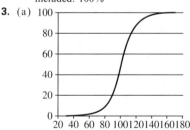

(b) (i) 101   (ii) 115   (iii) 149
   (c) (i) 2%   (ii) 88%   (iii) 54%

Exercise 19g – p. 310
1. 11.0
2. 208
3. 7.55

Exercise 19h – p. 312

1. 1.80
2. 13.1
3. 1.53
4. 1.65
5. (a) Higher mean weight with less variation.
   (b) Same mean weight but with less variation.
   (c) Lower mean weight and much less variation.
   (d) Higher mean weight but with greater variation.

Exercise 20a – p. 316

1. (a) $\frac{1}{2}$ (b) $\frac{1}{4}$
2. (a) $\frac{1}{52}$ (b) $\frac{1}{13}$ (c) $\frac{11}{13}$
3. (a) $\frac{1}{30}$ (b) $\frac{29}{30}$ (c) $\frac{11}{30}$
4. (a) (i) $\frac{1}{10}$ (ii) $\frac{2}{5}$ (iii) $\frac{9}{10}$
   (b) Any answer from 0.4 to 0.5 acceptable because the apple whose mass is listed as 200 g may be above or below 200 g as it lies in the range 119.5 up to 200.5 grams.
5. (a) $\frac{1}{5}$ (b) $\frac{10}{49}$
6. $\frac{1}{3}$

Exercise 20b – p. 318

1. (a) $\frac{11}{36}$ (b) $\frac{4}{9}$ (c) $\frac{5}{18}$

2.

	1	2	3	4
1	11	12	13	14
2	21	22	23	24
3	31	32	33	34
4	41	42	43	44

(a) $\frac{7}{16}$ (b) $\frac{1}{4}$

3.

	1	2	3	4	5	6
1	11	12	13	14	15	16
2	21	22	23	24	25	26
2	21	22	23	24	25	26
3	31	32	33	34	35	36
3	31	32	33	34	35	36
4	41	42	43	44	45	46

(a) $\frac{1}{6}$ (b) $\frac{3}{4}$ (c) $\frac{7}{18}$

4. $\frac{1}{6}$
5. Wrong. If two dice are tossed there are 36 possible scores, only one of which is 12 $\Rightarrow$ $P(12$ with two dice$) = 1/36$. If one die is tossed and the score doubled, there are six possibilities, one of which is 12 $\Rightarrow$ $P(12$ with one die$) = 1/6$.

6. (a) $1/18$ (b) 0

Exercise 20c – p. 320

M = Mutually exclusive,
I = Independent,
N = Neither

1. (a) M   (b) N
2. (a) I   (b) N
3. (a) N   (b) N
4. (a) M   (b) I   (c) I
5. N

Exercise 20d – p. 325

1. (a) $\frac{1}{8}$ (b) $\frac{7}{8}$ (c) $\frac{3}{8}$
2. (a) 4/25   (b) 0.384   (c) 0.5904
3. 0.107
4. (a) $\frac{1}{8}$ (b) 1/64   (c) 0.998
5. (a) 1/5   (b) 1/25   (c) 0.992
6. (a) $\frac{1}{8}$ (b) 1/16
7. 2/3 (0.667)
8. 0.330
9. (a) 0.9025   (b) 0.095   (c) 0.0025
10. 3.14×10^5

Exercise 20e – p. 331

1. (a) 0.6   (b) 0.167   (c) 0.3
2. (a) 0.000 181   (b) 0.0681   (c) 0.0769
3. 1/12
4. $P(A \cup B) > P(A) + P(B)$, $\therefore$ A and B are not mutually exclusive
   $P(A \cap B) = 0.2 < P(A)P(B)$, $\therefore$ A and B are not mutually independent.
5. 7/10, 11/14
6. 4/5, no because $P(A \cap B) \neq 0$
7. (a) 0.4   (b) 0.9   (c) 0.1
8. (a) 0.625   (b) 0.4
9. (a) 0.45   (b) 0.5   (c) 5/9
10. (a) $\frac{5}{7}$ (b) $\frac{13}{14}$ (c) $\frac{3}{4}$

Exercise 20f – p. 339

1. 0.588
2. (a) 0.313   (b) 0.6
3. 0.08
4. (a) 0.1   (b) 0.001
5. (a) 0.02   (b) 0.405
6. (a) 0.0329. A child is equally likely to be born on any one of the 365 days in a year.
   (b) 0.001 23
7. (a) 0.233   (b) 0.429

8. (a) 0.167   (b) 0.875   (c) 0.75
9. (a) 1/4   (b) 1/8

Multiple Choice Exercise E – p. 344

1. D
2. E
3. C
4. B
5. C
6. B
7. C
8. C
9. F
10. F
11. T
12. T
13. F
14. T
15. T

Miscellaneous Exercise E – p. 345

1. (a) 0.0902
   (b) A positive test result still means that the probability of having the disease is low.

2. 8.54, 0.0564, 0.237
3. (a) 0.3   (b) 3/7
4. (b) School register
5. (a) 1/8   (b) 3/11
6. (i) 2 squares
   (ii)

°F	50 – 60	60 – 70	70 – 75	75 – 80	80 – 90
f	10	30	35	25	20

   (iii) 72.3°; Actual readings not known.
7. (a) 29
   (b) (i) 8.6 minutes   (ii) 11.7 minutes
8. (i) 0.95
   (iii) $0.95p/(0.1 + 0.85p)$; 0.991
9. (i) (a) Class widths and frequencies too variable
      (b) Last class is open-ended so the midclass value is not known.
   (ii) 310
10. (a) (i) 0.599   (ii) 0.086
11. (a) 47, 12.5
    (b) They discount the few low values and the few high values..
12. (i) (a) 1/4   (ii) (b) 4/5

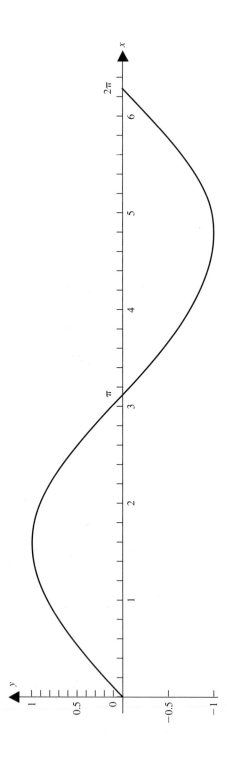

INDEX